Impact and Friction of Solids, Structures and Intelligent Machines

In Memoriam P. D. Panagiotopoulos (1950 - 1998)

SERIES ON STABILITY, VIBRATION AND CONTROL OF SYSTEMS

Series Editors: Ardéshir Guran & Daniel J. Inman

About the Series

Rapid developments in system dynamics and control, areas related to many other topics in applied mathematics, call for comprehensive presentations of current topics. This series contains textbooks, monographs, treatises, conference proceedings and a collection of thematically organized research or pedagogical articles addressing dynamical systems and control.

The material is ideal for a general scientific and engineering readership, and is also mathematically precise enough to be a useful reference for research specialists in mechanics and control, nonlinear dynamics, and in applied mathematics and physics.

Selected Forthcoming Volumes in Series B

Proceedings of the First International Congress on Dynamics and Control of Systems, Chateau Laurier, Ottawa, Canada, 5–7 August 1999
Editors: A. Guran, S. Biswas, L. Cacetta, C. Robach, K. Teo, and T. Vincent

Selected Topics in Structronics and Mechatronic Systems
Editors: A. Belyayev and A. Guran

Acoustic Interactions with Submerged Elastic Structures
Part II: Propagation, Ocean Acoustics and Scattering
Editors: A. Guran, G. Maugin and J. Engelbrecht

Part III: Acoustic Propagation and Scattering, Wavelets and Time-Frequency Analysis
Editors: A. Guran, A. de Hoop and D. Guicking

Part IV: Nondestructive Testing, Acoustic Wave Propagation and Scattering
Editors: A. Guran, A. Böstrom, and G. Maze

Dynamics with Friction: Modelling, Analysis and Experiments, Part II
Editors: A. Guran, F. Pfeiffer, and K. Popp

Selected Published Volumes in Series A

Vol. 1 Stability Theory of Elastic Rods
Author: T. Atanackovic

Vol. 2 Stability of Gyroscopic Systems
Authors: A. Guran, A. Bajaj, Y. Ishida, G. D'Eleuterio, N. Perkins, and C. Pierre

Vol. 3 Vibration Analysis of Plates by the Superposition Method
Author: Daniel J. Gorman

SERIES ON STABILITY, VIBRATION AND CONTROL OF SYSTEMS

Series B Volume 14

Series Editors: Ardéshir Guran & Daniel J Inman

Proceedings of the First International Symposium on

Impact and Friction of Solids, Structures and Intelligent Machines

In Memoriam P. D. Panagiotopoulos (1950 - 1998)

Ottawa Congress Centre, Ottawa, Canada 27–30 June 1998

Editor

Ardéshir Guran
Institute for Structronics, Canada

World Scientific
Singapore • New Jersey • London • Hong Kong

Published by

World Scientific Publishing Co. Pte. Ltd.
5 Toh Tuck Link, Singapore 596224
USA office: 27 Warren Street, Suite 401-402, Hackensack, NJ 07601
UK office: 57 Shelton Street, Covent Garden, London WC2H 9HE

British Library Cataloguing-in-Publication Data
A catalogue record for this book is available from the British Library.

PROCEEDINGS OF THE FIRST INTERNATIONAL SYMPOSIUM IMPACT AND FRICTION OF SOLIDS, STRUCTURES AND INTELLIGENT MACHINES

ISBN-13 978-981-02-4370-8
ISBN-10 981-02-4370-7

INTERNATIONAL SYMPOSIUM ON IMPACT AND FRICTION OF SOLIDS, STRUCTURES AND INTELLIGENT MACHINES

June 27-30, 1998

Program

Chairman:
Ardéshir Guran
Institute for Structronics, 275 Slater Street, Suite 900
Ottawa, Ontario, Canada K1P 5H9

Co-Chairmen:
Brian Feeny
Department of Mechanical Engineering, Michigan State University
East Langsing, MI 48824-1226 USA

Anders Klarbring
Division of Mechanics, Linköping University
S-581 83 Linköping, Sweden

Yukio Ishida
Department of Electronic-Mechanical Engineering, Nagoya University
Nagoya, Aichi, 464-8603, Japan

P.D. Panagiotopoulos
Department of Civil Engineering, Aristotle University
GB-54006 Thessaloniki, Greece; and
Faculty of Mathematics and Physics, RWTH
D-52056 Aachen, Germany

Professor P. D. Panagiotopoulos (1950–1998)

Proceedings of the First International Symposium on Impact and Friction of Solids, Structures, and Intelligent Structures

Editor: Ardeshir Guran

Preface

This volume of Stability, Vibration and Control of Systems contains the proceedings of the first symposium on Impact and Friction of Solids, Structures, and Intelligent Machines, that took place in Ottawa Congress Center, Ottawa, Canada, June 27-30, 1998. The purpose of the symposium was to bring together engineers and scientists concerned with modelling, analysis, measurement and control of mechanical systems in presence of impact and friction. It was also the aim of this workshop to stimulate cross-fertilization of different disciplines involved in the impact and frictional aspects of solids, structures and machines and to promote the fundamentals as well as industrial applications. A total of 79 participants from 23 countries played an active role during the lectures and sessions of contributed papers.

The book is about dynamics of mechanical systems in presence of impact and friction. Impact and friction are usually introduced and studied during elementary courses in physics. Thus, one may think these topics are simple and well understood. However, nothing could be more complicated and less mature than the study of impact and friction in mechanical systems. Contact mechanics has been studied by many of the brightest scientists, e.g., Newton, Coulomb, Reynolds, Hertz, and Sommerfeld. In spite of its importance from a practical view, it is of great theoretical interest and involves fundamental physics, e.g. questions related to the origin of irreversibility and adiabaticity, the role of self-organized criticality and, dynamical phase transitions in molecularly thin lubricant layers. Modern contact laws lead to mathematical models that are highly nonlinear, discontinuous and nonsmooth. Discontinuous and/or nonsmooth processes introduce problems with data processing techniques such as phase-space reconstructions, and analytical techniques such as perturbation methods and linearization in the vicinity of equilibria. Today due to great advances in computer technology and computational analysis (advances in calculating branching behavior of BVP's, nonsmooth convex analysis, differential inclusions, and the design of proper numerical solution methods for inequality constrained optimization problems), it is possible to solve complex contact problems. Furthermore, new experimental devices, such as atomic force microscope and quartz-crystal-microbalance probe allow other aspects of contact mechanics to have potentially important applications in nano-tribology. In short, the study of impact and friction, which is one of the oldest problems in physics, is in a phase of rapid and exciting development. These growing research breakthroughs have promoted the development of new technologies in the description and design of systems with impact and friction models, to understand nature, structures, machines, transportation systems, and other processes.

In this volume on Stability, Vibration and Control of Systems, research papers on certain selected topics have been compiled to form an integrated set. Being aware that it is not possible to do justice to each paper in this preface, we have attempted to group them under the following headings:

- Inelasticity, Hysteresis, Wear - Stability and Structures - Nonlinear Dynamics and Vibrations - Nonlinear Dynamics and Experiments - Analysis, Rigid Bodies, Multi-Body Systems, Mechanics and Contact Models - Applications, Rotating Systems - Modeling, Computations

The sponsors of the ISIFSM98 are cited below. Here we acknowledge dutifully their support.

-The Canadian Society for Mechanical Engineering -Institute for Structronics, Ottawa, Canada -Worcester Polytechnic Institute, USA -Linkoping University, Sweden -Stochastic Mechanics/Analysis Research Group, Worcester, USA

The International Program Committee of ISIFSM98 was composed of:

-Brian Feeny (Michigan State University, USA) -Ardeshir Guran (Institute for Structronics, Canada) -Yukio Ishida (Nagoya University, Japan) -Anders Klarbring (Linkoping University, Sweden) -Panagiotis Panagiotopoulos (Aristotle University, Greece)

All members of the IPC played a key role both in the the organization of the ISIFSM98 and in the editing of this book. It was with shock and deep sadness that we learned of the sudden passing of Professor Panagiotis Panagiotopoulos shortly after the ISIFSM98 meeting on 12 August, 1998. Dr.Panagiotopoulos was a prolific scholar in theoretical and applied mechanics. He produced excellent fundamental research while simultaneously promoting the practical applications of mechanics to industrial problems. The mechanics community has lost a great scholar, the profession has lost an energetic champion and many of us have lost a warm and charming friend. He will be always remembered as a true scholar, a thoughtful person, and a gentleman. He will be greatly missed. This modest volume is dedicated to his memory.

Ardeshir Guran
Ottawa, October 1998

Table of Contents

SESSION I – (A)

Inelasticity, Hysteresis, Wear

Constitutive Modeling of Polymeric Foams for Impact-Contact Problems

J. Zhang[1]
Z. Lin[2]
A. Wong[2]
N. Kikuchi[1]
V. C. Li[2]
A. F. Yee[3]
G. S. Nusholtz[4]

ABSTRACT In this study, mechanical properties of three types of polymeric foams (polypropylene (PP), polysyrene (PS) and polyurethane (PU) foams) are investigated for impact-contact simulation. Focus has been placed on the strain rate and temperature effects. Based on selected experimental results from uniaxial compression, hydrostatic compression, and simple shear tests, phenomenological hydrodynamic elastoplastic constitutive law is developed to model these polymeric foams for impact simulation of dummy models.

1 Introduction

Polymeric foams are excellent energy absorbers and have found a wide range of applications, especially in the automotive industry. New provisions in the Federal Motor Vehicle Safety Standards (FMVSS) require the use of Head Injury Criterion (HIC) for regulation of the interior of motor vehicles in order to reduce the mortality resulted from severe head injuries during traffic accidents. To satisfy such a criterion, polymeric foams are currently

[1]Department of Mechanical Engineering and Applied Mechanics, University of Michigan, Ann Arbor, MI 48109
[2]Department of Civil Engineering and Environmental Engineering, University of Michigan, Ann Arbor, MI 48109
[3]Department of Material Science and Engineering, University of Michigan, Ann Arbor, MI 48109
[4]Chrysler Corporation, Auburn Hills, MI 48326

being considered as viable vehicle interior padding materials.

Polymeric foams consist of large quantity of either open or closed cells. A typical stress-strain curve of a polymeric foam exhibits three regimes (Gibson and Ashby, 1988): a linear elastic regime (due to cellular wall bending) at low stresses followed by a long collapse plateau and a densification regime in which the stress rises steeply. The mechanisms associated with the collapse plateau are different, depending on the properties of the cell walls. For flexible foams, the collapse plateau is due to elastic buckling of cell walls. For rigid and brittle foams, plastic yielding and brittle crushing of the cell walls are the main failure mechanisms respectively. Under tension, on the other hand, the post-elasticity behavior of a polymeric foam is controlled by cell wall alignment towards the loading axis and the stress-strain curve is ultimately truncated by either ductile or brittle fracture. Extensive analytical and experimental work has been carried out to characterize the elastic and failure behaviors of polymeric foams. The strain-rate and temperature effects are two important aspects to be considered in the applications of polymeric foams. The microstructural features of polymeric foams (e.g., cell configuration, cell size, strut thickness, etc.) lead to more complexity in material property characterization as compared to solid polymers.

The purpose of this study is to investigate the rate and temperature dependent behaviors of polymeric foams used in the automotive industry and to develop a phenomenological constitutive model targeted for computer simulation of crash impacts, in which compression and shear are the dominant deformation modes while tension is less relevant. The model will be further implemented into a dynamic finite element analysis code such as LS-DYNA3D for crashworthiness analysis.

2 Experimental Investigations

Experiments, including uniaxial compression, hydrostatic compression and simple shear, were conducted on: (1) an electro-hydraulic Instron 1331 machine equipped with a temperature chamber, and (2) a pneumatically driven ICI instrumented impact machine (used for uniaxial compression only). The Instron 1331 testing machine can maintain constant cross-head speeds ranging from quasi-static (0.08 mm/sec, ASTM Standard D1621) to 250 mm/sec with a closed-loop servo-controlled system. The pneumatically driven ICI machine can achieve initial impact speed of up to 10 m/s. Although the impactor speed of the ICI impact machine is varying during the impact process, the recorded impactor velocity history showed a broad plateau which was taken to compute the average strain rate. Foam specimens are $50 \times 50 \times 50$ mm^3 in dimension for compression and tension tests and $50 \times 50 \times 100$ mm^3 for simple shear tests. Four different cross-head rates of 8.0×10^{-5} m/s, 4.0×10^{-3} m/s, 0.229 m/s, and 4.45 m/s (corresponding strain rates 1.6×10^{-3}, 0.08, 4.6, and 88/sec) were used in uniaxial compression tests.

For simple shear tests, foam specimens were glued in-between two L-shaped loading fixtures made of steel. The uniaxial tension fixture consists of two circular loading plates to which the top and the bottom of the test specimen are glued. During a hydrostatic compression test, foam specimens were wrapped by latex rubber and immersed into a specially-designed hydrostatic compression chamber filled with water. Air is allowed to escape from the specimens through an air vent on the lid of the chamber. All specimens were subjected to 80% volumetric strain in uniaxial and hydrostatic compression tests and loaded until fracture in tension and simple shear tests. The tests conducted under various temperatures are for uniaxial compression only. The temperature range of −20°C to 80°C was selected according to the actual temperature range that vehicle interior padding might experience.

3 Constitutive Modeling

3.1 Yield Criterion

Gibson et al (1989) derived a plastic yield surface for rigid foams

$$\frac{\sigma_{vm}}{\sigma_{ys}} = \pm\gamma\left(\frac{\rho^*}{\rho_s}\right)^{3/2}\left\{1 - \left[\frac{3p}{\sigma_{ys}(\rho^*/\rho_s)}\right]^2\right\} \qquad (1a)$$

where p is the hydrostatic stress, σ_{vm} is the effective stress, and σ_{ys} is the yield strength of solid polymer. $\frac{\rho^*}{\rho_s}$ is the relative foam density and γ is a material parameter to be determined from an uniaxial compression test. For the PP foam (3.06 pcf), the material parameters in Eq. (5a) are: $\sigma_{ys} = 30\text{MPa}$, $\frac{\rho^*}{\rho_s} = 0.05$, and $\gamma = 0.447$. Puso and Govindjee (1995) proposed following the formula to fit the theoretical buckling surface by Triantafillou et al (1989)

$$\sigma_{vm}^2 + \frac{1}{R^2}(p^2 - h^2) = 0 \qquad (1b)$$

where R and h can be determined from uniaxial and hydrostatic compression tests.

Due to volumetric compressibility, the yield surface of crushable foams should incorporate first and second stress invariants. In this study, the yield locus is defined by a single ellipse spanned on the plane of effective stress σ_{vm} and hydrostatic pressure p

$$\frac{[p - x_0(\varepsilon_{vp})]^2}{a(\varepsilon_{vp})} + \frac{\sigma_{vm}^2}{b(\varepsilon_{vp})} \leq 1 \qquad (2)$$

where $x_o(\varepsilon_{vp})$, $a(\varepsilon_{vp})$ and $b(\varepsilon_{vp})$ are three material parameters that define the center and the lengths of the major and minor axes of the yield ellipse. This ellipse is extendible in the $\sigma_{vm} - p$ space as a foam speciman densifies under loading, and can be uniquely defined by three types of tests: (1) uniaxial compression, (2) hydrostatic compression, and (3) simple shear. The initial yielding ellipse of the PP foam (3.06 pcf, 5% relative density) is plotted in Fig. 1(a) along with the yield surface proposed by Gibson et al (1989) and the buckling surface by Puso and Govindjee (1995). The buckling cap parameters calculated for the PP foam (3.06 pcf) from hydrostatic compression and uniaxial compression test results are: R=0.364 and h=0.074 MPa. It can be seen from Fig. 1(a) that the single-surface yield criterion Eq. (6) agrees reasonably well with the multi-surface criteria defined by Eqs. (1a) and (1b). Also, the single-surface criterion possesses no corner point and offers simplicity for numerical implementation. The evolution of the yield surface of the PP foam (3.06 pcf) with increasing plastic volumetric strain is depicted in Fig. 1(b).

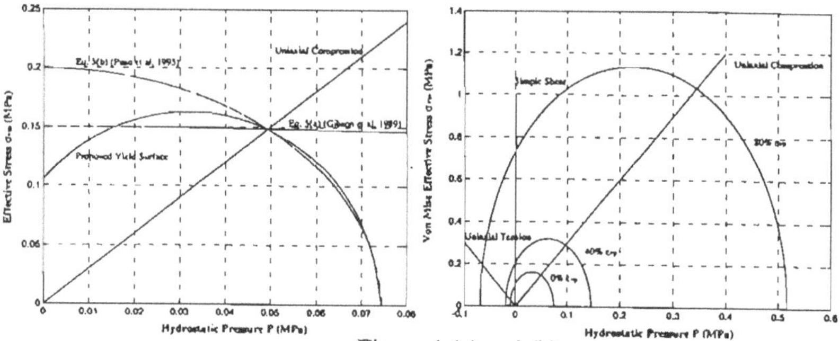

Figure 1 (a) and (b)

3.2 Flow Rule

For crushable foams with zero plastic Poisson's ratio, the flow potential can be written as (Bilkhu et al, 1993)

$$g = \sqrt{\frac{9}{2}p^2 + \sigma_{vm}^2} \tag{3}$$

3.3 Volumetric Hardening and Strain Rate Dependence

Strain-rate effect is resulted from the gas flow and viscoplasticity of the cell wall properties (Gibson and Ashby, 1988). At room temperature, it can be represented by

$$\dot{\bar{\varepsilon}}_p = D\left(\frac{f}{f_0}\right)^{\frac{1}{n}}, \qquad n = a + b\bar{\varepsilon}_p \tag{4}$$

where f and f_0 are the initial effective yield stresses under an arbitrary and a reference quasi-static strain rate respectively, and $\dot{\bar{\varepsilon}}_p$ is the effective plastic strain rate defined as $\dot{\bar{\varepsilon}}_p = \sqrt{\frac{2}{3}\dot{\varepsilon}^p_{ij}\dot{\varepsilon}^p_{ij}}$. For example, the calculated material parameters for the PP foam (3.06 pcf) are: $D = 0.768$, $a = 0.041$, and $b = 0.003$.

4 Numerical Implementation and Simulation

The above constitutive model was implemented into the finite element program LS-DYNA3D (Hallquist, 1994) as a user defined material subroutine. In this section, the numerical implementation procedures and applications of the current model will be presented.

4.1 Numerical Implementation

For transient analysis of large deformation impact problems, extreme small time steps need to be used. The stresses in each element are updated based on the constitutive law defined in last section at each time integration step.

The constitutive equations derived were implemented into LS-DYNA via a user defined material subroutine. Numerical simulations are conducted to validate the constitutive model under different loading conditions. The material parameters are calibrated from experimental results discussed in the previous sections. The initial yield locus at zero plastic strain can be obtained from: (1) uniaxial compressive strength σ_0, (2) hydrostatic compressive strength p_{c0}, and (3) simple shear strength τ_0. For the PP foam (3.06 pcf), these parameters are: $\sigma_0 = 0.15$ MPa, $p_{c0} = 0.074$ MPa, and $\tau_0 = 0.058$ MPa. The hardening behavior is defined by three tabular functions $x_0(\varepsilon_{vp})$, $a(\varepsilon_{vp})$ and $b(\varepsilon_{vp})$. The strain rate dependence is defined by Eq. (4).

Numerical simulations were performed to simulate a high speed indentation of a rigid hemisphere (ϕ127 mm, 22.2 kgm) is dropped onto a foam block (203 × 203 × 101 mm^3) with an initial speed of 4.5 m/sec (10 mph). Figure 2(a) shows the deformed configuration of the PP foam block (3.06 pcf) during the impact. Contact forces versus deflection curves are plotted in Fig. 2(b). Good agreement between the experimental result and the simulation result is found for the loading stage. Since the current model assumes elastic unloading, large discrepancy is shown for the unloading stage. Further research is needed to model the unloading behavior of the foam materials accurately.

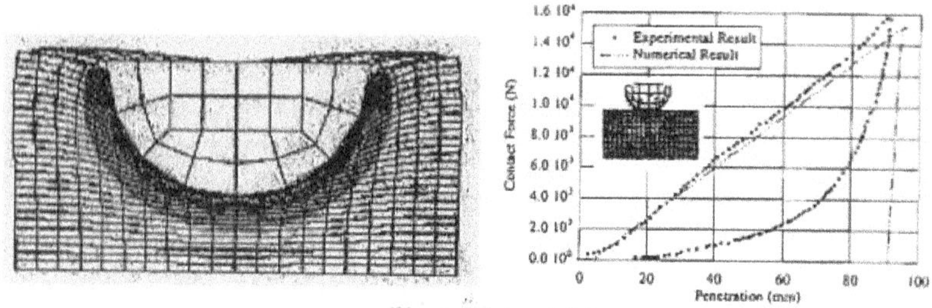

Figure 2(a) and (b)

5 Acknowledgements

This research has been supported by a grant from American Automobile Manufacturers Association. Special thanks to Livermore Software Technology Corporation for the use of LS-DYNA3D software in numerical investigations presented in this study. We are also grateful to two anonymous reviewers for their valuable comments and suggestions.

6 References

[1.] Bilkhu, S.S., Founnas, M. and Nusholtz, G.S., Material Modeling of Structure Foams in Finite Element Analysis Using Compressive Uniaxial and Triaxial Data, *SAE Technical Paper Series 930434*, International Congress and Exposition, Detroit, Michigan.

[2.] Gibson, L.J. and Ashby, M.F., *Cellular Solids-Structural and Properties*, Pergamon Press, New York, 1988.

[3.] Wallquist, J.O., *LS-DYNA3D Theoretical Manual*, LSTC Report 1018, 1994.

[4.] Puso, M.A. and Govindjee, A Phenomenological Constitutive Model for Polymeric Foam, *Mechanics of Plastics and Plastic Composites*, MD-Vol. 68/AMD-Vol. 215, ASME, pp. 159-176, 1995.

Shape Differential Equation for Viscoelastic Flows

N. Gomez[1]
J.-P. Zolésio[2]

ABSTRACT In this paper, we revisit the steady generalized Norton-Hoff problem on the rheology of visco-plastic flows arising in metal forming. Then we deal with maximizing the energy functional with respect to the domain. The velocity method yields the continuity of both the functional and the state. The so-called shape differential equation provides large virtual evolution of the shape that increase the energy and may provide stationary domains.

1 Steady Norton Hoff Flows

Given a bounded domain Ω of class C^k, for some $k > 0$, in $\mathbb{R}^3$ with non-empty boundary Γ, we consider the following penalization of Norton Hoff problem for visco-plastic flows under mixed boundary conditions:

$$\mathcal{P}\left\{\begin{array}{rcl} -\operatorname{div}\left[K|\varepsilon(u)|^{p-2}\varepsilon(u)\right] - \nabla\left[|\operatorname{div} u|^{p-2}\operatorname{div} u\right] & = & f \text{ on } \Omega \\ u & = & 0 \text{ on } \Gamma_D \subset \Gamma \\ K|\varepsilon(u)|^{p-2}\varepsilon(u.n + |\operatorname{div} u|^{p-2}(\operatorname{div} u)n & = & g \text{ on } \Gamma_N = \Gamma \setminus \Gamma_D \end{array}\right.$$

The coefficients p and K are positive and characterize the material. We are interested in the case $1 < p \leqslant 2$ corresponding to non-Newtonian fluids, and assume that $\overset{\circ}{\Gamma}_D$ has non-negative a $\mathcal{H}^2$-measure. It is natural to introduce the functional space

$$\mathcal{W}(\Omega) = \left\{ v \in \mathrm{W}^{1,p}(\Omega, \mathbb{R}^3) \;\middle|\; v_{|\Gamma_D} = 0 \right\}$$

We endow each space with its natural norm, unless otherwise specified. If we require that f belongs to the dual space $\mathcal{W}'$ and g is in $\mathrm{L}^{p'}(\Gamma_N, \mathbb{R}^3)$,

---

[1] CMA, Ecole des Mines de Paris - INRIA - 2004 route des Lucioles - B.P. 93 - 06902 Sophia Antipolis Cedex - France

[2] also CNRS, Institut Non Linéaire de Nice - 1361 rte des Lucioles - 06504 Sophia Antipolis, France

minimizing the functional

$$\begin{array}{rcl} \Phi : \mathcal{W}(\Omega) & \to & \mathbb{R} \\ v & \mapsto & \displaystyle\int_\Omega \frac{K}{p}|\varepsilon(v)|^p + \frac{\eta}{p}|\operatorname{div} v|^p - f.v - \int_\Gamma g.v \end{array} \tag{1}$$

provides a variationnal formulation of $\mathcal{P}$. This functional is obviously continuous and strictly convex. Moreover, Korn's inequality provides it is also coercive. Accordingly, the functional Φ has a unique minimum over $\mathcal{W}$, which will be named the *weak solution* of $\mathcal{P}$ in Ω.

Notice that from the properties of Φ, we deduce it is directional differentiable, with directional derivative at the point $u \in \mathcal{W}$ in direction $v \in \mathcal{W}$

$$\Phi'(u;v) = \int_\Omega \sigma(u)..\varepsilon(v) + \rho(u)\operatorname{div} v - f.v - \int_\Gamma g.\gamma v \tag{2}$$

Since Φ is convex, the necessary minimality condition is also sufficient:

$$\Phi(u) = \min\left\{ \Phi(v) \;\middle|\; v \in \mathcal{W} \right\} \iff \forall v \in \mathcal{W},\ \Phi'(u;v) = 0 \tag{3}$$

If we assume u is regular enough, this optimality condition may be rewritten *via* Green's formulae and u is a strong solution of $\mathcal{P}$. The use of Green's formula is licit when u lays in $W^{2,p}(\Omega, \mathbb{R}^3)$, to ensure the existence of the boundary expressions. We expect this extra-regularity to derive from assumptions on the domains or the data, as it does for instance for Laplace operator. As far as we know, there are no such results. However, if we just suppose $u \in \mathcal{W}(\Omega)$ the boundary integral may be given a weak-sense as a duality product.

Our main purpose will be the shape control of the energy functional

$$E(\Omega) = \min_{v \in \mathcal{W}(\Omega)} \Phi(v) \tag{4}$$

which equals, up to a constant, the compliance. We aim at maximizing it, and we will do so *via* a gradient-like method.

As usual in shape analysis (see [3] for instance), we suppose a smooth bounded hold-all $\mathfrak{D} \subset \mathbb{R}^3$ is given. The domain Ω undergoes the one-parameter deformation of the global flow T of a given speed-field V in

$$\mathbb{V}_k = \left\{ V \in C^0\left(\mathbb{R}^+, C^k(\bar{\mathfrak{D}}, \mathbb{R}^3)\right) \;\middle|\; V.\nu = 0 \text{ on } \partial\mathfrak{D} \right\} \cap \mathrm{L}^\infty\left(\mathbb{R}^+, C^k(\bar{\mathfrak{D}}, \mathbb{R}^3)\right)$$

where ν denotes the unit outward normal of $\partial\mathfrak{D}$, and $\mathbb{R}^+$ is the interval $[0,\tau]$ with $0 < \tau \leqslant \infty$. If, for $s \in \mathbb{R}^+$, we denote

$$\Omega_s = T(s)(\Omega)\ ,\quad \Gamma_s = T(s)(\Gamma)\ ,\quad \Gamma_{D,s} = T(s)(\Gamma_D)\ ,\quad \Gamma_{N,s} = T(s)(\Gamma_N)$$

and we assume the deformation is such that a non-negative $\mathcal{H}^2$-measure open subset Γ_f of Γ_D remains stable under the deformations. This is to say that we choose restricted speed-fields:

$$V \in \mathcal{V}_k = \left\{ V \in \mathbb{V}_k \;\middle|\; V = 0 \text{ on } \Gamma_f \right\}$$

We can regard Norton-Hoff weak problem in Ω_s as soon as we transport the boundary conditions.

Assumption 1 *Assume we are given a domain functional g such that, for any Lipschitz domain $\Omega \subset \mathcal{D}$, $g(\Omega)$ belongs to $\mathrm{W}^{1,p'}(\partial\Omega, \mathbb{R}^3)$, g is continuously shape-differentiable and both $g(\Omega)$ and its shape-gradient $g'(\Omega; .)$ are uniformly bounded with respect to Ω.*

Assume $f \in \mathrm{W}^{1,\infty}(\mathcal{D}, \mathbb{R}^3)$.

So we can set the transported problem $\wp_s = \wp(\Omega_s)$. This problem has a unique weak-solution u_s in $\mathcal{W}(\Omega_s)$. Using classical techniques of shape analysis (see [3]), the following result is derived in [1].

Theorem 1 *Under Assumption 1, the weak solution of $\mathcal{P}$ is shape continuous and the energy functional is shape differentiable. Its shape-gradient $dE(\Omega; V)$ in direction $V \in \mathcal{V}_k$ is*

$$\begin{aligned}&\int_\Omega -K|\varepsilon(u)|^{p-2}\varepsilon(u)..(\mathrm{D}u.\mathrm{D}V_0) - \eta|\mathrm{div}\,u|^{p-2}\mathrm{div}\,u\; tr\;(\mathrm{D}u.\mathrm{D}V_0)\\ &\quad - \int_\Omega \mathrm{D}f.V_0.u + \left[\frac{K}{p}|\varepsilon(u)|^p + \frac{\eta}{p}|\mathrm{div}\,u|^p - fu\right] \mathrm{div}\,V_0 \\ &\quad - \int_\Gamma g'(\Omega; V).u + g(\Omega).u\,[\mathrm{div}\,V_0 - \langle \mathrm{D}V_0.n\;,\;n\rangle]\end{aligned} \tag{5}$$

where u is the weak solution of $\mathcal{P}$ and $V_0 = V(0)$.

2 Shape Differential Equation

The existence of the shape gradient of the energy suggests a gradient-like method to maximize it. Intuitively, we would like to find a family of domains that follows the gradient. In [4], J.P. Zolésio gave sense to differential equations as $\partial_t\Omega_t = f(\Omega_t)$, named *shape differential equations*, where the term $\partial_t\Omega_t$ is a germ of speed-fields. The technique hereafter follows this work.

Using semi-group properties of the flow mapping, we derive from equation 5 that the expression of the gradient $dE(\Omega_s; V)$ of the energy for any s in $\mathbb{R}^+$ is linear with respect to $V(s)$. Thus it defines a continuous linear form on $\mathcal{C}^k(\mathcal{D}, \mathbb{R}^3)$. We denote it $\mathcal{G}(\Omega_s)$. So, for any s in $\mathbb{R}^+$,

$$dE(\Omega_s; V) = \langle \mathcal{G}(\Omega_s)\;,\; V(s)\rangle_{\mathcal{C}^k(\mathcal{D},\mathbb{R}^3)' \times \mathcal{C}^k(\mathcal{D},\mathbb{R}^3)} \tag{6}$$

But the form $\mathcal{G}(\Omega_s)$ depends on the speed-field V because Ω_s does. So we can consider the shape differential equation

$$\forall s > 0, \quad V(s) - A^{-1}\mathcal{G}(\Omega_s(V)) = 0 \tag{7}$$

where A is a duality operator. Thanks to boundedness properties we prove in [1] that:

Theorem 2 *Equation 7 has a solution in* $\mathcal{V}_k$.

3 Asymptotic Properties

Since for any s in $\mathbb{R}^+$

$$E(\Omega_s) = E(\Omega) + \int_0^s \langle \mathcal{G}(\Omega_t)\ ,\ V(t)\rangle_{C^k(D,\mathbf{R}^3)' \times C^k(D,\mathbf{R}^3)}\, dt \tag{8}$$

if V is a solution of equation 7, the properties of duality operators (see [2] for instance) yield

$$E(\Omega_s) = E(\Omega) + \int_0^s \|V(t)\|^2_{C^k(D,\mathbf{R}^3)}\, dt \tag{9}$$

Accordingly, the field V will provide a sequence of domains which increase the energy. The following result is proved in [1].

Theorem 3 *There exists a sequence* (t_n) *such that* $\mathcal{G}(\Omega_{t_n})$ *tends to zero in* $\mathcal{H}'$. *If* $V \in \mathrm{L}^1(\mathbb{R}^+, \mathcal{H})$ *then* $\Omega_s(V)$ *tends to* Ω^* *in* $\mathcal{O}_k$, *with*

$$\lim_{s\to\infty} E(\Omega_s) = E(\Omega^*) \text{ and } \mathcal{G}(\Omega^*) = 0$$

References

[1] N. Gomez and J.P. Zolésio. Shape differential equation in norton hoff flows. Preprint INLN 97.41.

[2] J.L. Lions. *Quelques méthodes de résolution des problèmes aux limites non linéaires.* Dunod, 1972.

[3] J. Sokolowski and J.P. Zolésio. *Introduction to Shape Optimisation: Shape sensitivity analysis*, volume 16 of *Computational Mathematics*. Springer-Verlag, New York, Berlin, Heidelberg, 1992.

[4] J.P. Zolésio. Identification de Domaines par Déformations. These de Doctorat d'Etat, Nice, 1979.

Analysis of Contact Damage Around Scratches in Glass

T. N. Farris[1]
S. Chandrasekar[1]

ABSTRACT An analytical model of the stress field caused by sliding microindentation of brittle materials is developed. The complete stress field is treated as the superposition of applied normal and tangential forces with a sliding blister approximation of the localized inelastic deformation occurring just underneath the indenter. It is shown that lateral cracking is produced by the sliding blister stress field and that median cracking is caused by the applied contact forces. The model is combined with measurements of the material displacement around an indentation to show that the relative magnitude of tensile stresses governing lateral crack and median crack growth varies with the magnitude of the applied load. The model also predicts a range of loads at which the lateral crack will grow only after the indenter is removed from the surface. These predictions are consistent with observations of the different regimes of cracking observed under a sliding pyramidal indenter in soda-lime glass and other brittle solids.

Microscopic observations of machined ceramic surfaces show that brittle fracture contributes substantially to material removal in abrasive machining processes. A careful examination of the fractures in machined ceramics reveals striking similarities with fractures about quasi-static and sliding indentations produced by sharp indenters in brittle solids (Figures 1-2). It was, therefore, decided to develop a model for characterizing the stress fields produced by sliding, sharp indenters in brittle solids [1]. This model is an extension of that proposed by Yoffe [2] to describe the stress fields and fractures generated under a quasi-static indentation by a pointed indenter in brittle materials.

A large body of experimental observations has shown that in brittle materials, the volume of material displaced by the penetration of a sharp indenter is accommodated by compaction or shear deformation, or both, and this may occur either uniformly or irregularly. In silicate glasses, which have a relatively open structure, the displaced material is most readily accommodated by compaction within a zone underneath the indenter. However, in soda-lime glasses and in most ceramics shear deformation dominates with some compaction also taking place underneath the indenter. The ex-

[1]Schools of Engineering, Purdue University, West Lafayette, IN 47907-1282

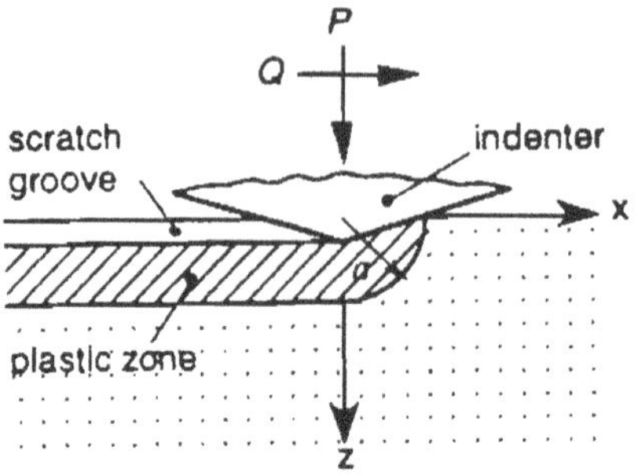

Figure 1. Schematic of indenter and inelastic zone. The inelastic zone is modeling using the sliding blister field

perimental model used by Yoffe to describe the deformation underneath a sharp conical indenter in soda-lime glass is based on experimental work that showed that the mean contact pressure under a sharp indenter remains constant and relatively independent of the indenter geometry. This model assumes that as a conical indenter is applied to the solid surface, yielding initially occurs in a hemi-spherical bowl under the indenter and continues until the yielded zone attains its stable preferred state for that pressure. As the indenter is loaded further, no additional flow occurs within this bowl, but fresh yielding takes place in an adjoining thin hemi-spherical shell. This process continues and "it is as if a set of nested hemi-spherical bowls separated by plastic filler were forced down in turn, one within the other, and become locked in the positions. [2]." The yielding process occurs through a combination of shear deformation and compaction within the yielded zone. Part of the material displaced by the indenter is accommodated by compaction within the yielded zone while the remaining material is pushed into the surroundings. At the end of the indentation process, the elastic half-space is left with an over-large hemi-spheroid fixed in a hemi-spherical cavity on its surface. This misfit leads to a residual stress in the region outside of the cavity. Consequently, as the material reacts to this residual stress, cracking may occur within the solid.

The "blister" field proposed by Yoffe [2] was for describing the development of cracks below a conical indenter acting on a brittle solid. The stress distribution in the surrounding elastic zone was obtained as the superposition of two stress fields—the Boussinesq field for the point force which idealized the indentation pressure distribution and a doublet force system which idealized the localized inelastic deformation. The doublet force system for the blister field consisted of superposition of a point center of expansion in an infinite elastic solid with an additional doublet to satisfy the free surface normal stress boundary condition. Such a doublet force system leads to a volume increase of the inelastic deformation zone

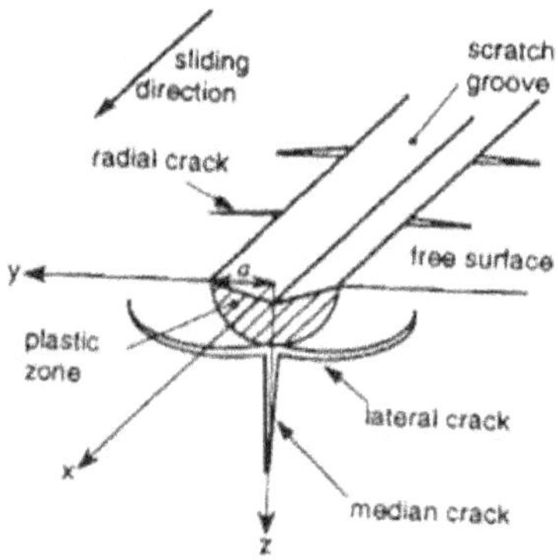

Figure 2. Schematic of cracking

Figure 3. Photograph of lateral and median cracks in glass A: extent of median crack; B: extent of lateral crack. The indenter load is 0.9 N

that must be taken up by compaction or elastic deformation of the remainder of the solid. The volume increase is a measure of the "strength" of the blister field. The strength of the blister field varies with the applied force; this variation is characterized by a material property which has thus far defied a precise estimation. Qualitatively, Yoffe's model has predicted the origin and growth of radial, median, and lateral crack systems quite well. In particular it is consistent with the observation that for a certain range of loads, the lateral crack forms and grows during unloading. It's only disadvantage is from a quantitative point of view as a procedure for determining the strength of the blister field for various indenter geometries is not yet available.

Models to describe the sliding indentation fracture process in ceramics have not yet evolved to a stage where the inelasticity under the indenter is accounted for completely. This is in large part due to the difficulty introduced by a lack of information concerning the constitutive behavior of ceramics in large hydrostatic compressive stress fields such as that existing underneath a microindentation. It seems that a reasonable first step for modeling the sliding indentation stress fields would be to extend Yoffe's work on static indentation to the sliding situation. This extension should include the load-history effects of inelastic deformation left behind by the sliding indenter and the elastic effects due to the frictional force.

We have developed an extension of Yoffe's blister field to model sliding microindentation stress fields called the sliding blister field [1]. The predictions of this theory have been compared to experimental observations of fractures around sliding indentations made through examination of the fractured specimens after testing. There is also brief discussion of recent in-situ, experimental observations of cracking [3] that are explained qualitatively by the sliding blister field model.

In-situ observations indicate that these cracks initiate at or close to the boundary between the inelastic deformation zone and the surrounding elastic solid (Figure 3). The subsequent growth of these cracks occurs in the elastic region. Bulsara [3] observes that for most loads at which the median and lateral cracks occur, the cracks propagate such that the crack front moves with the load. However, near loads of about 1N applied with a Vickers indenter in soda-lime glass the lateral crack does not appear until the indenter is lifted from the surface. Once the load is removed, the lateral crack forms beneath the end of the scratch and propagates rapidly along the entire length of the scratch and until the beginning of the scratch track where it stops.

The stress field produced by a sliding sharp indenter in brittle solids has been derived from an extension of Yoffe's model for static microindentation. Numerical results are illustrated and compared with experimental observations. The strength of the residual stress field has been evaluated as a function of the normal load has been discussed. It has been shown that the present analytical model provides reasonably good descriptions of the crack patterns observed around sliding indentations in brittle materials. In particular, the analysis accounts for the occurrence of lateral cracking under sliding indenters above an experimentally observed threshold load. Future work aims at inclusion of the blister field stresses in a fracture mechanics analysis of cracks observed during sliding microindentation.

The sliding blister field does not reveal any information about details of deformation in the inelastic zone near the indenter. However, the sliding blister field is a simple analytical tool that can be used to explain many interesting aspects of failure mechanisms in glass that occur near microindentations. Conclusions based on the sliding blister field are being used to develop polishing process models that predict surface finish.

Acknowledgment This work was supported in part by the National Science Foundation through grant CMS 97-14604,

[1] Y. Ahn, T. N. Farris, and S. Chandrasekar. Sliding microindentation fracture of brittle materials: Role of elastic stress fields. *Mechanics of Materials*, pages In–Press, 1998.

[2] E. H. Yoffe. Elastic Stress Fields Caused by Indenting Brittle Materials. *Philosophical Magazine A*, 46(4):617–628, 1982.

[3] V. H. Bulsara. *Scratch Formation in Brittle Solids and its Application to Polishing.* PhD thesis, Purdue University, 1997.

Remodeling Around Pressurized Holes: Prediction of Adaptive Elasticity

K. Firoozbakhsh[1]

ABSTRACT The problem of evolution of material inhomogeneity around pressurized holes in an adaptive elastic transversely isotropic homogeneous plate is considered. The internal remodeling theory of Cowin and Hegedus is modified and two cases of uniformly pressurized elliptical hole (vicinity of a force fitted rigid pin) and circular hole with localized pressure (vicinity of a loose rigid pin) are studied.

1. Introduction

To describe the functional adaptation behavior of certain smart materials the theory of adaptive elasticity -known as remodeling theories- was developed [1,2]. Remodeling theories have been used to explain the process of growth, reinforcement, and resorption in smart materials such as bones [2,3,5]. Adaptive elasticity theory predicts evolution of material inhomogeneity around the hole defects, which changes the classical values of stress concentration. This deviation from the classical theory of elasticity is important and should be observed in the formulation of such design problems. In this paper the behavior of an adaptive elastic material around two practically important pressurized holes is investigated.

2. Problem Formulation

The problem considered is an initially homogeneous, stress free, adaptive elastic transversely isotropic large plate which will be stressed at time t=0 by force fitting a rigid pin. Following [2] and [3] the material properties are assumed as

$$K_{ij}(e) = K_{ij}^{*}\,\kappa(e) \qquad (1)$$

[1] Department of Orthopaedic and Rehabilitation, Medical Center, The University of New Mexico, Albuquerque, NM 87131, USA

Where $K_{ij}(e)$ are the material coefficients after remodeling, $K^{\bullet}_{ij}$ are the initial material coefficients at equilibrium state, and their dependence upon the change in volume fraction e is given by $\kappa(e)$. Based on (1) it will be easy to show that for a transversely isotropic adaptive elastic material the constitutive relations will be simplified as

$$K_{11}(e) = K_{22}(e) = \frac{1}{E}\kappa(e) \quad ; \quad K_{66}(e) = \frac{2(1+\nu)}{E}\kappa(e) \tag{2a}$$

$$K_{16}(e) = K_{26}(e) = K_{36}(e) = 0 \tag{2b}$$

$$K_{12}(e) = K_{21}(e) = K_{13}(e) = K_{31}(e) = -\frac{\nu}{E}\kappa(e) \tag{2c}$$

Furthermore, the compatibility equations for the case of plane stress in terms of stress function U reads

$$\begin{aligned} &K_{22}(e)U_{,1111} - 2K_{16}(e)U_{,1112} + [2K_{12}(e) + K_{66}(e)]U_{,1122} \\ &- 2K_{26}(e)U_{,1222} + K_{11}(e)U_{,2222} = 0 \end{aligned} \tag{3}$$

Considering equation (2) this equation reduces to biharmonic equation for stress

$$U_{,1111} + U_{,1122} + U_{,2222} = 0 \tag{4}$$

2.1 *Case 1-Uniformly pressurized elliptical hole*

The classical solution may be written as [4]

$$trT^{\bullet}_{ij} = T^{\bullet}_{11} + T^{\bullet}_{22} = \frac{-4pmR}{K(m - \frac{1}{\xi^2})} \tag{5}$$

Where p is the uniform pressure around the hole, m=(a-b)/(a+b), and ξ <1 is referred to interior of unit circle so that we have

$$Z = K(\frac{1}{\xi} + m\xi) \; ; \; \xi = \rho e^{i\theta} \; ; \; Z = x + iy = r(a\cos\theta + ib\sin\theta) \tag{6}$$

And ρ will be calculated from

$$\rho = \frac{r \pm (r^2 - 1 + \frac{b^2}{a^2})}{1 \pm \frac{b}{a}} \tag{7}$$

Following the remodeling theory [1,2], the remodeling rate equation is simplified to read

$$e(x,t) = C_2 e^2(x,t) + C_1 e(x,t) + C_0 + A[1+e(x,t)]trE^{\bullet}_{ij}(x,t) =$$
$$\alpha[e^2(x,t) - 2b\,e(x,t) + c] = 0 \tag{8}$$

And $E^{\bullet}_{ij}(x,t)$ will be calculated from

$$tr\,E^{\bullet}_{ij}(x,t) = \frac{1}{E}\chi(e)\,tr\,T^{\bullet}_{ij}(x) - 2\frac{\vartheta}{E}tr\,T^{\bullet}_{ij}(x) \tag{9}$$

Knowing the tr $T^{\bullet}_{ij}(x)$, equation (8) leads to a nonlinear Ricatti differential equation whose solution will be in the form of

$$e_1, e_2 = b \pm \sqrt{b^2 - c} \tag{10}$$

And the remodeling around the hole will be calculated from

$$e(x,t) = \frac{e_1[e_0(x) + e_2[e_1 - e_0(x)]\exp[(sign\,\alpha)\frac{t}{T}]}{[e_0(x) - e_2] + [e_1 - e_0(x)]\exp[(sign\,\alpha)\frac{t}{T}]} \tag{11}$$

Where $T=1/[\alpha(e_1-e_2)]$ is the time constant.

2.2 Case 2- Locally pressurizes circular hole

For the case where the pin is loose and therefore the applied pressure is localized the loading boundary condition will be at r= R, p= X1+ iX2 where X1 and X2, respectively, represent the vertical and horizontal components of the localized pressure. Once again from the classical theory of elasticity we may write [4]

$$trT^{\bullet}_{ij} = T^{\bullet}_{11} + T^{\bullet}_{22} = 4\Re[\varphi(z)]\ ;\ \varphi(z) = \frac{-X_1 + iX_2}{8\pi(1-\nu)}.\frac{1}{z}\ ;\quad z \subset \Re \tag{12}$$

Where $\Re$ represents the real part and z belongs to the outside of pin region.

3. Numerical values

For numerical results we used the material coefficients given previously [3,5]:
$E=18.4^{Gpa}$; $\vartheta = 0.302$; $C_0 = 1.5X10^{-8}\ S^{-1}$; $C_1 = -1.5X10^{-7}\ S^{-1}$
$C_2 = 2.5X10^{-7}\ S^{-1}$; $A = -2.5X10^{-5}\ S^{-1}$: $\chi(e)=1-11[e-(e_2)_0]$
And the remodeling was calculated for the different pressures. For a uniformly pressurized circular hole results indicated that the theory of Cowin et al fails to predict any evolution around the hole. In the Cowin's theory the rate of remodeling will depend on the trace of strain tensor and, therefore, for cases in which this trace becomes zero remodeling will vanish. This seems to be a non-valid assumption and is suggested to be modified as follows. We propose that

the remodeling is dependent on the absolute trace value defined as the sum of absolute values of principal strains and redefine tr $T^{\bullet}_{ij}(x)$ in the form

$$trT^{\bullet}_{ij}(x) = |T^{\bullet}_{11}| + |T^{\bullet}_{22}| \qquad (13)$$

Based on elasticity theory the stresses outside the pressurized circular hole are

$$T_{rr} = T^{\bullet}_{11} = -p\frac{R^2}{r^2} \quad ; \quad T_{\theta\theta} = T^{\bullet}_{22} = p\frac{R^2}{r^2}$$

$$|T^{\bullet}_{11}| = |T^{\bullet}_{22}| = p\frac{R^2}{r^2} \qquad (14)$$

Knowing the stresses from (14), tr $T^{\bullet}_{ij}(x)$ will be known from (13) and hence coefficients α, b, c will be calculated and from the Ricatti equation the remodeling will be determined.

4. Results and Discussion

Our results indicate that the internal remodeling theory of Cowin is unable to predict any micro-structural evolution around the uniformly pressurized circular hole and accordingly the initially assumed density distribution remains unchanged. Based on the modified theory, however, there will be a distribution of material inhomogeneity around the hole. Our results indicated that the value of e raises from 0.1267 to 0.168 at the boundary R=a and decreases gradually as a function of ratio R/a, where a is the pin radius and R is the distance from the pin center. Based on the proposed theory the local compressive stress causes the adaptive material to evolve to one of the greater density, which in turn lowers the amount of stress concentration.

5. References

[1] Cowin, SC. and Hegedus, DH. Theory of Adaptive Elasticity J Elasticity 6; pp: 313-326, 1976.

[2] Cowin, SC. and Van Buskirk, WC. Internal Bone Remodeling Induced by a Medullary Pin, J Biomech. 11; pp:269-275, 1978.

[3] Firoozbakhsh, KK. and Cowin, SC. Devolution of Inhomogeneities in Bone Structure: Prediction of Adaptive Elasticity, ASME J Biomech Eng 10; pp: 287-293, 1980.

[4] Lekhnitski, SG. Theory of Elasticity of an Anisotropic Body, translated from Russian, Mir Publisher, Moscow, pp:67-72, 1981.

[5] Van Buskirk, WC. and Cowin, SC. and Ward, N. Ultrasonic Measurement of Orthotropic Elastic Constants of Bovine Femoral Bone, ASME J Biomech Eng 103; pp: 67-71, 1981

Evaluation of Fretting Stresses Through Full-Field Temperature Measurements

M. P. Szolwinski[1]
G. Harish[1]
T. N. Farris[1]
T. Sakagami[2]

ABSTRACT The near-surface stress field in fretting has long eluded experimental characterization due in large part to the fact that the friction coefficient in the slip zones associated with partial slip contacts cannot be evaluated from measured forces. However, newly available infrared technology is capable of resolving finely, both spatially and temporally, subsurface temperatures near the fretting contact. These temperatures are induced by both frictional heating due to the microslip and the coupled thermoelastic effect arising from strains in the material. This paper will present experimental results from an investigation into fretting of aluminum/aluminum contacts using this thermal imaging technique. Subsequent discussion focuses on the use of these data to validate a finite element model of the fretting contact.

A typical fretting fatigue test apparatus applies an oscillatory tangential load (ΔQ) to an indenter or fretting pad pressed into contact with the surface of a specimen by a normal force P. The tangential loading is usually either applied by a dedicated actuator or induced through the deformation of a fretting fixture or chassis attached to the pads by an in-phase cyclic bulk stress applied to the specimen. For complete details of the fretting test fixture and thermal imaging system, including specimen dimensions, contact surface preparation and a description of the data acquisition system, see other work by the authors.[1] (Note that a complete set of references for material presented subsequently can found there.)

A calibration between the intensity measured by the thermal imaging system and temperature was achieved by placing a thin aluminum plate outfitted with a thermocouple in front of the camera at the same focal distance used in the fretting experiments. A calibration curve between measured intensity and temperature was obtained by varying the plate temperature and matching thermal images with thermocouple readings. With calibration completed, an effort was made to detect (successfully) the thermoelastic effect arising from simple cyclic uniaxial tension and compression

[1] School of Aeronautics & Astronautics, Purdue University, West Lafayette, IN 47907
[2] Department of Mechanical Engineering & Systems, Graduate School of Engineering, Osaka University, 2-1, Yamadaoka, Suita, Osaka 565-0871, Japan

of the fatigue specimen and compare the resulting temperature changes predicted by thermoelastic theory, as shown in Figure 1.

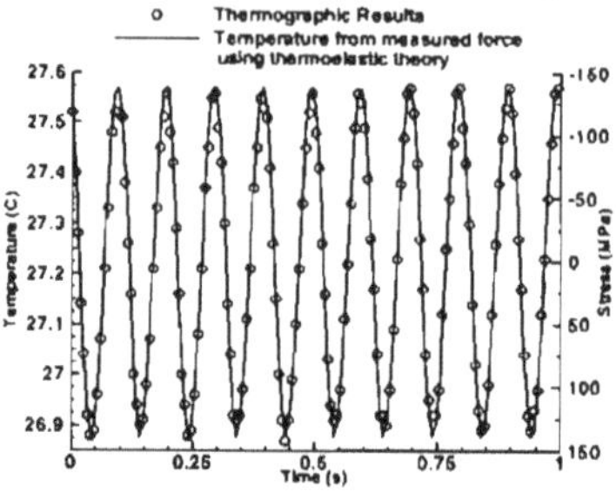

Figure 1. Comparison between measured temperatures during a uniaxial, fully-reversed cyclic loading of the fatigue specimen compared with those predicted by thermoelastic theory.

With a calibration between temperature and intensity determined and confidence in the ability of the system to measure accurately the small temperature changes induced by the thermoelastic effect established, a fretting experiment was conducted at a frequency of 2 Hz under fully-reversed (R − ratio $= L_{min}/L_{max} = -1$). Estimates of both the contact zone halfwidth, a and stick zone halfwidth, c, were be made by exercising available expressions for the mechanics of fretting contacts with applied loads and a value for coefficient of friction in the slip zones, μ. The evolution of μ to a steady-state value of 0.65 has been investigated in a separate series of experiments.

Figure 2 displays a sequence of temperature images from a 10.7 mm × 10.7 mm area around the right pad/specimen interface early in the fretting experiment (cycle 16). The specimen is to the left of the solid line with the pad to the right. The bright line (a mechanically-etched reference mark) on the pad projects through the nominal center of contact, with the Hertz contact length extending above and below this line. Focusing on the specimen portion of the images (left side) illustrates clearly the thermoelastic temperature changes induced by the cyclic bulk loading of the specimen. As the compressive bulk load on the specimen increases (top row of images in each block, progressing left to right), so does the bulk specimen temperature.

Conversely, decreases in the bulk specimen temperature are associated with the tensile portion of the bulk waveform (lower row of both sequences). This temperature cycle occurs at the same frequency as that of the applied bulk load waveform. Note that the thermoelastic effect is more pronounced below the centerline of the contact. This occurs since the bulk load magnitude at the bottom of the specimen is greater than at the top, as part of the applied bulk load is taken up from the portion of the specimen above the contact by the tangential loads on the pads.

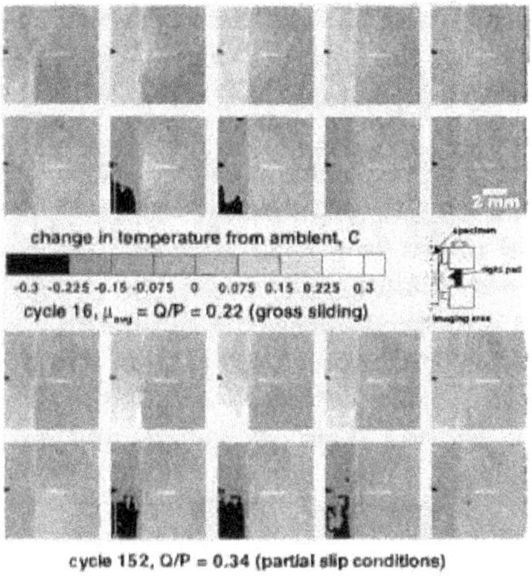

Figure 2. A sequence of thermal images taken at cycles 16 and 152 in the experiment. The upper sequence was imaged during gross sliding conditions, while the lower sequence was generated during conditions of partial slip.

The absence of the thermoelastic effects due to a cyclic bulk load in the fretting pad elucidates a second critical feature of the first block of images: a localized near-surface patch of heating spanning the width of the apparent area of contact. The frequency of this heating—twice the frequency of the applied bulk load waveform—provides insight into its source: frictional heating from gross sliding at the contact interface. Preliminary efforts to characterize the evolution of interfacial friction under the influence of fretting have revealed a sharp rise in friction coefficient over the first 50 cycles of similar fretting tests, correspondingly roughly to the transition from sliding to the establishment of partial slip conditions.

The lower sequence in Figure 2 presents a series of thermographic images acquired 152 cycles into the fretting test. While a similar temperature range is noted in the sequence of images, the nature of the field is significantly different. Absent is the patch of near-surface heating across the entire contact area, evidence of the transition from gross sliding to partial slip conditions at the interface. Indeed no distinct frictional effect is observed in the images, in spite of the microslip regions in the contact area.

Instead, the thermoelastic effect resulting from the cyclic contact stresses dominates the temperature field, providing excellent insight into the nature of the contact stress field arising from the fretting shear tractions. Of particular interest is the zone of tensile stress acting parallel to the contact surface, demarcated by a localized decrease in temperature at the trailing edge of contact (lower edge) in the specimen during the tensile portion of the applied waveform (second block, lower row of images). The surface tangential stress distribution as determined both analytically from Westergaard functions and an approximate approach to include the bulk stress effect; and numerically with a finite element approach using friction elements resolve accurately the peak in tangential stress at the trailing edge of contact. Also note that the tangential stress in the pad at this location is compressive.

Researchers have observed a characteristic fatigue-cracking behavior at this trailing edge location in an exhaustive series of fretting fatigue tests. These experimental results have been correlated with predictions for fretting fatigue crack nucleation based on using this tangential stress peak in conjunction with a multiaxial fatigue life parameter.

Using the boundary conditions inferred from friction tests, a commercial finite element code was modified to include the coupled-thermoelastic effect. Comparisons between these calculations and the measurements are shown in Figure 3.

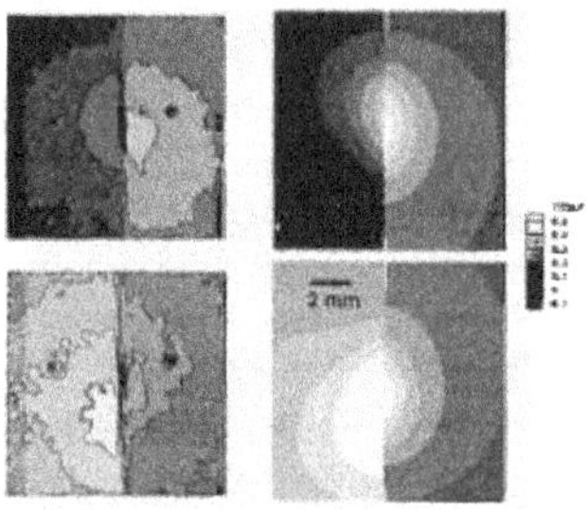

Figure 3. Comparison of FEM and experiment for a 10 Hz loading at cycle 7. For the case considered, the bulk stress amplitude, $\sigma_o = 100$ MPa, maximum Hertzian pressure, $p_o = 180$ MPa, and $\mu_{avg} = Q/P = 0.25$. (Temperatures contours in C)

Thus the successful comparison between the finite element results and experimental thermal images validates both the finite element model and the ancillary experiments that have estimated the evolution of friction coefficient under the influence of fretting. Furthermore, the finite element calculations also offer validated values for the sharp gradients in near-surface contact stresses and strains, and surface microslip amplitudes responsible for wear and nucleation of fatigue damage at the surface. This technique is being exercised currently on titanium contacts representative of those in turbomachinery components. With the low thermal conductivity of titanium comes the potential for higher temperatures generated during conditions of gross sliding and related degradation of near-surface material properties.

Acknowledgements: This research was supported in part by AFOSR through contract #F49620-93-1-0377 and a National Defense Science and Engineering Graduate (NDSEG) fellowship for M. P. Szolwinski.

[1] M. P. Szolwinski, G. Harish, T. N. Farris, and T. Sakagami. In-situ measurement of near-surface fretting contact temperatures in an aluminum alloy. *Journal of Tribology*, (in press), 1998.

Non-Gaussian Equivalent Linearization for Nonstationary Random Vibration of a 2-DOF Hysteretic System

K. Kimura[1]
S. Tamura[2]

ABSTRACT In the previous paper an approximate analytical method was developed for calculating the nonstationary responses of a single DOF system with bilinear hysteresis subjected to amplitude modulated nonwhite random excitation. The method consists of the non-Gaussian equivalent linearization technique and the use of the moment equations approach. In the present paper the method is extended and applied to random vibration analysis of a two DOF hysteretic system. The mean square responses and the mean dissipated hysteretic energy are obtained and compared with the corresponding digital simulation results.

1 Introduction

The study of the dynamic behavior of a hysteretic system subjected to random excitation is of great importance in safety analysis and reliability assessment of actual structures. The response distribution of the system with strong nonlinearity such as hysteresis exhibits the non-Gaussian property.

The objective of the present paper is to calculate the stochastic responses of a two DOF system with bilinear hysteresis subjected to nonwhite random excitation by employing the previously proposed method [1].

[1]Department of Mechanical and Environmental Informatics, Tokyo Institute of Technology, Ookayama, Meguro-ku, Tokyo 152, JAPAN

[2]Department of Mechanical Engineering Science, Tokyo Institute of Technology

2 Problem Statement

Consider a two DOF hysteretic system as shown in Figures 1 and 2. The equation of motion is written in terms of state variables u_1–u_6 (u_1, u_4: relative displacement, u_2, u_5: relative velocity, u_3, u_6: additional variables for hysteresis). The random forcing function is assumed to be in the form $f(t) = e(t)n(t)$, where $e(t)$ is a deterministic envelope function and $n(t)$ is a Gaussian stationary nonwhite noise with zero mean.

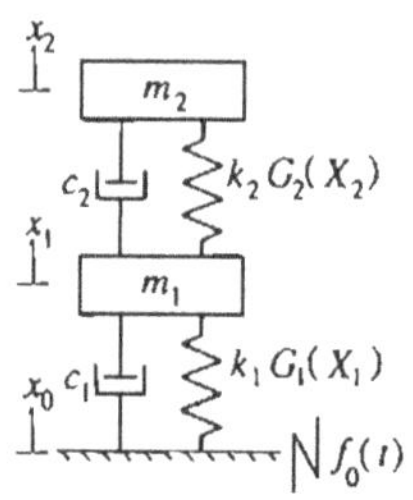

Figure 1: 2–DOF hysteretic system

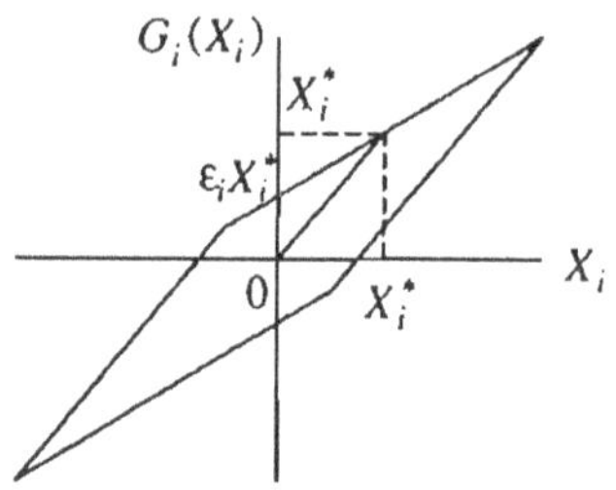

Figure 2: Bilinear hysteretic characteristic

3 Response Analysis

The approximate solution of the system can be obtained from the equivalent linearized equation. The probability density function(pdf) of the responses is necessary for determining the equivalent linear coefficients. Since the response distribution of the system with strong nonlinearity is widely different from the Gaussian distribution, the non-Gaussian pdf of the responses has to be introduced. Figures 3 and 4 respectively shows the non-Gaussian pdf of u_3 and joint pdf of u_2 and u_3, which consists a truncated Gaussian probability density function and delta functions.

The second order moment equations are derived from the equivalent linearized equation. The nonstationary mean square relative displacement and velocity responses are obtained by solving the moment equations numerically [2]. The mean square absolute acceleration responses and the mean dissipated hysteretic energy responses can be also obtained by using the solutions of the moment equations.

4 Illustrative Examples

Noise with an exponentially decaying harmonic correlation function is considered as a typical example of nonwhite noise. A unit step function is used

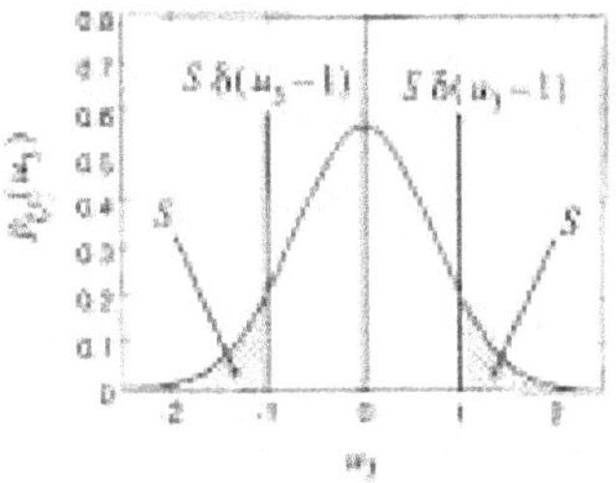

Figure 3. Probability density function of u_3

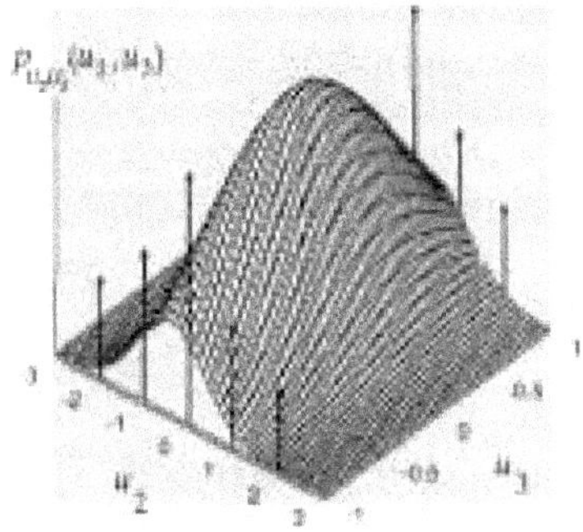

Figure 4: Joint prob. density function of u_2 and u_3

as an envelope function.

Figures 5-7 respectively show the transient rms relative displacement and absolute acceleration responses and also the transient mean dissipated energy response, where the dominant frequency of the random excitation coincides with the first coupled natural frequency of the system. The solid lines and diamonds represent the results obtained by the present method and digital simulation, respectively. A good agreement is found between both results.

5 References

[1] Kimura, K., Yasumuro, H. and Sakata, M., Non-Gaussian equivalent Linearization for Nonstationary Random Vibration of Hysteretic System, *Probabilistic Engineering Mechanics*, 9, 15-22, 1994

[2] Kimura, K, Yagasaki, K. and Sakata, M., Non-stationary Responses of a System with Bilinear Hysteresis Subjected to Non-white Random Excitation, *Journal of Sound and Vibration*, 91, 181-194, 1984

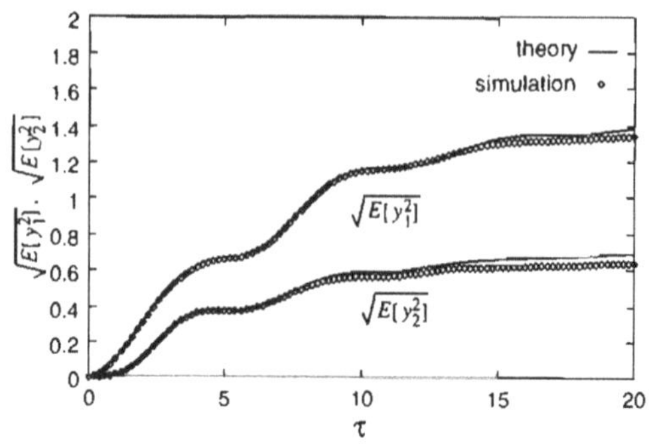

Figure 5: Transient rms relative displacement response

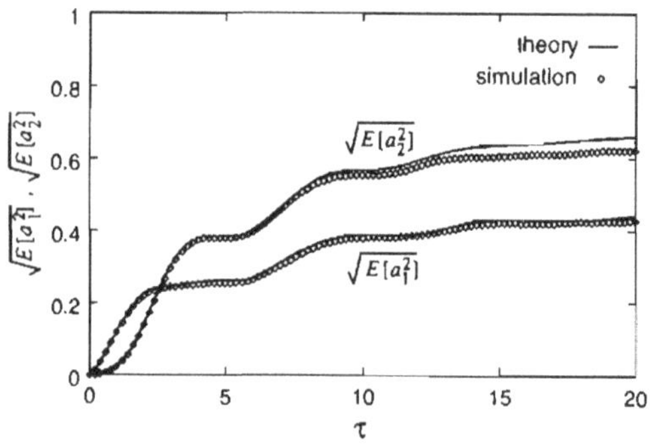

Figure 6: Transient rms absolute acceleration response

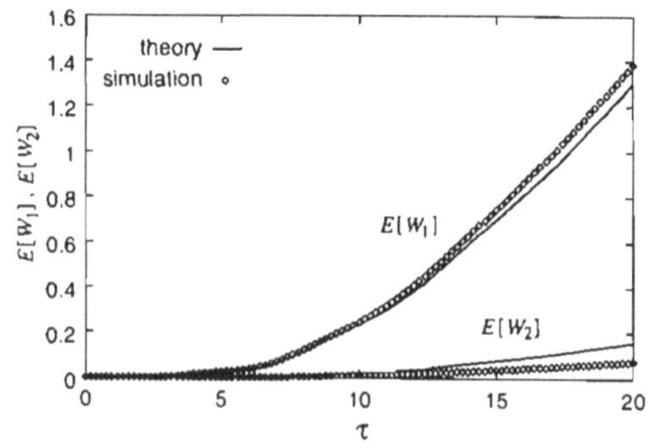

Figure 7: Transient mean dissipated energy response

"Soft" Reliability Assessment of Existing Transmission Lines

Ibrahim Hathout[1]

ABSTRACT This paper introduces a part of a system progressively developed by the author since 1990 to assess the damage state, reliability, and remaining life of existing transmission lines. The model presented here deals with the reliability evaluations in the presence of structural deterioration. The probability of failure is derived as a function of the cumulative distribution of the standardized safety margin. Cognitive uncertainty inherent in the visual inspection and assessment of old structures, is included using a fuzzy strength multiplier function. This fuzzy function is obtained by combining the condition ratings of the major components in the line using a fuzzy weighted averages technique.

1 Introduction

It is generally recognized that safety of a transmission line must be statistically evaluated, and reliability or probability of failure has been proposed as a "reasonable" criterion for transmission line safety. Existing transmission lines are designed using experience based deterministic methods with safety factors. The most fundamental problem for design engineers assessing the safety of existing transmission lines is relating the design safety factor to the failure probability in the presence of structural deterioration. The safety factor, S_F, is defined as:

$$S_F = R_{min}/Q_{max} = (\Phi\ \mu_R)/(\gamma\ \mu_Q) = (\Phi/\gamma)\ S_{FC} \qquad (1)$$

where R_{min}, Q_{max} are the resistance and load in the deterministic design, μ_R, μ_Q are the mean values of R and Q, Φ and γ are constant design parameters, and S_{FC}

[1] Ontario Hydro-H6 G26, 700 University Ave, Toronto, Ontario, Canada, M5G 1X6, E-mail: i.hathout@hydro.on.ca

is the central factor of safety. If R and Q are independent random variables, then failure probability, P_f, is given by:

$$P_f = P[Z < 0] = P[\frac{Z - \mu_Z}{\sigma_Z} < -\frac{\mu_Z}{\sigma_Z}] = F_{ZS}(-\frac{1}{V_Z}) \quad (2)$$

where F_{ZS} is the cumulative distribution function (cdf) of the standardized safety margin $Z_S = (Z-\mu_Z)/\sigma_Z$, and V_Z is the coefficient of variation of the safety margin.

The quantitative probability of failure defined by equation 2, states that "failure occurs when the line loads, transgress its tolerance limits." This conventional definition of failure assumes "perfect" knowledge of the relevant line loads and the associated tolerance limits of the line structures and makes no distinction between "sudden" and "drift" failures. Failures of existing transmission structures usually occur by slow drift (fuzzy or soft failure) of some relevant performance (i.e. gradual increase of corrosion, deformation, differential settlement, deflection etc.). The ordinary meaning of failure is qualitative; it implies that the line becomes unfit to accomplish a specific function, which is power transfer. The notion of failure in reliability theory differs from the ordinary meaning by the rigor of its definition, which lends it a somewhat artificial flavor. The artificiality of the reliability model is common to all mathematical models because they are based on a quantitative definition of a qualitative concept. The failure in this sense is fuzzy or soft and not crisp or hard. In this study, the conventional (crisp or hard) probability of failure, Eq. 2, is fuzzified (softened) by introducing a fuzzy strength condition function to account for cognitive type uncertainty inherent in the visual inspection and damage assessment process.

2 Fuzzification of Failure Probability

To accommodate cognitive type uncertainty inherent in the visual inspection and assessment of existing transmission lines, a fuzzy condition rating function "C" is introduced such that $P_f = P[C.R < Q]$. The condition rating of a transmission tower is obtained using the fuzzy logic expert system "FESTA" [2]. FESTA uses a fuzzy weighted averages technique and its knowledge base to evaluate the condition rating and maps it to one of the five standards rating functions of Fig.(1) using the Euclidean distance technique [2].

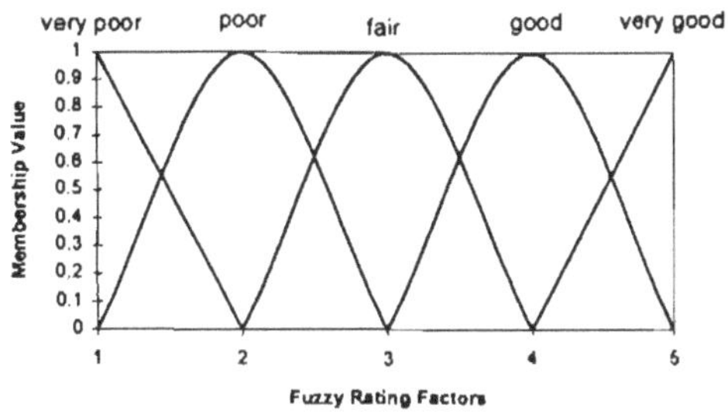

Figure 1: Fuzzy Condition Rating Functions

Once the condition rating of each structure in a transmission line is obtained, the overall rating of this line is calculated by combining the ratings of all structures or components using the following fuzzy weighted averages formula:

$$C = \frac{\sum_{1}^{n} W_i c_i}{\sum_{1}^{n} W_i} \tag{3}$$

where c_i is the overall rating of the i-th tower and W_i is a fuzzy weight factor reflecting the site conditions (wind direction, proximity of pollution sources, public safety etc.). The solution of Eq. 3 is based on a fuzzy computational technique called resolution identity of fuzzy sets. The procedure is based on the decomposition of the fuzzy sets (Wi and c_i) into non-fuzzy level-sets by slicing the fuzzy sets at different membership levels and then combining the lower boundaries and upper boundaries at these levels. The fuzzy weight factors W_i may range from "small" to "large" and should be assigned by a senior engineer.

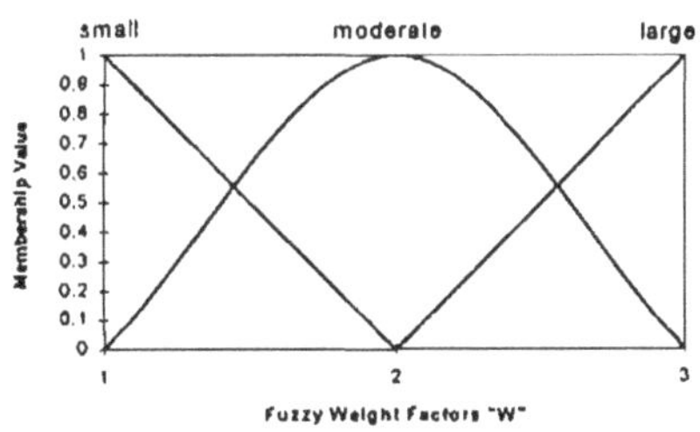

Figure 2. Fuzzy Weigh Functions, W_i

If we assume that C, R, and Q are independent random variables, then the first four moments μ_{z1}, μ_{z2}, μ_{z3}, and μ_{z4} of the safety margin Z=CR-Q can be calculated. Using equation 1, the relationship between P_f and S_{FC} may be obtained by re-writing equation 2 as follows:

$$P_f = F_{zs}\left(-\frac{\mu_{c1} - S_{FC}}{\sqrt{\mu_{c1}^2 V_R^2 + \mu_{c2} + V_Q^2 S_{FC}^2}}\right) \tag{4}$$

Since equation (4) has no exact solution, an approximate solution using Edgeworth series is proposed. The procedure is to expand Edgeworth's generalized law using Taylor's series; expressing the expansion in term of the Tchebycheff-Hermite polynomial; using the orthognality characteristics of the Tchebycheff-Hermite polynomial; replacing the cumulants with the moments of the distribution; and then integrating to determine the cumulative distribution function (cdf). It can be shown that the probability of failure expressed in term of β_1 "skewness" & β_2 "kurtosis" (for standard normal $\mu_2=1$; $\mu_3{}^2=\beta_1$ "skewness"; $\mu_4=\beta_2$ "kurtosis") is given by:

$$P_f = \left(1 - \frac{\sqrt{\beta_{Z1}}}{3!} H_2\left(-\frac{1}{V_Z}\right) - \frac{\beta_{Z2}}{4!} H_3\left(-\frac{1}{V_Z}\right) - 10\frac{\beta_{Z1}}{6!} H_5\left(-\frac{1}{V_Z}\right)\right) \Phi\left(-\frac{1}{V_Z}\right) \tag{5}$$

where β_{z1} and β_{z2} are given by,

$$\sqrt{\beta_{Z1}} = \frac{\sqrt{\beta_{R1}}\left(\frac{1}{V_C S_{FC}}\right)^3 \left(\frac{V_r}{V_Q}\right)^3 + \sqrt{\beta_{C1}}\left(\frac{1}{V_Q S_{FC}}\right)^3 - \sqrt{\beta_{Q1}}\left(\frac{1}{\sqrt{\mu_{C2}}}\right)^3 + 6\left(\frac{V_R}{V_Q S_{FC}}\right)^3 \left(\frac{1}{V_R V_C}\right)}{\left(\frac{1}{\mu_{C2}} + \left(\frac{1}{S_{FC} V_Q}\right)^2 + \left(\frac{V_R}{V_Q}\right)^2 \left(\frac{1}{V_C S_{FC}}\right)^2\right)^{3/2}}$$

$$\beta_{Z2} = \frac{\beta_{R2}\left(\frac{V_R}{V_Q V_C S_{FC}}\right)^4 + \beta_{Q2}\left(\frac{1}{\mu_{C2}}\right)^2 + \beta_{C2}\left(\frac{1}{V_Q S_{FC}}\right)^4 + 6\left(\left(\frac{1}{\mu_{C2}^{1/2} V_Q S_{FC}}\right)^2 + \left(\frac{V_R}{V_C \mu_C^{1/2} V_Q S_{FC}}\right)^2 + \left(\frac{1}{V_C V_R}\right)^2 \left(\frac{V_R}{V_Q S_{FC}}\right)^4\right)}{\left(\frac{1}{\mu_{C2}} + \left(\frac{1}{V_Q S_{FC}}\right)^2 + \left(\frac{V_R}{V_Q V_C S_{FC}}\right)^2\right)^2}$$

A computer program "SRA" has been written to relate safety factors of existing transmission lines to failure probabilities in the presence of structural deterioration. The program is being tested on a number of existing transmission lines in southern Ontario.

3 Numerical Example: Reliability of a Double Circuit, 115 kV Transmission Line

The reliability of a double circuit, 115 kV line built in Toronto some 70 years ago was analyzed. The line was built using a deterministic method with a factor of safety of about 1.5. Visual inspection has indicated that some sections of the line have suffered major symptoms of failure. The worst section is 2.2 km, spans between towers 6 and 18. For simplicity, reliability assessment is limited to that 2.2 km section.

Symptoms of failures: Visual inspection has revealed that corrosion and deformation are the predominant symptoms of failures. The towers were found to be suffering advanced corrosion state (that is predominantly dark red). Most of the bracing members were, visually, rated "poor" and the leg members are rated "poor-fair." Deformation of the diagonal braces in the lower panels is found to be a chronic problem in this section. One of the legs of tower 16 suffered large deformation (rated "poor") and another leg on tower 11 also suffered lesser deformation (rated "fair"). Figure 3, shows a magnified cross-section of a member suffering from pitting corrosion and a nut suffering corrosion crack.

The condition rating of each tower was obtained with the aid of the fuzzy logic expert system [2]. The condition rating of the entire line section (between towers 6 and 18) was then obtained using equation 3. The resultant condition rating of the line section is mapped to the standard condition ratings of Fig.(1). The overall rating of the line section is found to be about "poor" with the following characteristics:

$$\mu_{C1} = 0.4505, \qquad \mu_{C2} = 0.1400, \qquad \beta_{C1} = 1.1327, \qquad \beta_{C2} = 1.1744$$

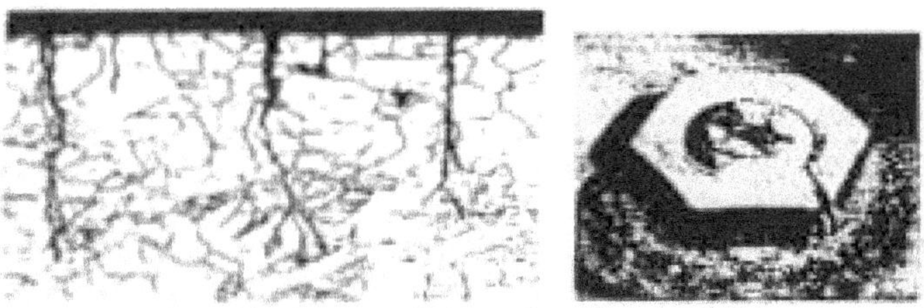

Figure 3 Corrosion Cracks of an Angles and Nut

3.1 Reliability

Wind Load: The nearest weather station to the line is that at the Toronto International Airport. The recorded wind speeds at the Airport were converted to a wind speed at line location using the following transfer function:

$T_f = (C_h/1000)^{\alpha 1}(1000/33)^{\alpha 2}$, where C_h is the conductor height, α_1, α_2 are the terrain factors (α_1=0.21; residential suburbs), and (α=0.14; airport open flat terrain).

Resistance of the structures: Detailed structural analysis was carried out to determine the capacities of the towers (6 to 18) in term of wind speed. The wind resistance of each tower was calculated from actual wind spans assuming as "new" condition.

Results: Using the wind data and wind resistance of the towers, the computer program "SRA" is used to calculate the probability of failure of the line section for two cases: 1. Using actual condition rating "about poor", and 2. As "new" condition.

The results may be summarized as follows:

Case 1: P_f (actual condition) = 0.103866
Case 2: P_f (as new condition) = 0.001144

The result has indicated that the introduction of the damage state of the line has increased the probability of failure order of magnitude. The result has also indicated that ignoring the damage state of the line will yield unrealistic results.

4 Conclusions

Conventional (hard or crisp) reliability is not practical when dealing with existing structures where damage assessment is based on visually inspection. The fuzzification of failure probability is essential for realistic assessment of line reliability. The cognitive uncertainty may be minimized if a "reasonably accurate" condition assessment is made. Computerizing this method will to enable transmission engineers with little or no background in reliability and fuzzy set theories to assess the safety of existing transmission lines. Testing the methodology and the associated software on selected transmission lines, is underway.

5 References

1. Murotsu, Y. Nakayasu, H., Mori, K. and Kase, S. "New Method of Relating Safety Factor to Failure Probability in Structural Design", Advances in Reliability and Stress Analysis, ASME Annual Meeting, San Francisco, CA, December 1978.

Soil-Foundation Interaction Literature Review

H. R. Hamidzadeh[1]
M. Dehghani[2]

ABSTRACT Research on dynamics of soils and foundations has yielded several fundamental methods for formulation of interaction problems. This paper is intended to survey the development of the current state-of-practice for design and analysis of dynamically loaded rectangular foundations. This survey is confined to the reported linear mathematical models for foundation resting on a surface of an elastic half space medium.

1 Introduction

The possible occurrence of extreme dynamic excitation, either natural or manmade, has a major influence on the design of buildings and machine foundations. A primary concern in designing foundations is the knowledge of how they are expected to respond when subjected to dynamic loadings. The validity of the mathematical analysis depends entirely on how well the mathematical model simulates the behavior of the real foundation. In most of these analyses it is common to assume that the footing is rigid and the medium is a homogeneous elastic half-space.

This paper will discuss some of the issues of dynamics of soils and foundations from a practical point of view. Since this topic is quite broad, a brief description of methodology will be outlined. One of the main objectives of this review paper is to survey evolution of techniques for solving the dynamic response of foundations when subjected to harmonic loadings.

[1]Department of Mechanical Engineering, South Dakota State University, Brookings, SD 57007.

[2]Applied Research Engineering Division, Lawrence livermore National Laboratory, Livermore, CA 94550.

2 Dynamics of Rectangular Foundations

The first attempt to solve the dynamic response of rectangular base was made by Thomson and Kobori [1] and Kobori et al [2]. They provided computational results for components of the complex displacement functions by assuming a uniform stress distribution on the contact area of the base and medium.

Several investigations have been conducted to extend the static solution to the corresponding dynamic problem by considering mixed boundary value problem. Awojobi and Grootenhuis [3] used the Hankel transform to present the complete dynamic problem by dual integral equations. They gave an analytical solution for the vertical and rocking modes of an infinitely long strip foundation. Karasudhi et al [4] treated the vertical, horizontal and rocking oscillations of a rigid strip footing, on an elastic half-space, by reducing the dual integral equations into the Fredholm integral equations. Luco and Westmann [5] solved the mixed boundary value problems by considering massless rigid strip footings. Their procedure reduced the resulting dual integral equations to the Fredholm integral equations.

The above investigations were based on infinitely long strip foundations. Few investigators have paid attention to the dynamics of a rectangular foundation, due to the difficulty of the asymmetric problem. Elorduy et at [6] introduced a numerical technique based on the uniform displacements for a number of points on the contact surface of the rectangular footings. They gave complex displacement functions for vertical and rocking modes for different rectangular footings. Awojobi and Tabiowo [7] gave a solution to this problem by extending solution for circular foundations to an equivalent rectangular base. Wong and Luco [8] solved this problem for the three modes of vertical, horizontal and rocking vibrations. They used the approach reported by Kobori et al [2] to provide an approximate solution to a footing which was divided into a number of square subregions. They assumed that the stress distribution for each subregion is uniform with unknown magnitude, while all the subregions experienced uniform displacements. Their solution considered the coupling effect for viscoelastic medium and the complex stiffness coefficients were tabulated [9] for different loss factors. Hamidzadeh [10] and Hamidzadeh and Grootenhuis [11] presented an improved version of Elorduy's method to obtain the dynamic responses for three modes of vertical, horizontal and rocking vibration for rectangular foundations. In their analysis a uniform displacements for all nodes of rectangular subregions were assumed. They utilized the impedance matching technique to formulate the dimensionless response for foundations. An approximate method for computation of compliance functions of rigid plates resting on an elastic or viscoelastic half-space excited in all directions was reported by Rucker [12]. The reported method provides compliances for vertical, horizontal, rocking and torsional motion of foundations. Another approximate solution for harmonic response of an arbitrary shape foundation on an elastic half-space was reported by Chow [13].

In his analysis the contact area was divided into square subregions and the influence of the square subregion was approximated by that of an equivalent circular base. He then compared his results with those of Wong and Luco [9] and Hamidzadeh and Grootenhuis [11]. The dynamic stiffness of a rigid rectangular foundation on the half-space was considered by Triantafyllidis [14], who provided solutions for the mixed-boundary value problem. The problem was formulated in terms of coupled Fredholm integral equations of the first kind.

Lumped parameter models to represent the soil structure interaction for embedded foundations were developed by Wolf [15]. Dobry and Gazetas [16 and 17] presented a method to determine the effective dynamic stiffness and damping coefficient of a rigid footing by considering variations of foundation shape, soil type and length ratio of the base.

3 Conclusions

A historical literature review on the dynamic response of rigid foundations subjected to loadings is presented. Significant progress has been established in simulation of the embedded foundations and foundations on elastic half-space and layered media. Techniques described here can be utilized with some degree of confidence to estimate dynamic responses for a number of problems in the field of soil-foundation interaction.

Further advances in development of mathematical models are needed for flexible foundations, structures supported by piles, interaction between foundations and the development of a low-frequency measuring system for in-situ testing.

4 References

[1] Thomson, W.T. and Kobori, T., "Dynamical compliance of rectangular foundations on an elastic half-space." J. Appl. Mech, Trans. ASME, 579-584, 1963.

[2] Kobori, T., Minai, R., Suzuki, T. and Kusakabe, K., "Dynamical ground compliance of rectangular foundations." Proc. 16th Japan Nat. Congress Appl. Mech. Eng., 301-315, 1966.

[3] Awojobi, A.O. and Grootenhuis, P., "Vibration of rigid bodies on semi-infinite elastic media." Proc. Roy. Soc., 287, Series A, 27-63, 1965.

[4] Karasudhi, P., Keer, L.M. and Lee, S.L., "Vibratory motion of a body on an

elastic half-space." J. Appl. Mech., 35, 697-705, 1968.

[5] Luco, J.E. and Westmann, R.A., "Dynamic response of a rigid footing bonded to an elastic half-space." J. Appl. Mech., ASME, 39, 527-534, 1972.

[6] Elorduy, J., Nieto, J.A. and Szekely, E.M., "Dynamic response of bases of arbitrary shape subjected to periodic vertical loading." Proc. Int. Symp. On Wave Propagation and Dynamic Properties of Earth Materials, Albuquerque, University of New Mexico, 105-123, 1967.

[7] Awojobi, A.O. and Tabiowo, P.H., "Vertical vibration of rigid bodies with rectangular bases on elastic media." Int. J. of Earthquake Eng. and Struct. Dyn., 4, 439-454, 1976.

[8] Wong, H.L. and Luco, J.E., "Dynamic response of rigid foundations of arbitrary shape." Int. J. Earthquake Eng. Struct. Dyn. 4, 579-587, 1976.

[9] Wong, H.L. and Luco, J.E., "Table of impedance functions and input motions for rectangular foundations," Report CE 78-15, USC, CA, 1978.

[10] Hamidzadeh, H. R. "Dynamics of rigid foundations on the surface of an elastic half-space," Ph.D. Dissertation, University of London, 1978.

[11] Hamidzadeh, H.R. and Grootenhuis, G., "The dynamics of a rigid foundation on the surface of an elastic half-space," Int. J. Earthquake Eng. Struct. Dyn., 9, 501-515, 1981.

[12] Rucker, W., "Dynamic behavior of rigid foundations of arbitrary shape on a half-space," Int. J. Earthquake Eng. Struct. Dyn., 10, 675- , 1982.

[13] Chow, Y.K., "Vertical vibration of three-dimensional rigid foundations on layered media," Int. J. Earthquake Eng. Struct. Dyn., 15, 585-594, 1987.

[14] Triantafyllidis, T., "Dynamic stiffness of rigid rectangular foundations on the half-space," Int. J. Earthquake Eng. Struct. Dyn., 14, 391-411, 1986.

[15] Wolf, J.P. and Somaini, D.R., "Approximate dynamic model of embedded foundation in time domain," Int. J. Earthquake Eng. Struct. Dyn., 14, 683-703, 1986.

[16] Dobry, R. and Gazetas, G., "Dynamic response of arbitrary shaped foundation," J. Geotech. Eng, ASCE, 112, 109-135, 1986.

[17] Dobry, R., Gazetas, G., and Strohoe, K.H., "Dynamic response of arbitrary shaped foundation II," J. Geotech. Eng, ASCE, 112, 136-154, 1986.

Moment-Rotation Hysteresis Behavior of Semi-Rigid Steel Building Frame Connections

Anant R. Kukreti[1]
Ali Abolmaali[2]

ABSTRACT This paper presents a procedure to obtain a family of mathematical models to predict the moment-rotation hysteresis behavior of steel frame connections.

Introduction

Most steel beam-to-column connections are semi-rigid in nature, and their behavior has been described by many researchers by the moment-rotation (M-θ) curve of the connection, where the rotation (θ) represents the change in the angle between the beam and the column from its original configuration. The moment-rotation curves of commonly used connections when subjected to static loads are well established [1]. Few studies [2, 3] have been conducted to evaluate their behavior under cyclic loads. Semi-rigid connections are well suited for seismically induced loads because of their high ductility and energy dissipation capacities. In this paper, the procedure to experimentally obtain the moment-rotation hysteresis loops of semi-rigid steel connections is described, and typical moment-rotation plots are presented. The parameters which describe the hysteresis behavior are identified, and analytical hysteresis models suggested. The procedure to obtain prediction equations for these parameters is also described.

---

[1]Acting Director/Samuel Roberts Noble Foundation Presidential Professor, School of Civil Engineering and Environmental Science, University of Oklahoma, Norman, OK 73019-1024

[2]Graduate Teaching Assistant, School of Civil Engineering and Environmental Science, University of Oklahoma, Norman, OK 73019-1024

Testing Procedure and Results

The setup consists of a cantilever beam specimen bolted to a column specimen (selected to remain elastic) using the desired connection, and the column is bolted to a reaction frame. A concentrated vertical cyclic force, with a specified cyclic load history, is applied to the free-end of the beam. Thus, the test setup consists of: (1) an actuator to apply the force; (2) beam of a reaction frame to support the actuator; and (3) column of a reaction frame to support the column in the test specimen. Lateral braces are provided at the beam-end connected to the actuator swivel to prevent out-of-plane buckling of the test specimen. The instrumentation consists of two linear variable displacement transducers (LVDTs) to calculate the relative connection rotation, two wire potentiometer to measure beam displacements at two points, and strain gaged bolts to measure bolt forces. The actuator has a load cell and displacement transducers installed in it to measure the cyclic load applied to the beam-end and the actuator stroke (displacement), respectively. The two LVDTs must be placed such that their tips touch the connection column flange at the level of the beam flanges, as close as possible. Thus, the relative displacements of the two LVDTs divided by the vertical distance between the two LVDT tips gives the local rotation of the connection. The moment on the connection is calculated by multiplying the actuator force by the distance from the centerline of the actuator to the face of the column. The cyclic load is applied until either bolt failure occurs or excessive connection rotation is noticed. Typical experimental moment-rotation hysteresis loops for a semi-rigid steel connection are presented in Fig. 1. Failure of the connection is usually observed to occur due to either bolt failure (rupture) in tension or excessive rotation of the connection. The excessive or ultimate rotation (θ_u) is defined to be when excessive rotation causes significant loss in stiffness of the connection due to excessive yielding of the connection elements (end-plate or angle legs). Most of the semi-rigid steel connections exhibit "pinching" at higher load levels, as shown in Fig 1(b). From Fig. 1 it can be seen that the connection behavior is characterized by three parameters, the initial stiffness, K_i, ultimate moment capacity, M_u, and the ultimate rotation, θ_u.

Moment-Rotation Models and Development of Parametric Equations

To idealize the actual observed moment-rotation hysteresis behavior following four types of analytical models can be considered: elasto-plastic, bilinear, Ramberg-Osgood, and modified bilinear, which are presented in Fig. 2. They will vary in the levels of complexity. The bilinear model is defined by four parameters, which are K_i, M_u, θ_u, and a transition moment, M_t. The transition moment M_t is defined as the moment where the tangent line drawn from the point with coordinates (M_u, θ_u) and touching the enveloping moment-rotation curve intersects the initial stiffness slope line drawn at the origin, as shown in Fig. 1. The elasto-plastic model is a special case of the bilinear model, and it is described by only three parameters, K_i, M_u, and θ_u. The Ramberg-Osgood model consists of three nonlinear portions, a loading, an unloading, and a reloading path. Ramberg and Osgood [4] have proposed equations for these curves, which involve three parameters, characteristic moment, M_y, characteristic rotation θ_y, and a rigidity parameter, r. In the bilinear model M_y is chosen instead of M_t.

The test cases are chosen such that pertinent geometric variables are varied, all possible combinations are considered, and sufficient number of practical cases chosen in which the values of K_i and M_u, computed using the monotonic (static) moment-rotation relationships [1], are varied to incorporate very stiff to very ductile connections. The moment-rotation hysteresis test results obtained for these test cases are regressed to obtain the prediction equations for K_i, M_u, θ_u, M_t, M_y, θ_y. A prediction equation is usually acceptable only if it has a value of $R^2 > 0.90$, where R^2 is the coefficient of multiple regression.

References

[1] Kishi, N., and Chen, W.F. (1986). "Steel Construction Data Bank Program," *Structural Engineering Report No. CE-STR-86-18*, School of Civil Engineering, Purdue University, West Lafayette, Indiana.

[2] Nader, M.N., and Astaneh, A. (1996). "Shaking Table Tests of Rigid, Semi-Rigid, and Flexible Steel Frames, *Journal of Structural Engineering*, ASCE, Vol. 122, No. 6, June, pp. 589-596.

[3] Kukreti, A.R., and Biswas, P. (1997). "Finite Element Analysis to Predict the Cyclic Hysteretic Behavior and Failure of End-Plate Connections," *International Journal of Computers and Structures*, Vol. 65, No. 1, July 1997, pp.127-147.

[4] Ramberg, W., and Osgood, W.R., (1943). "Description of Stress-Strain Curves by Three Parameters," *Monograph No. 4*, Publicazione Italsider, Nuova Italsider, Genova.

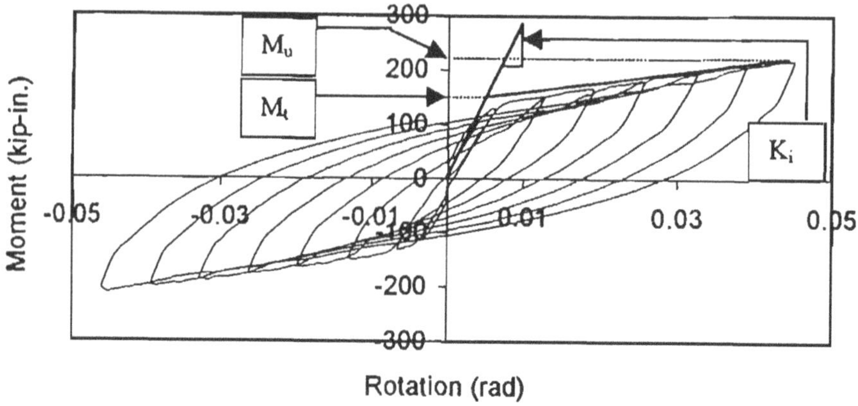

(a) Without Pinching

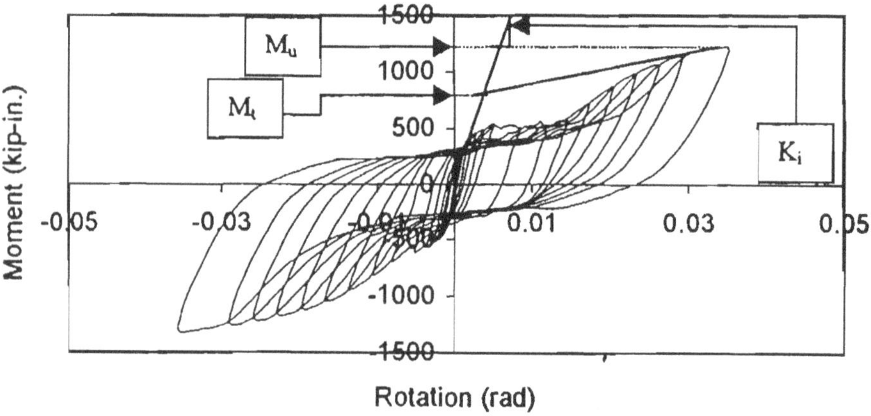

(b) With Pinching

Fig. 1. Typical Moment-Rotation Hysteresis Loops for Semirigid Steel Frame Connections

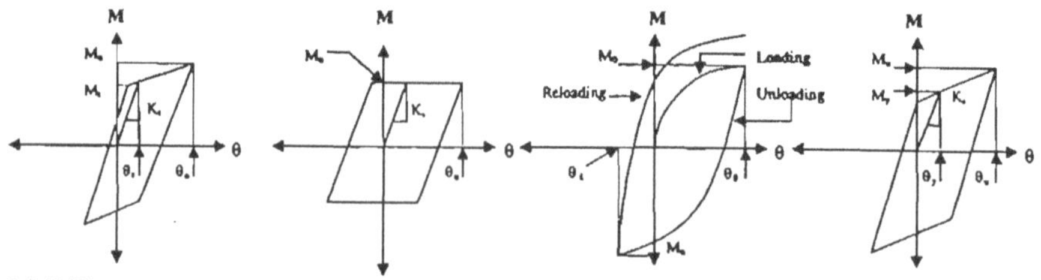

(a) Bilinear (b) Elasto-Plastic (c) Ramberg-Osgood (d) Modified Bilinear

Fig. 2. Moment-Rotation Analytical Models

SESSION I – (B)
Stability and Structures

Chairmen: Prof. T.N. Farris, Purdue University, USA
Prof. L. Johansson, Linköping University, SWEDEN

Stability of finite dimensional systems with unilateral contact and friction: theoretical results

J.A.C. Martins[1]
A. Pinto da Costa[1]

ABSTRACT This paper addresses some questions in the general area of the instability of finite dimensional elastic systems in unilateral frictional contact with rigid obstacles. Some of the results obtained by Martins et al. [1] are generalized here to encompass the effects of the system non-linear elastic behavior and of the obstacle curvature.

1 Dynamic and static contact problems with friction

We consider a plane holonomic and scleronomic finite dimensional system whose configuration at each time $t \geq 0$ is described by the vector $\mathbf{X}(t) \in \mathbb{R}^N$ of independent generalized coordinates. A finite number n_C of particles (p) of that mechanical system is subjected to unilateral frictional contact with fixed obstacles defined by equations $\varphi^p(\mathbf{x}) = 0$, with $\varphi^p: \mathbb{R}^2 \to \mathbb{R}$ sufficiently smooth. At each point on or sufficiently close to an obstacle, an orthonormal basis $(\mathbf{n}^p(\mathbf{x}), \mathbf{t}^p(\mathbf{x}))$ is defined such that $\mathbf{n}^p(\mathbf{x})$ has the direction of $\partial\varphi^p/\partial\mathbf{x}$ $(\neq \mathbf{0})$. The position of each particle p is identified by the vector $\mathbf{x}^p(t) = \mathbf{x}^p(\mathbf{X}(t)) \in \mathbb{R}^2$. The vector of the normal and tangential components of the velocities of all the particles p is denoted by $\mathbf{v}(t) \in \mathbb{R}^{2n_C}$, and is related to the generalized velocities $\dot{\mathbf{X}}(t)$ by

$$\mathbf{v}(t) = \mathbf{G}(\mathbf{X}(t))\, \dot{\mathbf{X}}(t). \tag{1}$$

The column vector that groups the normal and tangential components of the reaction forces that act at some time $t \geq 0$ on all contact candidate particles is denoted by $\mathbf{r}(t) \in \mathbb{R}^{2n_C}$, and is related to the corresponding vector of generalized reactions $\mathbf{R}(t) \in \mathbb{R}^N$ by

[1] Dep. Eng. Civil and ICIST, Instituto Superior Técnico. Av. Rovisco Pais, 1096 Lisboa Codex, Portugal

$$\mathbf{R}(t) = \mathbf{G}^T(\mathbf{X}(t))\, \mathbf{r}(t). \tag{2}$$

In the following it will be always assumed that:

the lines of the $2n_C \times N$ *matrix* $\mathbf{G}$ *are linearly independent.* (3)

In these circumstances the linear map $\mathbf{G}^T: \mathbb{R}^{2n_C} \to \mathbb{R}^N$ has a left inverse, which can be represented by the $2n_C \times N$ matrix $\mathbf{G}^\dagger = (\mathbf{G}\,\mathbf{G}^T)^{-1}\,\mathbf{G}$; this means that, for some vector of generalized reactions $\mathbf{R}(t) \in Rg(\mathbf{G}^T)$ the corresponding column vector $\mathbf{r}(t)$ of the particle reactions can be computed by

$$\mathbf{r}(t) = \mathbf{G}^\dagger(\mathbf{X}(t))\, \mathbf{R}(t). \tag{4}$$

The mechanical system is assumed to be non-linear elastic and is acted upon by constant external applied forces. We denote by $\mathbf{M} = \mathbf{M}(\mathbf{X})$ the symmetric, positive definite mass matrix. Along portions of the system trajectory where $\mathbf{X}(t)$ is twice continuously differentiable and $\mathbf{r}(t)$ is continuous, the motion is governed by the N *Lagrange equations*

$$\mathbf{M}(\mathbf{X}(t))\, \ddot{\mathbf{X}}(t) + \mathbf{B}(\mathbf{X}(t), \dot{\mathbf{X}}(t)) = \mathbf{G}^T(\mathbf{X}(t))\, \mathbf{r}(t), \tag{5}$$

where $\mathbf{B}(\mathbf{X}(t), \dot{\mathbf{X}}(t))$ groups the contributions from the internal and external forces, as well as inertia terms that depend quadratically on the generalized velocities. In addition, the *classical unilateral contact conditions* and the *friction law of Coulomb*

$$\phi^p(\mathbf{X}(t)) \overset{\text{def}}{=} \varphi^p(\mathbf{x}^p(\mathbf{X}(t)) \le 0, \quad r_n^p(t) \le 0, \quad \phi^p(\mathbf{X}(t))\, r_n^p(t) = 0, \tag{6}$$

$$r_t^p(t) \in \mu\, r_n^p(t)\, \sigma[v_t^p(t)] \tag{7}$$

have to be satisfied for all contact candidate particles p; $\mu \ge 0$ is the coefficient of friction, and $\sigma[.]$ denotes the multi-valued application such that, for each $x \in \mathbb{R}$, $\sigma[x] = x/|x|$, if $x \ne 0$, $\sigma[x] = [-1, 1]$, if $x = 0$.

The *static equilibrium* of the same mechanical system is characterized by time independent vectors of generalized coordinates and of contact reactions, $\mathbf{X}^0$ and $\mathbf{r}^0$, such that the governing equations (5) are satisfied with $\mathbf{X} = \mathbf{X}^0$, $\mathbf{r} = \mathbf{r}^0$, $\dot{\mathbf{X}} = \ddot{\mathbf{X}} = \mathbf{0}$, together with the form $\phi^p(\mathbf{X}^0) \le 0$, $r_n^{0p} \le 0$, $\phi^p(\mathbf{X}^0)\, r_n^{0p} = 0$, $|r_t^{0p}| \le -\mu\, r_n^{0p}$, of the unilateral frictional contact conditions.

2 Initial acceleration and reaction discontinuities

At the static equilibrium configuration $\mathbf{X}^0$, for each $(\mathbf{V}, \mathbf{V}') \in \mathbb{R}^N \times \mathbb{R}^N$ and each $(\mathbf{r}, \mathbf{r}') \in \mathbb{R}^{2n_C} \times \mathbb{R}^{2n_C}$, we define the functionals

$$m^*(\mathbf{V}, \mathbf{V}') \stackrel{\text{def}}{=} \mathbf{M}\,\mathbf{V} \cdot \mathbf{V}' - \sum_{p \in P_c} \mu\, \mathbf{G}_n^{\dagger p} \mathbf{M}\,\mathbf{V}\, |\mathbf{G}_t^p \mathbf{V}'|, \tag{9}$$

$$m^{\#}(\mathbf{r}, \mathbf{r}') \stackrel{\text{def}}{=} \mathbf{G}\,\mathbf{M}^{-1}\,\mathbf{G}^T \mathbf{r} \cdot \mathbf{r}' - \sum_{p \in P_c} \mu\, |\mathbf{G}_t^p \mathbf{M}^{-1} \mathbf{G}^T \mathbf{r}|\, r_n', \tag{10}$$

where $P_c = P_c(\mathbf{X}^0)$ is the set of particles p that are in contact at the equilibrium configuration $\mathbf{X}^0$. We also define the sets of admissible reactions $K_\mathbf{r}(\mathbf{X}^0)$ and right accelerations $K_\mathbf{A}(\mathbf{X}^0, \mathbf{0})$ at the equilibrium state $\mathbf{X} = \mathbf{X}^0$, $\dot{\mathbf{X}} = \mathbf{0}$ [2].

Then a dynamic solution with initial conditions $\mathbf{X}(\tau) = \mathbf{X}^0$, $\dot{\mathbf{X}}(\tau) = \mathbf{0}$ may be initiated with an acceleration discontinuity *only if*:

$$\exists\, \mathbf{A} \in K_\mathbf{A}(\mathbf{X}^0, \mathbf{0}),\ \mathbf{A} \neq \mathbf{0}:\ m^*(\mathbf{A}, \mathbf{A}') - m^*(\mathbf{A}, \mathbf{A}) + [\mathbf{r}^0 \cdot \mathbf{G}(\mathbf{A}' - \mathbf{A}) - \sum_{p \in P_c} \mu\, r_n^0\, (|\mathbf{G}_t^p \mathbf{A}'| - |\mathbf{G}_t^p \mathbf{A}|)] \geq 0,\ \forall\, \mathbf{A}' \in K_\mathbf{A}(\mathbf{X}^0, \mathbf{0}); \tag{11}$$

$$\exists\, \Delta\mathbf{r} \in K_\mathbf{r}(\mathbf{X}^0) - \mathbf{r}^0,\ \Delta\mathbf{r} \neq \mathbf{0}:\ m^{\#}(\Delta\mathbf{r}, \Delta\mathbf{r}' - \Delta\mathbf{r}) \geq 0,\ \ \forall\, \Delta\mathbf{r}' \in K_\mathbf{r}(\mathbf{X}^0) - \mathbf{r}^0. \tag{12}$$

The following conditions are *sufficient* to guarantee that such intial discontinuity will actually be followed by a smooth dynamic evolution:
(i) the necessary conditions (11), (12) hold;
(ii) the acceleration and reaction jump vectors $\mathbf{A}$ and $\Delta\mathbf{r}$ [$\mathbf{M}\,\mathbf{A} = \mathbf{G}^T \Delta\mathbf{r} \neq \mathbf{0}$, cf. (11), (12)] give an unambiguous information on the near future states of the particles that are in contact at the equilibrium configuration (strictly free states, strictly stick states, or strictly slip states in the near future);
(iii) the mass and friction dependent *effective mass matrix* for such near future dynamic evolution of the system is non-singular;
(iv) $\mathbf{M}(\mathbf{X})$, $\mathbf{B}(\mathbf{X}, \dot{\mathbf{X}})$, $\mathbf{G}(\mathbf{X})$ and $\partial\mathbf{G}(\mathbf{X})/\partial\mathbf{X}$ are bounded and Lipschitz continuous in $(\mathbf{X}, \dot{\mathbf{X}})$.

3 Divergence of smooth dynamic solutions with perturbed initial conditions

In the neighbourhood of the equilibrium state, and for admissible directions of the increments of the generalized coordinates and the contact reactions the equations of motion (5) have the following linearized form

$$\mathbf{M}(\mathbf{X}^0)\, \delta\ddot{\mathbf{X}}(t) + \mathbf{K}(\mathbf{X}^0, \mathbf{r}^0)\, \delta\mathbf{X}(t) = \mathbf{G}^T(\mathbf{X}^0)\, \delta\mathbf{r}(t), \tag{13}$$

where the N × N *tangent stiffness matrix* $\mathbf{K}(\mathbf{X}^0, \mathbf{r}^0)$ has contributions from the internal and the external forces, as well as from the contact related nonlinearities.

For $t \geq \tau$ we consider perturbed dynamic solutions of the linearized system (13) in the form

$$\delta\mathbf{X}(t) = \alpha(t)\,\mathbf{V} \in K_V(\mathbf{X}^0, \mathbf{r}^0), \qquad \delta\mathbf{r}(t) = \beta(t)\,\mathbf{w} \in K_w(\mathbf{X}^0, \mathbf{r}^0, \mathbf{V}); \quad (14)$$

the function α is twice continuously differentiable, α and $\dot{\alpha}$ are non-negative and non-decreasing; the function β is continuous, non-negative and non-decreasing; the initial values $\alpha(\tau) \geq 0$ and $\dot{\alpha}(\tau) \geq 0$ are arbitrarily small.

The sets $K_V(\mathbf{X}^0, \mathbf{r}^0)$ and $K_w(\mathbf{X}^0, \mathbf{r}^0, \mathbf{V})$ contain the directions of the admissible right velocities and reaction rates at the equilibrium state $(\mathbf{X}^0, \mathbf{r}^0)$ [2]. For each $(\mathbf{V}, \mathbf{V}') \in \mathbb{R}^N \times \mathbb{R}^N$ we continue to use the notation (9) and we define:

$$a^*(\mathbf{V}, \mathbf{V}') \stackrel{\text{def}}{=} \mathbf{K}\,\mathbf{V} \cdot \mathbf{V}' - \sum_{p \in P_c} \mu\, \mathbf{G}_n^{\dagger p}\, \mathbf{K}\,\mathbf{V}\, |\mathbf{G}_t^p\, \mathbf{V}'|. \quad (15)$$

Then:

$$\exists\, \lambda^2 \geq 0 \text{ and } \mathbf{V} \in K_V(\mathbf{X}^0, \mathbf{r}^0),\ \mathbf{V} \neq \mathbf{0}:$$
$$\lambda^2\,[m^*(\mathbf{V},\mathbf{V}') - m^*(\mathbf{V},\mathbf{V})] + [a^*(\mathbf{V},\mathbf{V}') - a^*(\mathbf{V},\mathbf{V})] \geq 0,\ \forall \mathbf{V}' \in K_V(\mathbf{X}^0, \mathbf{r}^0) \quad (16)$$

if and only if the linearized system (13) admits a solution of the form (14). In these circumstances, the equilibrium state $(\mathbf{X}^0, \mathbf{r}^0)$ of the linearized system (13) is dinamically unstable by divergence.

Under some additional conditions we can conclude that the equilibrium state of the given non-linear system is also unstable by divergence. We assume that:
(i) condition (16) holds for a strictly positive λ^2, and, in the specific region of directional linearization that contains the vectors $\mathbf{V}$ and $\mathbf{w}$ obtained from (16), all other λ^2 that solve (16) are non-positive real numbers;
(ii) the vectors $\mathbf{V}$ and $\mathbf{w}$ obtained from (16) give an unambiguous information on the near future states of the particles that are in contact at the equilibrium configuration;
(iii) the mass and friction dependent *effective mass matrix* for such near future dynamic evolution of the system is non-singular;
(iv) $\mathbf{M}(\mathbf{X})$, $\mathbf{B}(\mathbf{X}, \dot{\mathbf{X}})$, $\mathbf{G}(\mathbf{X})$ and $\partial\mathbf{G}(\mathbf{X})/\partial\mathbf{X}$ are continuously differentiable in the neighbourhood of $(\mathbf{X}, \dot{\mathbf{X}}) = (\mathbf{X}^0, \mathbf{0})$.
Then the equilibrium state of the actual non-linear system corresponding to $\mathbf{X}^0$ and $\mathbf{r}^0$ is dynamically unstable (a divergence instability).

4 References

See J.A.C. Martins, A. Pinto da Costa, Stability of finite dimensional systems with unilateral contact and friction: illustrative examples of small dimension.

Stability of finite dimensional systems with unilateral contact and friction: illustrative examples of small dimension

J.A.C. Martins[1]
A. Pinto da Costa[1]

ABSTRACT Following the theoretical developments in the first part of this paper, we present here some illustrative examples relative to: (i) the occurrence of dynamic solutions initiating at an equilibrium position with an initial acceleration and reaction discontinuity; (ii) the occurrence of quasi-static bifurcations and divergence instabilities.

1 Initial acceleration and reaction discontinuities

We consider a rigid rod of length L and mass m with plane motion that may establish contact with a flat obstacle at the extremity A of the rod (see Figure 1). The generalized coordinates are grouped in the vector $\mathbf{X} = [\theta\ X\ Y]^T$. Constant forces F_X, F_Y and moment M_Z are applied at the rod center of mass.

Numerous studies have considered this or related systems [3–5]. Our interest here is to study the occurrence of dynamic solutions initiating at some equilibrium states of the rod with an initial acceleration and reaction discontinuity; vanishing initial velocities are considered in the present study.

We consider equilibrium states in contact such that $\mathbf{r}^0 = [r_n^0\ \ r_t^0]^T = [-1\ \ \alpha]^T$, with $-\mu < \alpha \leq 0$: a static reaction in the interior of the friction cone. The analysis of the other relevant case ($\alpha = -\mu$) can be found in [2]. The initial jump equations are then

$$\mathrm{M}\ \ddot{\mathbf{X}}^+ = \mathrm{G}^T \Delta\mathbf{r}, \tag{1}$$

[1] Dep. Eng. Civil and ICIST, Instituto Superior Técnico. Av. Rovisco Pais, 1096 Lisboa Codex, Portugal

where $\ddot{\mathbf{X}}^+ = [\ddot{\theta}^+\ \ddot{X}^+\ \ddot{Y}^+]^T$, $\mathbf{M} = \mathrm{diag}(mL^2/12, m, m)$, $\mathbf{G}_n^A = [-\ L/2\ \sin\theta\ \ 0\ \ -1]$, $\mathbf{G}_t^A = [L/2\ \cos\theta\ \ 1\ \ 0]$, $\Delta\mathbf{r} = [r_n^+ - r_n^0\ \ r_t^+ - r_t^0]^T$. Here we only discuss jumps towards a contact state with non-vanishing reaction and possible slip to the right: $a_t^+ > 0$, $a_n^+ = 0$, $r_n^+ < 0$, $r_t^+ = \mu\, r_n^+$ (see [2] for other discussions).

The expressions of a_t^+ and r_n^+ obtained from (1) yield $1/3 + (\sin\theta^0 + \alpha\cos\theta^0)\sin\theta^0 < 0$ as a necessary condition for such a jump, which occurs only if $\alpha < 0$. The consideration of the equations of motion for a smooth near future sliding to the right yields an equation for θ where the multiplier of $\ddot{\theta}$, $(1/3 + (\sin\theta - \mu\cos\theta)\sin\theta) < 0$, is equal, apart from a positive constant factor, to the determinant of the *effective mass matrix* mentioned in condition (iii) of Section 3 of the first part of this paper. In addition, the condition (ii) (the reaction and acceleration jumps strictly indicate near future sliding states) as well as the remaining conditions (iii), (iv) are also satisfied. Then a smooth evolution exists after that initial acceleration jump. A stroboscopic representation of that evolution for a suitable set of data is shown in Figure 1. Immediately after the initial jump, the extremity A begins sliding towards the right with increasing contact reactions. After some time, its horizontal velocity vanishes. At that instant a reaction jump towards the vertex of the friction cone occurs and the bar looses contact.

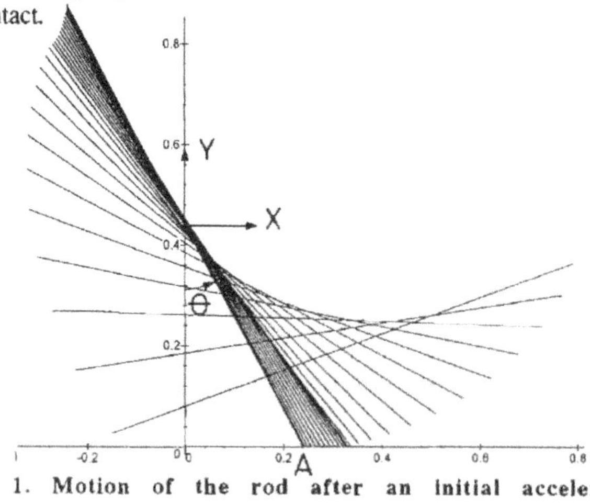

FIGURE 1. Motion of the rod after an initial acceleration and reaction discontinuity. Data: L = 1, m = 1, μ = 2, $\mathbf{X}^0$ = [0.5 0 0.439]^T, $\mathbf{r}^0$ = [-1 -1.8]^T.

2 A "non-associative Shanley column"

This example deals with the rigid column AB of lenght L and mass M that is rigidly connected at point B to a rigid massless bar orthogonal to AB (Figure 2). This system is a modified version of an example presented by Klarbring [6]. Point B has no horizontal displacement. The column is supported by four springs of stiffness K which remain vertical. Particles C and D have mass M_p

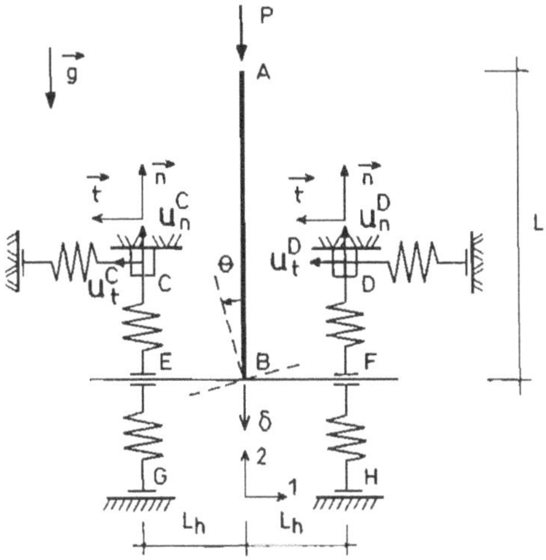

FIGURE 2. A column with frictional contact supports.

and may establish unilateral frictional contact with horizontal obstacles. The initial distance of the vertical springs to the axis of symmetry is L_h. An initial vertical downward displacement $\bar{u}_v$ is applied to both the horizontal obstacles. The extremities of the horizontal springs opposite to particles C and D have both an initial horizontal prescribed displacement $\bar{u}_h$ which corresponds to an initial compresssion of those springs.

The vector of generalized coordinates is $\mathbf{X} = [\delta\ \theta\ u_n^C\ u_n^D\ u_t^C\ u_t^D]^T$ and P denotes the external vertical force. Here we consider equilibrium states such that C and D are in a state of impending slip: $P_A \leq P < P_B$, where P_A is the load at which a state of impending slip is first attained and P_B is the load corresponding to geometric contact with no reaction. Other cases are studied in [2].

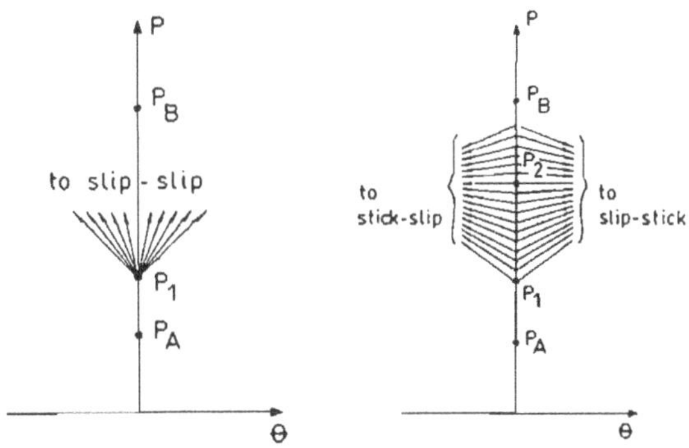

FIGURE 3. Bifurcation diagram for $P_A \leq P < P_B$ (case $P_A < P_1$ and $P_2 < P_B$). Rate solutions leading to: (a) SLIP-SLIP frictional contact states ($P = P_1$); (b) STICK-SLIP or SLIP-STICK states.

Several possibilities for the near future evolution of the system exist: for instance, STICK-SLIP means that particle C sticks and particle D slides. Figure 3 represents the general aspect of the bifurcation diagram of the fundamental trajectory. The rate problem may have infinitely many SLIP-SLIP solutions for a particular load P_l (Figure 3a). Above P_B there exists a continuous range of bifurcation points into a STICK-SLIP or a SLIP-STICK solution with $\dot{P} > 0$. The upper load (P_2) of this segment is given by condition $\dot{P} = 0$ (Figure 3b). For $P_2 \geq P_B$, the segment of bifurcation points into STICK-SLIP or SLIP-STICK solutions ends at the load P_B. On the contrary, if $P_2 < P_B$ (the case of Figure 3b) there are also bifurcations with a load decrease ($\dot{P} < 0$).

The study of the divergence instability of the equilibrium states of the fundamental path is done by solving *a set* of linear eigensystems, one for each possible near future evolution of the system, which is equivalent to the solution of (16) in the first part of the paper. For instance, the linear eigenproblem corresponding to STICK-SLIP near future states is

$$\begin{bmatrix} \lambda^2 + 4 & 0 & 0 \\ 0 & \frac{\lambda^2}{3} + a & \frac{P}{2} \\ -\mu & b & \lambda^2 m + 1 \end{bmatrix} \begin{bmatrix} V_\delta \\ V_\theta \\ V_t^D \end{bmatrix} = \begin{bmatrix} 0 \\ 0 \\ 0 \end{bmatrix}.$$

The coefficient matrix is $(\lambda^2\mathbf{M}^* + \mathbf{K}^*)$ where $\mathbf{M}^*$ and $\mathbf{K}^*$ are the effective mass and stiffness matrices, and V_δ, V_θ and V_t^D are generalized displacement rates. From the latter system we conclude that divergence instability towards STICK-SLIP states may occur for $P \geq P_2$. From the previously mentioned set of eigensystems we conclude that a divergence instability of this kind does not occur in the continuous range of bifurcation points $[P_1, P_2]$.

3 References

[1] Martins, J.A.C., Barbarin, S., Raous, M. and Pinto da Costa, A., Dynamic stability of finite dimensional linearly elastic systems with unilateral contact and Coulomb friction (1998)(submitted).

[2] Martins, J.A.C. and Pinto da Costa, A., Dynamic stability of finite dimensional nonlinear elastic systems with unilateral contact and friction (1988)(submitted).

[3] Lötstedt, P.,Coulomb friction in two-dimensional rigid body systems. *ZAMM*, **61**, 605-615, 1981.

[4] Stewart, D. E. and Trinkle, J. C., An implicit time-stepping scheme for rigid body dynamics with inelastic collisions and Coulomb friction. *Int. J. Num. Meth. Engrg.*, **39**, 2673-2691, 1996.

[5] Moreau, J.J., CISM courses and lectures **302**, 1-81, 1988.

[6] Klarbring, A., On discrete and discretized non-linear elastic structures in unilateral contact (stability, uniqueness and variational principles). *Int. J. Solids Structures*, **24**, 5, 459-479, 1988.

Response of a Vibroimpact System With Secondary Structure to a White-Noise Excitation: Case of Inelastic Impacts

M. F. Dimentberg[1]
H. G. Haenisch[1,2]
D. V. Iourtchenko[1]

ABSTRACT A system consists of a primary mass, impacting at a one-sided barrier, and a secondary structure. It is excited by a white-noise random force. Mean square response of a secondary structure is studied via "cascade" approximation, whereby the problem is decomposed into a nonlinear one for a SDOF primary structure and a MDOF linear one for a secondary structure under base excitation. Exact and asymptotic analytical solutions for the cases of zero and small impact losses are presented for a special case of a zero offset between barrier and system's equilibrium position. These solutions are used as benchmarks for Monte-Carlo simulation studies for the cases where impact losses aren't small (restitution factor is not close to unity). The results clearly demonstrate possibility for incorporating impact losses into random vibration analyses using previously derived formula for the equivalent viscous damping factor.

A primary SDOF system is considered, which has a rigid barrier at its static equilibrium position $y = 0$. The system's displacement $y(t)$ under stationary zero-mean white-noise random excitation $\zeta(t)$ with intensity D is governed by the following equation of motion for $y > 0$, i.e. between impacts

$$\ddot{y} + 2\alpha\dot{y} + \Omega^2 y = \zeta(t) \tag{1}$$

This equation is supplemented by the following impact/rebound condition

[1]Mechanical Engineering Department, Worcester Polytechnic Institute, 100 Institute Rd., Worcester, MA 01609

[2]Presently with Raytheon Co., 1001 Boston Post Rd., Marlborough, MA 01752

at impact instants t_*, $y(t_*) = 0$, where r is restitution factor

$$\dot{y}_+ = -r\dot{y}_-, \ \dot{y}_\pm = \dot{y}(t_* \pm 0), \ 0 < r \leq 1 \tag{2}$$

Consider now the primary mass' response $y(t)$ as a base excitation for a linear secondary structure as attached to the primary mass. Such a "cascade" approximation for the whole MDOF (nonlinear) vibroimpact system would be adequate at least in the case of small secondary/primary mass ratio. Whilst an arbitrary structure can be treated in this way, the specific case of a single secondary mass will be studied here. Then the relative displacement $z(t)$ of the mass is governed by equation

$$\ddot{z} + 2\alpha_s\dot{z} + \Omega_s^2 z = -\ddot{y} \tag{3}$$

This approach permits to obtain analytical solution for the case of zero impact losses ($r = 1$) through the use of the following exact solution for autocorrelation function $K_{yy}(\tau)$ of stationary response $y(t)$ as obtained in [1]

$$\begin{aligned} K_{yy}(\tau) &= (D/2\alpha\pi\Omega^2)[R(\tau)\sin^{-1}R(\tau) + \sqrt{1-R^2(\tau)}] \\ R(\tau) &= \exp(-\alpha|\tau|)[\cos\omega_d\tau + (\alpha/\omega_d)\sin\omega_d|\tau|] \\ \omega_d &= \sqrt{\Omega^2 - \alpha^2} \end{aligned} \tag{4}$$

The corresponding quadrature expression for power spectral density (PSD) $\Phi_{yy}(\omega)$ of $y(t)$ has been studied both numerically and analytically in [2] and used for predicting mean square response of a secondary structure; in particular, peaks of this PSD at $\omega = 2n\Omega$, $n = 1, 2, \ldots$ were identified. Moreover, successive integration by parts in this expression yields the following asymptotic formula for the acceleration PSD at high frequencies:

$$\begin{aligned} \lim_{\omega\to\infty}\Phi_{\ddot{y}\ddot{y}}(\omega) &= \lim_{\omega\to\infty}\omega^4\Phi_{yy}(\omega) = (D/2\pi)Q \\ Q &= 1 + 2\Omega/\pi\alpha \end{aligned} \tag{5}$$

This limiting value of the PSD of base excitation is directly applicable in case of large Ω_s/Ω, with corresponding mean square response of the secondary mass being

$$\sigma_{z\infty}^2 = DQ/4\alpha_s^2\Omega_s^2 \tag{6}$$

Convergence rate to this high-frequency limit increases with primary damping ratio α/Ω [2]. The case of inelastic impacts, or $r < 1$, will be addressed now. It should be noted, first of all, that the case of vanishingly small impact losses can be handled by asymptotic averaging over the period [3]. Specifically, if value of $1-r$ is proportional to a small parameter (is much smaller than unity), then these losses are found to be equivalent to those due to viscous damping with the "asymptotically equivalent" factor $\alpha_{as} = (1-r)\Omega/\pi$ [3]. Of course, the resulting solution is approximate rather than exact. This equivalent viscous damping may be regarded as a "universal" one, valid for free vibration also. However, for the random vibration problem at hand another formula for equivalent damping may be more relevant, as derived in [4] through a direct analytical solution of the vibroimpact problem (1), (2) (with zero viscous damping) by the method of moments:

$$\alpha_{eq} = [(1 - r^2)/(1 + r^2)](\Omega/\pi) \tag{7}$$

The equivalency has been established in [4] by considering mean square value of $y(t)$. And whilst the equivalent damping factor (7) clearly approaches the asymptotic one, or α_{as}, with r approaching unity, it should provide more accurate response predictions for not-very-small $1 - r$. This conclusion was confirmed by Monte-Carlo simulations [4].

For the response of secondary structure the PSD of $y(t)$ is more relevant than its mean square value. Therefore, applicability of the improved formula (7) for incorporating the impact losses should be verified through Monte-Carlo simulations. The simulations were based on simultaneous numerical solution of the stochastic ODE (1) with $\alpha = 0$ together with impact condition (2) and the following ODE for absolute displacement of secondary mass $x(t) = z(t) + y(t)$, as derived from the ODE (3):

$$\ddot{x} + 2\alpha_s \dot{x} + \Omega_s^2 x = -2\alpha_s \dot{y} - \Omega_s^2 y \tag{8}$$

To compare the results with predictions via formulae (5), (6) and with those presented in [2], the assigned values of the restitution factor r were calculated according to the relation (7). Damping ratio of the secondary mass has been $\alpha_s/\Omega_2 = 0.01$ throughout all simulation runs.

Figure 1 illustrates high accuracy of the asymptotic expression (6) with "impact magnification factor" Q as calculated via formula (5) by using expression (7) for equivalent damping, down to rather low value of the restitution factor $r = 0.72$. The limiting condition for Monte-Carlo simulations used to be established from analytical data as presented in [2]; for this specific case the corresponding natural frequencies ratio was $\Omega_s/\Omega = 12$.

In Figures 2 and 3, the present Monte-Carlo simulation results for $\sigma_z/\sigma_{z\infty}$ vs. Ω_s/Ω are compared with the curves, which are reproduced from [2] and are based on analytical solution (4) for $\alpha/\Omega = 0.01$ and 0.1 respectively. Good correlation is seen once again between simulation results and the analytical ones, based on the equivalent viscous damping (7). The latter is seen therefore as providing a viable simplified approach for incorporating impact losses into random vibration analyses.

Acknowledgements

This work was sponsored by the NSF, Grant CMS-9610363. This support is most highly appreciated.

References

[1.] Dimentberg, M.F., Hou, Z. and Noori, M., Spectral Density of a Non-Linear Single-Degree-of-Freedom System's Response to a White-Noise Excitation: A Unique Case of an Exact Solution, *International Journal of Non-Linear Mechanics*, vol. 30, #5, 673-676, 1995.

[2.] Dimentberg, M.F. and Haenisch, H.G., Pseudolinear Vibroimpact System with a Secondary Structure: Response to a White-Noise Excitation, *Transactions of ASME, Journal of Applied Mechanics* (to appear).

[3.] Dimentberg, M.F., *Statistical Dynamics of Nonlinear and Time-Varying Systems*, Research Studies Press, Taunton, England, 1988.

[4.] Dimentberg, M.F. and Iourtchenko, D.V., Incorporating Impact Losses into Random Vibration Analyses, *submitted to Probabilistic Engineering Mechanics.*

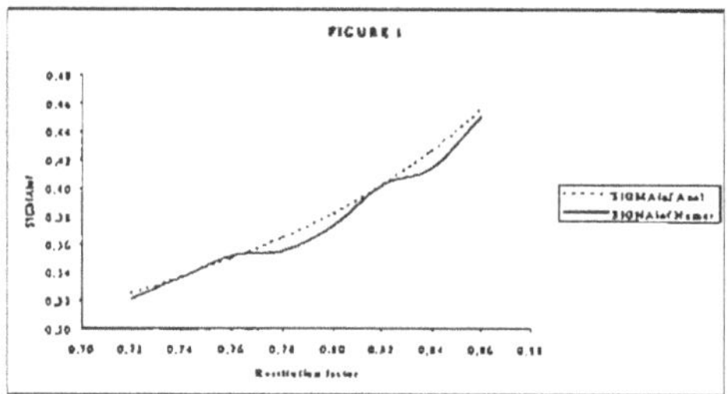

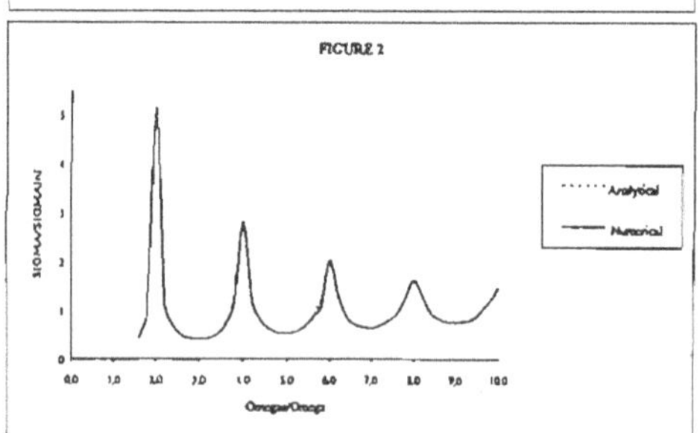

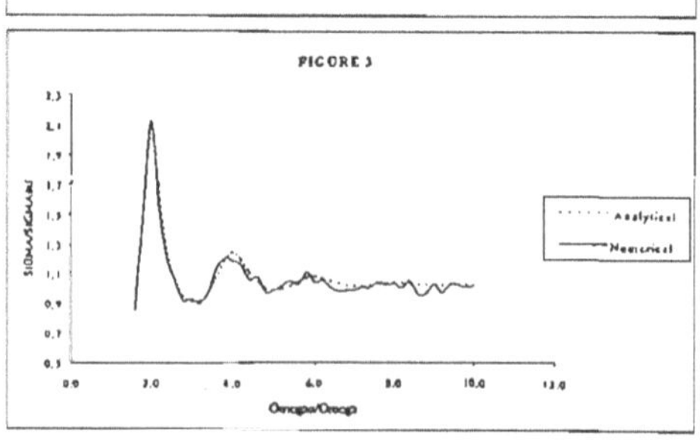

Assessment of Linear Joint Effects on Space Structures

A. El-Saadani[1]
I. I. Orabi[1]

ABSTRACT In this work, the dynamics of linear joint-dominated space structures is studied. The linear analysis deals primarily with the study of linear viscous damping in the joints. Undamped jointed modeshapes as well as damped modeshapes are presented. The effect of linear joint stiffness and damping on the mode shapes of the system is studied. A finite element technique using ANSYS is used in this analysis. The variation of eigenvalues resulting from increasing joint damping is also considered. The linear analysis are particularly useful in determining how damping mechanisms located in the joints distribute their effect to provide modal damping, especially to those modes which exercise the joint most actively. It is also possible to determine the level of joint damping that introduces a maximum amount of global damping into the system, for any given mode.

1 Introduction

A new era of space development has begun in which very large structures will be constructed in space. These include satellites, telescopes, antennas, and most prominently, NASA's manned Space Station. These structures will be lightweight and will include complex joints to allow for on orbit deployment or assembly. This process is particularly difficult when dealing with structures that contain a large number of joints that exhibit nonlinear, hysteretic behavior. For example, this may be the case for prefabricated truss structures that are designed to collapse into a dense package for transportation to orbit. The truss is then expanded in space by

---

[1] Mechanical Engineering Department, University of New Haven, West Haven, CT 06516

using joints that are especially designed to rotate and lock into place. The dynamic response of such structures is said to be joint-dominated if the amount of damping or the stiffness of the system is strongly affected by the joints' behavior. If damping in the system is primarily due to joint hysteresis, then joint behavior controls the amplitude of the steady-state response as well as the rate at which transients decay. Additionally, in some cases, joint flexibility can significantly reduce the stiffness of the structure, thus reducing its natural frequencies and altering the associated mode shapes. The presence of joints can strongly affect the dynamics of space structures in weightlessness, especially if the joints are nonlinear. When joints have linear characteristics, their presence is manifested by distinctive mode shapes and damping properties.

2 Undamped Modeshapes Analysis

In this analysis, four joint model is considered with the following characteristics: massless joints, distributed mass and stiffness in beam elements, free -free end conditions, all joints are identical and elements are modeled as beam in bending.

The effect of joints on modeshapes can be easily illustrated when comparison are made between undamped jointed modeshapes and undamped continuous beam modeshapes. Figure 1 shows symmetric jointed and continuous modeshapes and Fig. 2 illustrates antisymmetric jointed and continuous beam modeshapes. These figures show that the joints represent a local minimum in bending stiffness. It is also clear that the frequencies of the jointed model are less than the corresponding frequencies of the continues model and the joints are sharply bent in the higher modes. It is obvious that the presence of joints makes the model more flexible and accordingly lowers the stiffness in all modes. It is also noted that as the joint stiffness is varied from zero to infinity, the modeshapes vary from those of a pinned structure to those of a single continuous beam. At low stiffness and for higher modes, the modeshapes of the system become cusped at the joints due to the joint effects.

3 Damped Modeshapes Analysis

The most important source of passive damping in space structures is the joints damping. Unlike standard undamped modal vibrations, the modeshape changes its spatial distribution as well as its magnitude during each cycle. Figure 3 shows damped modes of the four joint model. It can be seen from this figure that when joint damping is increased from zero to infinity, the modeshapes show the same variation. The reason for that because joint damping tends to tighten up the joint and ultimately lock it, in the same way as an increase in joint stiffness would.

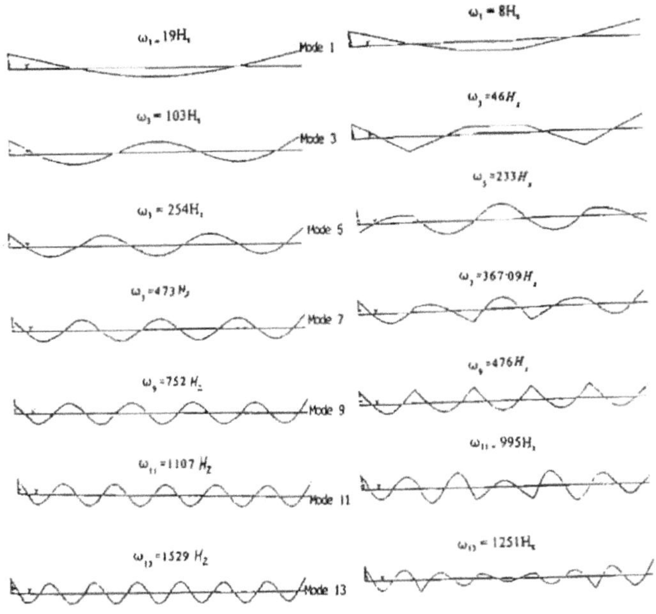

Fig. (1): Jointed and Continuous Beam Modeshapes (symmetric)

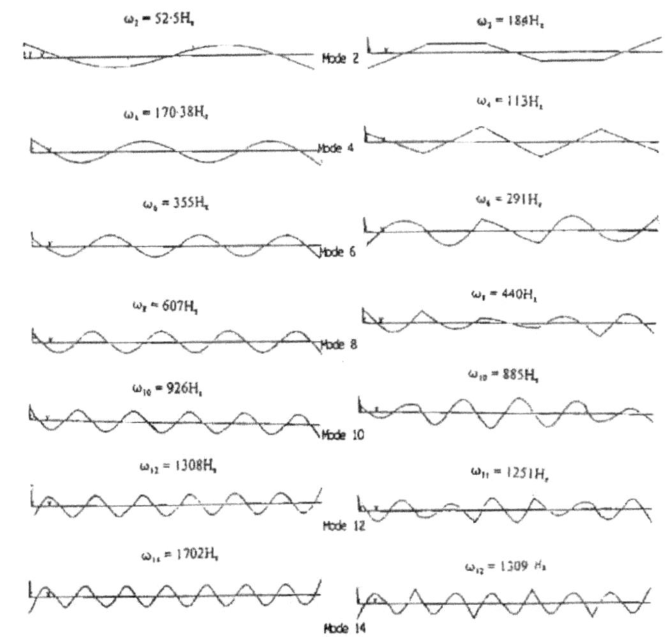

4 Effects of Joint Damping

Figures 4 and 5 show the root locus diagram of antisymmetric and symmetric modes for four joint model. It is observed that the peak at which the maximum modal damping occurs corresponds to a different value of joint damping for each mode. The lower modes require a large amount of damping to reach the maximum deflection. It is also observed that the joints participation at lower modes are much smaller compare to the higher modes. In the higher modes, in which the joints are very active, it takes much less joint damping to reach to the peak deflection.Note that it takes more joint damping to reach peak modal damping for higher joint stiffness,. It is clear that the less a joint is exercised, the more joint damping is needed to achieve maximum modal damping.

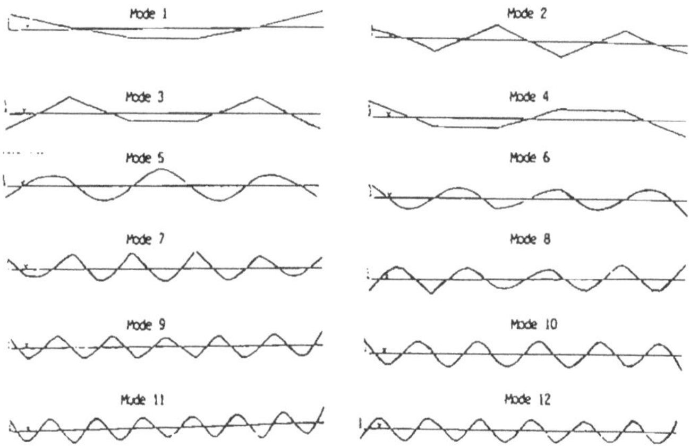

Fig. 3: Damped Mode Shapes

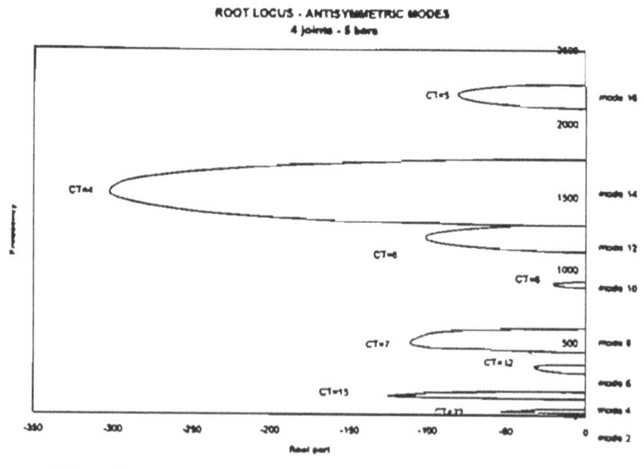

Fig. 4: Antisymmetric Modes of 4 Joints

5 Conclusions

Increasing passive damping into a system by increasing damping in the joints works only to a certain level. Increasing joint damping tends to decrease the mobility of the joint so much that it effectively inhibits the global damping effect.

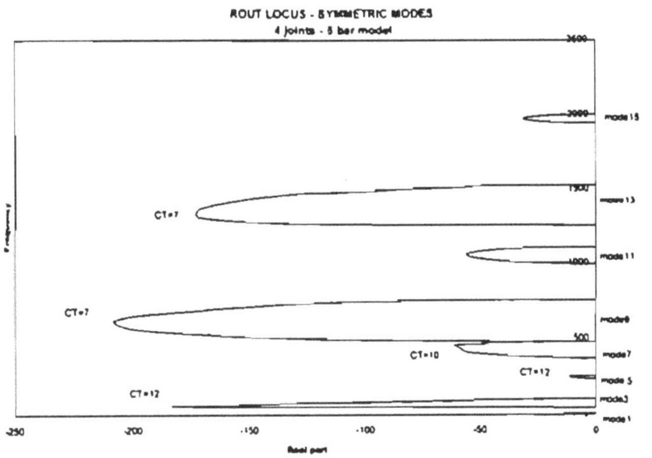

Fig. 5: Symmetric Modes of 4 Joints

The maximum amount of modal damping achievable is different for each mode, is higher for low stiffness joints, and is roughly proportional to the number of joints active in that modeshapes. In a realistic structure, the joints should be near the peaks of the modes that are critical to damping.

Acknowledgment

The authors wish to acknowledge the Yankee Ingenuity Initiative of the State of Connecticut for supporting this work.

6 References

1. Crawley, E.F. and O Donnell, K.J., "A Procedure for Calculating the Damping in Multi-Element Space Structures", Acta Astronautica, Vol. 15, No. 12, pp. 987-996, 1987.
2. Hertz, T.J., and Crawley, "Damping in Space Structure Joints", AIAA. 84-1039, 1984.
3. Nafeh, A.H., Mook, D.T., and Nayfeha, J.F., "Some Aspects of Model Interactions in the Response of Beams", AIAA 87-0777, 1987.

The Incorporation of Hysteretic Damping in Structural Analysis and Design

F.C. Nelson[1]

ABSTRACT The most frequently used model for structural damping is linear viscous proportional damping. Yet there are important areas of structural dynamics for which this damping model is inappropriate. Prominent among these is the use of shear-action rubber isolators at the base of large structures, such as buildings and bridges, to provide protection against earthquakes. These isolators dominate the damping of the low frequency modes and demand that a hysteretic damping model be used in the formulation of the equations of motion.

1. Viscous vs. Hysteretic Damping in 1DOF Systems

For 1DOF linear passive systems, the choice of damping model is usually between linear viscosity and linear hysteresis. If one chooses viscosity, the damping force is assumed proportional to velocity and in anti-phase with velocity, i.e.

$$F = -c\dot{x} \tag{1}$$

where c is the viscous damping coefficient. If one chooses hysteresis, the damping force is assumed proportional to displacement and in anti-phase with velocity, i.e.

$$\underline{F} = -ik\eta\underline{x} \tag{2}$$

where $\underline{\ \ }$ denotes a complex quantity and $i = (-1)^{1/2}$. The constant of proportionality in (2) is the product of the stiffness, k, and the material loss factor, η. Equation (1) has general validity but (2) is limited to harmonic response.

[1] Department of Mechanical Engineering, Tufts University, Medford MA 02155, USA.

2. Proportional Damping in NDOF Systems

For viscous damping, the multi-degree-of-freedom equations of motion have the form

$$[\mathrm{M}]\{\ddot{x}\} + [\mathrm{C}]\{\dot{x}\} + [\mathrm{K}]\{x\} = \{\mathrm{F(t)}\} \tag{3}$$

For an N degree-of-freedom system, [M], [C], and [K] are the NxN mass, damping, and stiffness matrices and $\{x\}$, $\{\dot{x}\}$, and $\{\ddot{x}\}$ are the Nx1 displacement, velocity, and acceleration vectors.
Let $[\Phi]$ be the modal matrix of the reduced system

$$[\mathrm{M}]\{\ddot{x}\} + [\mathrm{K}]\{x\} = 0 \tag{4}$$

where the columns of $[\Phi]$ are the eigenvectors of (4). Then (3) can be written

$$[\mathrm{m}]\{\ddot{q}\} + [\hat{\mathrm{c}}]\{\dot{q}\} + [\mathrm{k}]\{q\} = [\Phi]^T\{\mathrm{F}\} \tag{5}$$

where $[\mathrm{m}] = [\Phi]^T[\mathrm{M}]\,[\Phi]$ and is diagonal; $[\mathrm{k}] = [\Phi]^T[\mathrm{K}]\,[\Phi]$ and is diagonal; and the modal coordinates $\{q\}$ are defined by $\{x\}=[\Phi]\{q\}$. The diagonal elements of [m] and [k] are, respectively, the modal masses m_i, and the modal stiffnesses, k_i. The matrix $[\hat{\mathrm{c}}]=[\Phi]^T[\mathrm{C}]\,[\Phi]$ is diagonal, if and only if

$$[\mathrm{C}][\mathrm{M}]^{-1}[\mathrm{K}]=[\mathrm{K}]\,[\mathrm{M}]^{-1}[\mathrm{C}]. \tag{6}$$

If (6) is satisfied, (5) may be solved sequentially for the $\{q\}$ and then the physical coordinates recovered from $\{x\}=[\Phi]\{q\}$.

Lacking the condition of (6), diagonalization can be achieved by assuming that the system possesses a specific type of damping that has been designated as "proportional" because it assumes that [C] is proportional to [M] or [K], or both, i.e.

$$[\mathrm{C}] = \mathrm{a}_0[\mathrm{M}] + \mathrm{a}_1[\mathrm{K}]. \tag{7}$$

This is also called Rayleigh damping.

3. Hysteretic Damping in NDOF Systems

The analog to (3) for hysteretic damping is

$$[M]\{\ddot{\mathbf{x}}\} + i[H]\{\mathbf{x}\} + [K]\{\mathbf{x}\} = \{\underline{F}\}e^{i\Omega t} \quad (8)$$

where bold face type indicates a phasor $\{\mathbf{x}\} = \{\underline{x}\}e^{i\Omega t}$. Hysteretic damping is provided by the matrix [H]. The eigenvalue problem associated with (8) yields complex eigenvalues $\underline{\lambda}_i^2 = \lambda_i^2(1 + i\eta_i)$ and complex eigenvectors, $\{\underline{\varphi}_i\}$ with the properties

$$\{\underline{\varphi}_i\}^T[M]\{\underline{\varphi}_j\} = m_i\delta_{ij} \quad (9)$$

$$\{\underline{\varphi}_i\}^T([K] + i[H])\{\underline{\varphi}_j\} = \underline{k}\delta_{ij} \quad (10)$$

where δ_{ij} is the Kronecker delta. Notice that no special restrictions need to be placed on [H] in order to achieve completely uncoupled equations of motion in terms of modal coordinates given by $\{\underline{q}\} = [\underline{\Phi}]^{-1}\{\underline{x}\}$. In this sense, hysteretic damping is a "simpler" model than viscous damping. One may also argue that hysteretic damping is a "better" model of the damping of metals and polymers than is viscous damping, since measurements of the energy dissipated per cycle for metals and polymers at high frequencies is almost constant, consistent with the hysteretic damping model, whereas the viscous damping model shows a linear dependence.

One can propose proportional hysteretic damping, i.e.

$$[H] = a_0[M] + a_1[K]. \quad (11)$$

This has the advantage of making the mode shapes real and, moreover, equal to the undamped mode shapes of (4). The eigenvalues remain complex but simplify somewhat to $\underline{\lambda}_i^2 = \omega_i^2(1 + i\eta_i)$ where ω_i^2 are the eigenvalues of (4). Thus, for the important case of harmonic forcing, the proportional hysteretic damping model has much to recommend it, being "simpler" and "better" than proportional viscous damping. It is therefore surprising how seldom it is used.

4. Seismic Applications

The idea of using discrete, flexible connections (i.e. isolators) between a machine or a structure and its supporting foundation to reduce the transmission of vibration or force, or both, has a long history and a highly developed technology, see [1]. The related concept of using isolators at the base of large structures and buildings to protect them against earthquakes is more recent but of growing popularity, see [2].

The concept can be illustrated using the 3-story, steel-frame, shear building model depicted in Fig. 1.

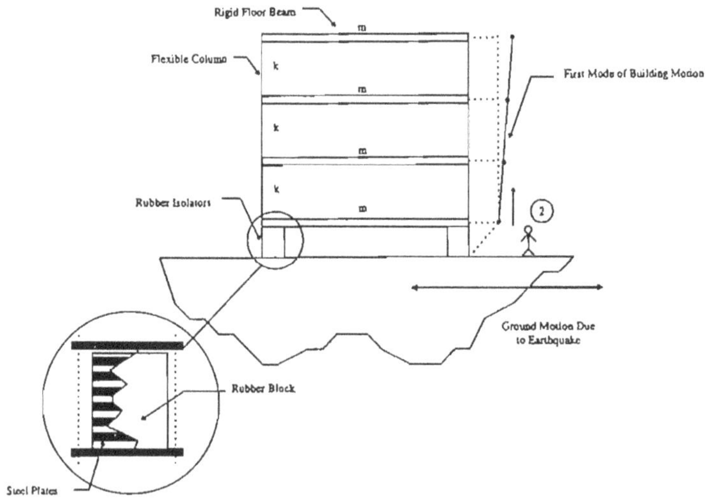

Fig. 1: Model of a steel-frame shear building with base isolation.

The rubber blocks that serve as base isolators typically must support high axial loads (e.g. 1,000 tons) while permitting a low frequency in shear (e.g. 1Hz). A design of growing acceptance is the use of a highly filled natural rubber block (rectangular or circular in cross-section) with embedded horizontal steel plates to enhance vertical stiffness; see Fig. 1 for a simplified depiction; design details and performance results can be found in [3]. Of course, such rubber isolators are the quintessential example of a system with hysteretic damping and their participation in any mode will likely dominate the damping of that mode. For example, the first (lowest frequency) mode shown in Fig. 1. should exhibit almost pure hysteretic damping.

For the model shown in Fig 1, the distribution of earthquake-induced inertia forces as seen by observer 2 will be uniform from the top to the bottom story. This force distribution thus has almost the same shape as the first mode. Higher modes will involve increasing amounts of frame deformation but will likely still be hysteretically damped. From the above discussion, all these damped modes will be mutually orthogonal. In particular, all modes higher than the first will then be very nearly orthogonal to the distribution of inertia forces and, as such, they will absorb only a minimal amount of energy from the action of these forces and therefore exhibit only a minimal vibration. In other words, the higher modes will be isolated from the earthquake ground motion. This is the essence of base isolation in physical terms and for a simple model. The mathematical explanation, which is easier to generalize to more complex models, can be found in [3].

In [4] it has been suggested that discrete viscous dampers distributed throughout a structure can achieve levels of seismic protection comparable to base isolation but at significantly lower cost. One way of accomplishing this show in Fig. 2.

Discrete viscous dampers imply a nondiagonal modal damping matrix. The assumption of proportional damping, as per (7), is difficult to justify since in this case the mass and stiffness are distributive but the damping is discrete.

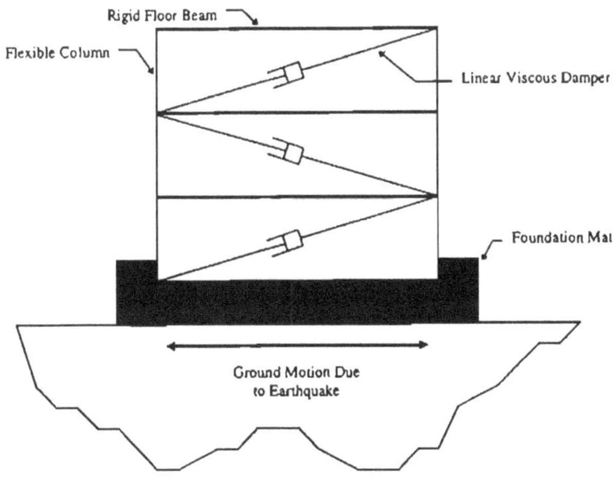

Fig. 2: Model of a steel-frame shear building with inter-story viscous dampers.

As a result, the modes will not be orthogonal and the seismic energy of an earthquake applied to the structure of Fig. 2 will be distributed among many modes rather than being concentrated in one mode as in the case with the structure of Fig. 1. We see then, that the approach of Fig. 2 is not based on isolation but is based on the augmentation of modal damping; a different approach, but one that has been used with success, see [4]. It would be interesting to investigate the model of Fig. 2 with the discrete viscous dampers replaced with discrete viscoelastic shear dampers, such as those in [5]. Modal orthogonality should be recovered and the seismic energy should become more concentrated in the fundamental mode.

5. References

[1] Nelson, F.C. 1994 Vibration Isolation: A Review, I. Sinusoidal and Random Excitation *Shock and Vibration*, 1, 485-493.

[2] Buckle, I.G. and Mayes, R.L. 1990 Seismic Isolation: History, Application, a Performance -- A World View *Earthquake Spectra*, 6, 161-201

[3] Kelly, J.M. 1993 *Earthquake-Resistant Design with Rubber*, Springer-Verlag, London

[4] Haskell, G. and Lee, D. 1996 Fluid Viscous Damping as an Alternate to Base Isolation *Seismic, Shock, and Vibration Isolation*, ASME, PVP - vol 341

[5] Housner, G.W. et al., 1997, Structural Control: Past, Present, and Future, Special Issue of the *Journal of Engr. Mech.* ASCE, Section 2.3 (Viscoelastic Dampers)

Multi-Linear and Smooth Hysteretic Models for Degrading Structures

M. V. Sivaselvan[1]
K. Skliros[1]
A. M. Reinhorn[2]

1 Introduction

Monotonic load-deformation characterization is not sufficient for the inelastic seismic analysis of structures, since when a structure is seismically excited, it is loaded cyclically and is taken to several excursions into the inelastic range of behavior. In order to analyze a structure close to collapse and beyond under such loading, an accurate modeling of its hysteretic behavior is required.

Structural elements subjected to cyclic loading undergo deterioration of strength and stiffness when they are stressed beyond the elastic limit. Besides, certain elements such as reinforced concrete beams and columns and bolted moment resisting steel connections undergo the phenomenon of pinching or slip. These have to be accurately modeled for computational analysis of structures in the inelastic range of behavior. At the same time, the model should be kept reasonably simple to be computationally tractable and efficient.

Here, two hysteretic models – one multi-linear and the other continuous - have been developed to fulfill these requirements.

2 The Multi-Linear Model

The state of the structural element, whose hysteresis is being modeled, is completely defined by a set of *control points* with coordinates calculated by functions of *database variables*. These functions are summarized in Table 1. Lines between these points are called *branches* and represent the path along the hysteresis loop (Fig. 1). The transitions between such branches are predefined by a set of rules (Table. 2).

The major advantage of the model is its versatility. It can be driven using either force or displacement control, with only minor changes in the implementation. Since both trilinear and bilinear envelopes are integrated, it can model the behavior of many materials such as steel and reinforced concrete. Special behavior such as Bauschinger effects can also be captured. Finally the model can be incorporated as a "black box" in any inelastic analysis program since it operates only on the database variables that come to it as parameters. Some examples of the capabilities of this model are shown in Fig. 3

[1] Graduate Student and [2] Professor and Chair, Dept. of Civil, Structural and Environmental Eng., University at Buffalo (SUNY), NY 14260, USA

POINT		FORMULA: φ	FORMULA: M	FORMULA: Other Terms Used
1		M^+_{cr}/K_0	M^+_{cr}	-
2		M^-_{cr}/K_0	M^-_{cr}	-
3		ϕ^+_y	M^+_y	-
4		ϕ^-_y	M^-_y	-
5		ϕ^+_{max}	M^+_{max}	-
6		ϕ^-_{min}	M^-_{min}	-
7		ϕ^+_c	$M^+_y + K^+_u(\phi^+_c - \phi^+_y)$	$K^+_u = a^+ K_0$
8		ϕ^-_c	$M^-_y + K^-_u(\phi^-_c - \phi^-_y)$	$K^-_u = a^- K_0$
9	9'	Point of intersection of 1. Line joining point 4 and point 8 2. Line passing through point $(M^+_{max}, \phi^+_{max})$ and having slope $R^+_k K_0$		$R^+_k = \frac{M^+_{max} + \alpha M^+_y}{K_0\phi^+_{max} + \alpha M^+_y}$
	9	$\phi^+_{max} - \frac{M^+_{max}}{R^+_k K_0}$	0	
10	10'	Point of intersection of 1. Line joining point 3 and point 7 2. Line passing through point $(M^-_{min}, \phi^-_{min})$ and having slope $R^-_k K_0$		$R^-_k = \frac{M^-_{min} + \alpha M^-_y}{K_0\phi^-_{min} + \alpha M_y}$
	10	$\phi^-_{min} - \frac{M^-_{min}}{R^-_k K_0}$	0	
11		Point of intersection of 1. Line joining point 5 and point 13 2. Branch 1		-
12		Point of intersection of 1. Line joining point 10 and point 14 2. Branch 1		-
13	13'	Point of intersection of 1. Line joining point 6 and point 4 2. Branch 1		-
	13	$WF\phi^-_p + (1-WF)\phi^-_w$	γM^-_y	$R^-_{k,min} = \frac{M^-_{min} + \alpha M^-_y}{K_0\phi^-_{min} + \alpha M^-_y}$ $\phi^-_p = \phi^-_{min} - \frac{M^-_{min} - \gamma M^-_y}{R^-_{k,min} K_0}$ $\phi^-_y = M^-_y / K_0$ $\phi^-_w = \gamma\phi^-_y$ Weighting Factor, $WF = \gamma$
		If $M^-_{min} > M^-_y$ and $M < M^-_w$ then		
		M^-_w / K_0	M^-_w	
14	14'	Point of intersection of 1. Line joining point 7 and point 3 2. Branch 1		-
	14	$WF\phi^+_p + (1-WF)\phi^+_w$	γM^+_y	$R^+_{k,max} = \frac{M^+_{max} + \alpha M^+_y}{K_0\phi^+_{max} + \alpha M^+_y}$ $\phi^+_p = \phi^+_{max} - \frac{M^+_{max} - \gamma M^+_y}{R^+_{k,max} K_0}$ $\phi^+_y = M^+_y / K_0$ $\phi^+_w = \gamma\phi^+_y$ Weighting Factor, $WF = \gamma$
		If $M^+_{max} < M^+_y$ and $M > M^+_w$ then		
		M^+_w / K_0	M^+_w	
15		ϕ^+_{max}	M^+_{max}	-
16		ϕ^-_{min}	M^-_{min}	-
17	A	Point of intersection of 1. Line joining point 1 and point 3 2. Line passing through current point and having slope $R^+_k K_0$		$R^+_k = \frac{M_{max} + \alpha M^+_y}{K_0\phi^+_{max} + \alpha M^+_y}$
	B	Point of intersection of 1. Line joining point 3 and point 5 2. Line passing through current point and having slope $R^+_k K_0$		
	C	Point of intersection of 1. Line joining point 10 and point 12 2. Line passing through current point and having slope $R^+_k K_0$		
	D	Point of intersection of 1. Line joining point 14 and point 15 2. Line passing through current point and having slope $R^+_k K_0$		
18	A	Point of intersection of 1. Line joining point 2 and point 4 2. Line passing through current point and having slope $R^-_k K_0$		$R^-_k = \frac{M_{min} + \alpha M^-_y}{K_0\phi_{min} + \alpha M^-_y}$
	B	Point of intersection of 1. Line joining point 4 and point 6 2. Line passing through current point and having slope $R^-_k K_0$		
	C	Point of intersection of 1. Line joining point 9 and point 13 2. Line passing through current point and having slope $R^-_k K_0$		
	D	Point of intersection of 1. Line joining point 3 and point 13 2. Line passing through current point and having slope $R^-_k K_0$		
19		Point of intersection of 1. Line joining point 11 and point 15 2. Line passing through current point and having slope $R^+_k K_0$		$R^+_k = \frac{M_{max} + \alpha M^+_y}{K_0\phi^+_{max} + \alpha M^+_y}$
20		Point of intersection of 1. Line joining point 10 and point 14 2. Line passing through current point and having slope $R^-_k K_0$		$R^-_k = \frac{M_{min} + \alpha M^-_y}{K_0\phi_{min} + \alpha M^-_y}$
21		$\phi_{current}$	$M_{current}$	

Table 1 Point Formulas

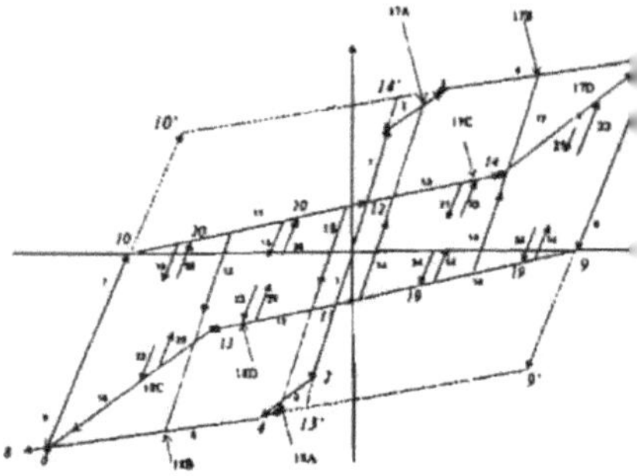

Fig. 1 Points and Branches

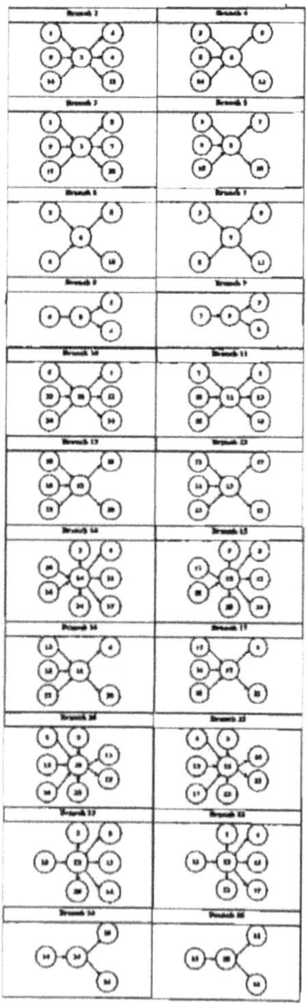

Table 2 Branch Map

3 The Smooth Model

Alternatively, the continuous model developed here is a slight modification of the popular Wen-Bouc model [1]. The concept is shown in Fig. 2. The Wen-Bouc model uses a hysteretic parameter 'Z' and defines the stiffness at any given force and deformation levels implicitly. Thus implementation of this model involves solving a differential equation. On the other hand, the present model defines the stiffness explicitly as a function of the current force and deformation levels as shown below. Examples of the hysteretic behavior of this model are shown in Fig. 4.

$$\frac{dF}{d\mu} = \alpha K_0 + \frac{1}{f_1 + f_2}$$

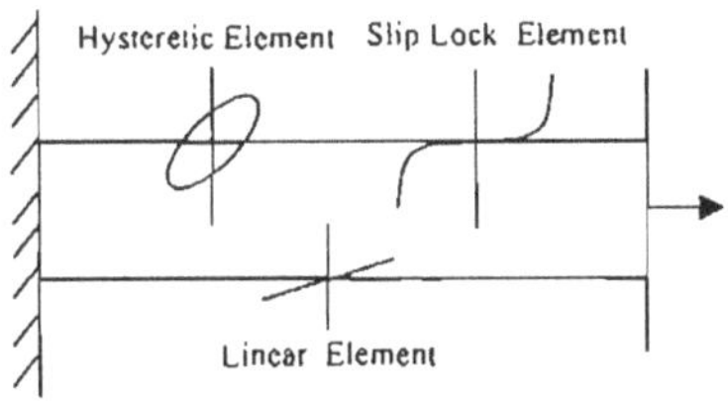

Fig. 2 Concept of the Smooth Model

where,

K_0 = Initial Elastic Stiffness

μ = ductility = $\dfrac{u}{u_y}$

α = Ratio of Post-yield stiffness to initial elastic stiffness

f_1 = Flexibility of the Hysteretic Element =

$$= \frac{1}{(1-\alpha)K_0\left[1-\left|\mu_{cur}-\mu_0\right|^n\left\{\eta_1 \operatorname{sgn}\left(\dot{F}_{cur}\Delta\mu\right)+\eta_2\right\}\right]\left[1-U\left(\left|\mu_{cur}-\mu_0\right|-1\right)\left\{\eta_1 \operatorname{sgn}\left(\dot{F}_{cur}\Delta\mu\right)+\eta_2\right\}\right]}$$

where $\dot{F}_{cur} = F_{cur} - \alpha K_0 u_{cur}$

η_1 and η_2 are parameters that determine the shape of the unloading curve;

the sum of these is always one, to satisfy the causality condition

$$\mu_0 = \mu_{cur} - \frac{\dot{F}_{cur}}{(1-\alpha)K_0 u_y} + \frac{1}{n+1}\operatorname{sgn}(\Delta\mu)$$

and U *is* the Heaviside Step Function

f_2 = Flexibility of the Slip Lock Element = $\sqrt{\dfrac{2}{\pi}}\dfrac{a}{\sigma}\exp\left[-\dfrac{(F_{cur}-\bar{F})^2}{2\sigma^2}\right]$

where $2a$ = Slip Length

$\bar{F}$ = the average force level about which the slip occurs

and σ governs the rate of slip, i.e. in what force interval the slip occurs

4 Conclusions

Two hysteretic models were developed. The polygonal (multi-linear), an extension of a previous version (IDARC 2D [2]), was motivated by the need of force/displacement control, while the smooth model is developed due its computational simplicity. The models were incorporated in the nonlinear analysis program IDARC2D [3] with similar performance.

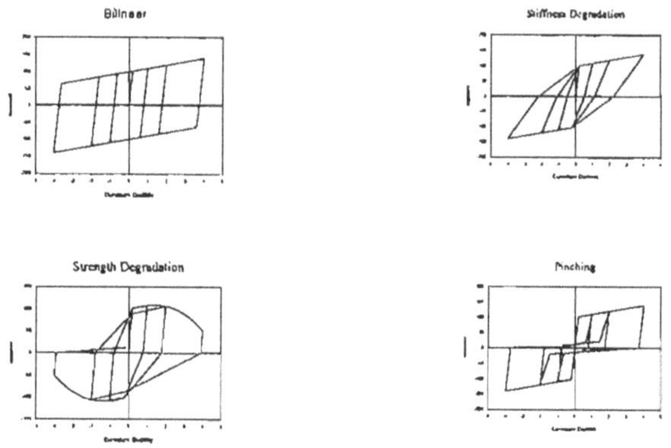

Fig. 3 Capabilities of Multi-linear Hysteretic Element

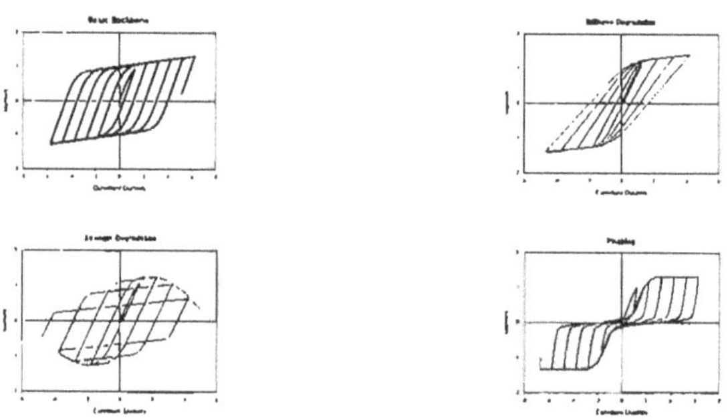

Fig. 4 Capabilities of the Smooth Model

References

1. Baber, T.T. and Noori, M.N., Random Vibration of Degrading, Pinching Systems, *J. Eng. Mech., ASCE,* Vol. 111, No. 8, August, 1985
2. Valles, R.E., Reinhorn, A.M., Kunnath, S.K., Li, C. and Madan, A., IDARC2D Version 4.0: A Computer Program for Inelastic Damage Analysis of Buildings. *Technical Report NCEER-96-0010,* June, 1996
3. Skliros, K., Reinhorn, A.M., Sivaselvan, M.V. and Kunnath, S.K., IDARC2D Version 5.0: A Computer Program for Inelastic Damage Analysis of Buildings. *Technical Report NCEER-98-xxxx*
4. Sivaselvan, M.V., Skliros, K, Reinhorn, A.M. Multi-linear and Smooth Hysteretic Models for Degrading Structures, Technical Report, University at Buffalo (SUNY), April 1998

Disk Brake Squeal
(by a Model of a Rotating Disk and a Beam)

Masaru Suganami[1], Mikio Nakai[1], Masayuki Yokoi[2]
Toshifumi Miyahara[3], Hiromitsu Matsui[4],
Ichirou Yamazaki[4]

ABSTRACT *This paper describes squeals caused by the in-plane vibration in the circumferential direction of the rotating disk when the beam was rubbed against it.*

1 Introduction

Disk brake squeals are frequently generated by the axial vibration of a disk. Only the axial vibration of the disk has therefore been noted for squeal analysis or elimination of squeal. However, squeals are also generated by the in-plane vibration in the circumferential direction of the disk (Here, we simply call it the circumferential vibration), and comprehension of these squeals is far from complete. In this paper, the possibility of occurring the squealing sound due to the circumferential vibration is pursued.

2 Experimental Procedures

Figure 1 shows an experimental apparatus composed of only a rotating disk and a beam. A beam with a resin serving as a brake pad was pressed against the side face of a rotating disk in its axial direction. The disk was clamped centrally with a free periphery, and three disks of 0.75, 1.6 and 2.0 cm in thickness (5.0 cm in inner diameter, 30 cm in outer diameter) were used. A squeal noise was generated when the disk was rotated at 40 min^{-1} under a contact load which was applied to the beam through a wire by a weight. A beam can move freely in the direction of the load through pins. In order to produce a squeal due to the circumferential vibration, the beam was inclined horizontally and vertically with clamps and shaft. A squeal due to the circumferential vibration was detected through two procedures: (1) Vibrational modes in the circumferential direction of the disk were examined by means of strain gages bonded on the disk circumference. (2) Two accelerations were measured by two accelerometers mounted outward on both sides of disk as shown in Fig.3(b). Since the circumferential

[1]Department of Precision Engineering, Kyoto University, Kyoto, 606-8501 Japan, [2]Osaka Sangyo University,[3]Matsushita Electric Co., Ltd, [4]Nissan Motor Co., Ltd.

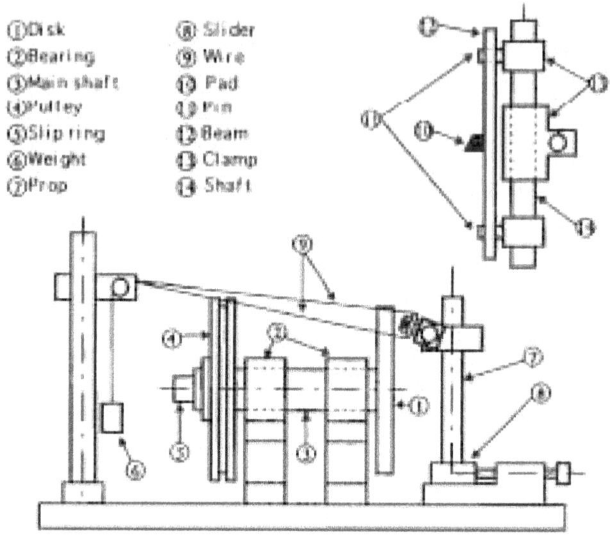

Figure 1: Experimental setup

vibration of the disk produces a wave of condensation and rarefaction on the disk in the presence of a squeal, the disk becomes thick on condensation and thin on rarefaction. Two accelerations, therefore, have the same phase for a squeal due to the circumferential vibration. Sound pressure was measured at a distance of about 5 cm from the contact part of the disk and the pad. A used beam is 3.2 cm in width, 0.56 cm in height and 30 cm in length. Applied contact loads are 78.4∼176.4 N.

3 Experimental Results and Discussion

In close relation to squeal frequency, natural frequencies in the transverse vibration of the beam and those in the axial and circumferential vibrations of the disk (2.0cm in thickness) are shown in Figs.2 (a),(b) and (c). In these figures, (n-s) and [n-s] indicate natural frequencies in the axial and circumferential vibrations of the disk, where n is the number of nodal diameters and s is that of nodal circles. Table 1 also shows calculated natural

Table 1: Calculated natural frequencies in the circumferential vibration of a disk (2.0 cm in thickness) and mode shape for [3-1]

1645Hz	[0-0]		
4922	[1-0]	11878Hz	[1-1]
8512	[2-0]	14869	[2-1]
12290	[3-0]	19706	[3-1]
15794	[4-0]	25057	[4-1]
19223	[5-0]		

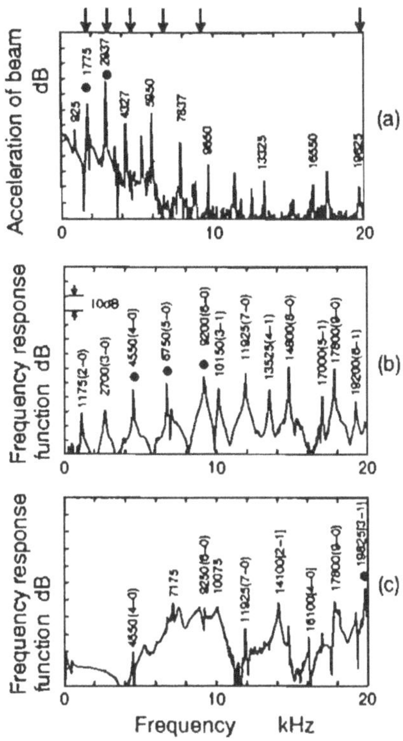

Figure 2: Natural frequencies (a): beam, (b): axial direction of the disk, (c): circumferential direction of the disk (2.0 cm in thickness)

frequencies in the circumferential vibration of the disk. As an example, the mode shape of [3-1] is illustrated, where the length and the direction of an arrow indicate the magnitude and the direction of displacement. The results for each disk thickness are as follows :

(1) Disk of 0.75 cm in thickness: Three kinds of squeals occurred at 1575, 4350 and 9650 Hz. We could, however, not verify if their squeals were generated by the circumferential vibration.

(2) Disk of 2.0 cm in thickness: Squeal frequencies are plotted in Fig.2 by ●. Squeals at 1700 Hz and 3050 Hz are generated by the mode coupling of only the beam. Squeals at 4600, 6750 and 9200 Hz are dominated by the axial vibrations of the (4-0), (5-0) and (6-0) modes, respectively. A squeal might occur at 19750 Hz by the circumferential vibration. Figure 3(a) presents time histories of sound pressure, axial and circumferential accelerations of the disk, circumferential strain of the disk and acceleration of the beam for the squeal at 19750 Hz. Figure 3(b) shows enlarged time histories of axial accelerations measured by two accelerometers mounted

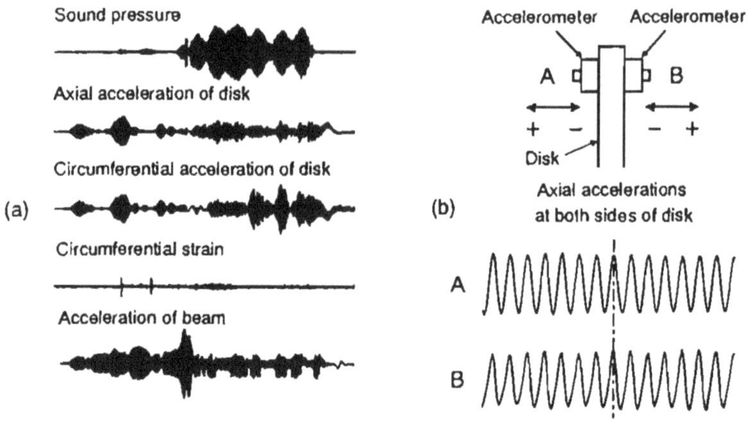

Figure 3: Time histories at 19750 Hz (a): sound pressure, accelerations and strain of the disk and acceleration of the beam, (b): axial accelerations

outward on both sides of disk. It can be seen that there is no phase difference between the two time histories as shown by a chain line in Fig.3(b). This demonstrates that the disk vibrates in the circumferential direction. Furthermore, we measured phase differences in time histories of displacements by strain gages bonded on each position of the disk circumference. As a result, it was found that this squeal was dominated by the [3-1] mode of the circumferential vibration. Furthermore, the squeal of 25375 Hz occurred in the [4-1] mode of the circumferential vibration. Both squeals due to the circumferential vibration occurred under high contact load 176.4 N.
(3) Disk of 1.6 cm in thickness: The squeal of the [3-1] mode at 19750 Hz was also generated by the circumferential vibration of the disk.

The above experimental results provided support of linear analysis on the system of the disk and the beam.

4 Conclusions

The experiments show frequent occurrences of squeals whose frequencies are close to the natural frequencies in the circumferential vibration of a disk. In addition, the squeal due to the circumferential vibration occurs in the vibrational mode having only one nodal circle and it is readily generated for thicker disk and higher contact load.

Reference

[1] Matsui H., *et al.*, Analysis of Disc Brake Squeal, SAE Paper 920553, 1992, p15.

Modelling and Stability Analysis of an Airplane Engine with a 4-Blade Propeller

D. Boulahbal[1], F. Golnaraghi[1], A. Khajepour[1], T. Haskett[1] and T. Zelenka[1]

ABSTRACT Dynamic balancing of turboprop engines is very difficult to achieve because of both the large size of the propellers' blades and their inherent flexibility. This latter aspect causes the propeller's mass center to whirl at a rate different from the rotation of the propeller. Soft rubber mounts are utilized to reduce the vibration transmission from the engine to the airplane's structure. Very resilient mounts would maximize the vibration isolation but lower the onset speed of instability of the engine/propeller assembly. In this paper, a ten degrees-of-freedom model of the engine is developed, and later used to investigate the stability issues.

Introduction

Turboprop engines consume less fuel than turbojet engines. Thus, they are more economical for short haul flights. A major setback is increased subsonic vibration levels in the airplane cabin. This vibration has its origin in the flexibility of the propeller blades which gives rise to a rotating unbalance, not necessarily synchronous with the propeller rotation, that travels through the wings to the airplane fuselage. In this undertaking, we develop a dynamic model for the engine/propeller assembly, and use it to investigate the stability aspects. Of primary interest are the effects of blade and mounts stiffness on the instability behavior of the dynamic system.

Assumptions and Mathematical Modelling

A schematic depiction of a typical turboprop engine/propeller is given in Figure 1. The engine block is assumed perfectly rigid and attached to the wing structure through five resilient mounts. The wing is assumed immobile in this study and its dynamics can be easily incorporated through the Impedance Coupling Method. Each flexible blade is modeled as a pendulum pinned to the shaft's hub. The length of the blades corresponds to the radius of gyration of the actual physical model. Six coordinates are required for the engine block and two for each of the four blades. The coordinates for each blade are coupled through a non-holonomic constraint that involves

[1]Department of Mechanical Engineering, University of Waterloo, Waterloo, Ontario, Canada, N2L 3G1

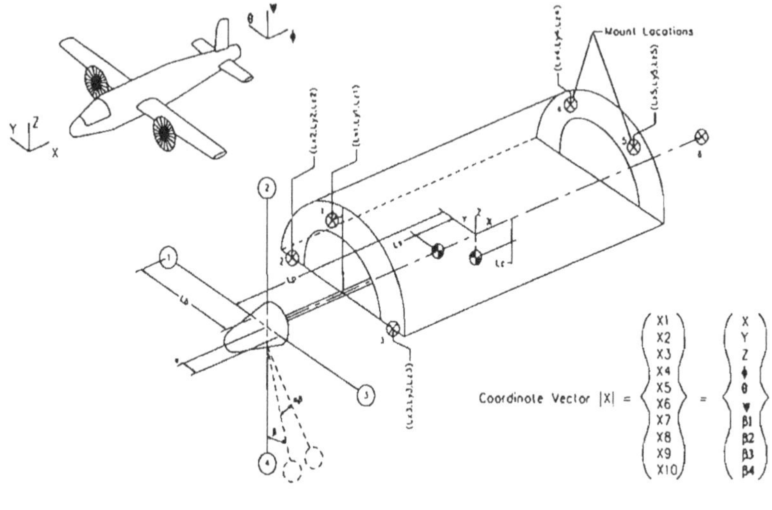

FIGURE 1. Engine/Propeller Degrees-of-Freedom

the blade's angle of attack. Thus we have a total of ten degrees-of-freedom. The rotation of the shaft is specified and it is an ignorable coordinate. The aerodynamic loading on the blades balances the torque provided by the engine and thus both loadings are ignored. The equations of motion are obtained via Lagrange's technique and can be cast in the matrix form.

$$[\mathrm{M}(t)]\{\ddot{\mathrm{x}}\} + [\Omega \mathrm{G}(t)]\{\dot{\mathrm{x}}\} + \left[\mathrm{K} + \Omega^2 \mathrm{E}(t)\right]\{\mathrm{x}\} = 0 \tag{1}$$

Because of the rotation rate Ω of the engine shaft, all matrices but K have time-dependent entries. This aspect complicates the stability analysis and to eliminate it, we use a Multiblade Coordinate Transformation (MCT) as in [2], which then yields a new set of equations with *constant* coefficients.

$$[\mathrm{M}]\{\ddot{\mathrm{x}}\} + [\Omega \mathrm{G}]\{\dot{\mathrm{x}}\} + \left[\mathrm{K} + \Omega^2 \mathrm{E}\right]\{\mathrm{x}\} = 0 \tag{2}$$

Though we have kept the same generic notation, the four new matrices in (2) are different from those of (1). In the present study, we have made use of the Buckingham Π theorem to non-dimensionalize (1) to enable comparison with an experimental test-rig currently being constructed, and make it applicable to similar dynamic applications. Dynamic systems described by (2) are usually rotordynamic in nature and have several potential instability sources, which we address next.

Stability Analysis and Discussion

Imposing harmonic time dependence $\mathbf{x}(t) = \mathbf{X}e^{st}$ in (2), yields

$$\left[s^2\mathbf{M} + s\left(\Omega\mathbf{G}\right) + \left(\mathbf{K} + \Omega^2\mathbf{E}\right)\right]\{\mathbf{X}\} = 0 \tag{3}$$

The eigenvalue problem (4) is solved for a range of values of Ω and the resulting eigenvalues s are displayed in the form of Campbell diagrams. Two cases are presented: in Figure 2 the propeller blades are assumed absolutely rigid whereas in Figure 3 they are given some flexibility. In both cases the siffnesses of the five mounts are kept at nominal values.

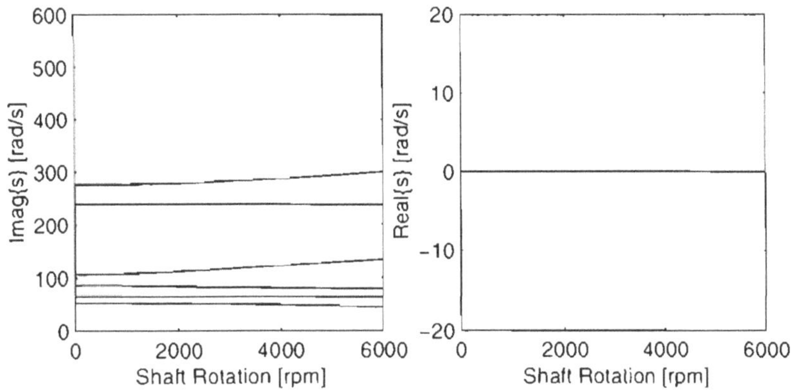

FIGURE 2. Campbell Diagram for Engine with Rigid Propeller

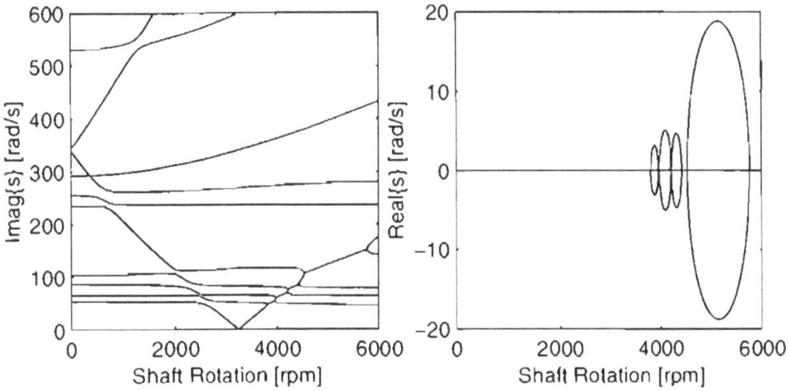

FIGURE 3. Campbell Diagram for Engine with Flexible Propeller

From the graph for $\Re\{s\}$ in Figure 2, operation of the engine is stable for all speeds. Unstable regions due to nonconservative coupling appear as soon as the blades are allowed some flexibility, Figure 3. Operation of the engine should of course be below the lowest onset speed of instability. The location and width of the unstable regions can be altered by changing both the mounts stiffness and the blade stiffness, though the second option is rather unrealistic. Figure 4 shows the effect the stiffness of the mounts has upon the unstable regions of the engine.

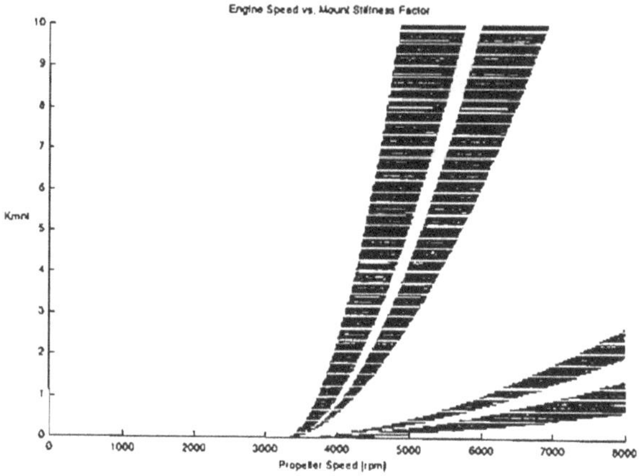

FIGURE 4. Effects of Mount Stiffness on Stability Boundaries

The immediate conclusion is that further reduction of the mounts stiffness, for better vibration isolation, is not a sound approach because of the stability issue. The alternative is to use an active vibration control approach, and this will be addressed in the future work.

Acknowledgments

We acknowledge the financial support provided by deHavilland Inc., and the Natural Sciences and Engineering Research Council of Canada. We also thank Mr Barry Leigh of deHavilland Inc., for providing valuable advice.

References

[1] D. Boulahbal, M.F. Golnaraghi, A. Khajepour, T. Haskett, T. Zelenka, and L. Toth, The deHalvilland Engine Propeller Investigation. *Interim Company Report*, Waterloo, Ontario, February 1998.

[2] Crandall, S.H. & Dugundji, J., Resonant whirling of airplane propeller-engine systems. *ASME J. of Applied Mechanics*, 48:929–935, 1981.

Motion of Beams with Unilateral Contact Constraints and Wear

L. Johansson[1]

ABSTRACT In the present paper the motion of a beam subjected to a stochastic load and constrained by unilateral contact conditions, which are allowed to change due to wear, is studied.

1 Introduction

In this paper the motion of a beam excited by random forces, characterized by their spectral density, is considered. The beam is constrained by unilateral contact against springs at contact sites along the beam. The problem is strongly nonlinear, since there might be initial gaps at the contact sites, since the contact sites are allowed to wear down and since Coulomb friction is included. The corresponding problem for a single DOF system without friction and wear has, of course, been studied previously, see e.g. Shaw and Holmes [1] and references therein. This Paper concentrates on the behaviour with several degrees of freedom when wear is allowed and the load is random. There is also an analysis of the best choice of parameters in Newmark's integration scheme in order to represent important frequencies accurately.

2 The Computational Algorithm

In the algorithm used, the beam equation of motion is discretized in space by the finite element method and in time by Newmark's method. The contact forces are computed from the displacements using a point contact algorithm with Coulomb friction, and wear is calculated using Archard's law.

[1]Department of Mech. Eng., Linköping University, S-581 83 Linköping, Sweden

To calculate the time history of the beam displacements, an iterative procedure in two levels is adopted. Thus, it is first assumed that the contact forces $\boldsymbol{F}_{cont}$ are known and the corresponding displacements $\boldsymbol{x}$ are computed from the discretized beam equation of motion, $\boldsymbol{M}\ddot{\boldsymbol{x}}^{n+1} + \boldsymbol{C}\dot{\boldsymbol{x}}^{n+1} + \boldsymbol{K}\boldsymbol{x}^{n+1} = \boldsymbol{F}^{n+1}_{appl} + \boldsymbol{F}^{n+1}_{cont}$, where $\boldsymbol{F}_{appl}$ is the applied force which is assumed to be known. Next the displacements are used as input to the contact algorithm to compute the contact forces. The process is repeated until convergence is reached in the contact forces. The gaps at the contact sites are then adjusted for wear, and the process is repeated until convergence is also reached in the forces with respect to wear. The algorithm used is described in detail in Johansson [2].

2.2 *Simulation of stochastic loading*

In order to simulate a stochastic load, sample time histories tailored to mimic specified spectral densities are generated. For this purpose trigonometric functions are used:

$$F_{appl}(t) = \sum_{k=1}^{m} \sqrt{2}\sigma_k \cos(\omega_k t + \phi_k) \tag{1}$$

where ϕ_k are independent random variables distributed between 0 and 2π.

The corresponding spectral density is, Soong and Grigoriu [3],

$$S(\omega) = \frac{1}{2}\sum_{k=1}^{m} \sigma_k^2 \delta(\omega - \omega_k) \tag{2}$$

so that σ_k and ω_k can be chosen to obtain a desired spectral density for F_{appl}.

2.3 *Time discretization*

Using Newmark's method for the time discretization, Newmark [4], we introduce

$$\dot{x}^{n+1} = \dot{x}^n + [(1 - \gamma)\ddot{x}^n + \gamma\ddot{x}^{n+1}]\Delta t \tag{3b}$$

$$x^{n+1} = x^n + \dot{x}^n\Delta t + [(\frac{1}{2} - \beta)\ddot{x}^n + \beta\ddot{x}^{n+1}]\Delta t^2 \tag{3a}$$

Since an important application of this work is to apply a broad band noise to structures and analyze the resulting displacement to identify dominant frequencies in the response, the parameters γ and β are chosen so as to give as accurate as possible values for the eigenfrequency of a single degree of freedom linear example. The value γ=0.5 is chosen in order to avoid numerical dissipation.

To study the influence of β on the forced responce, consider a single degree of freedom linear oscillator with harmonic excicitation:

$$\ddot{x} + \omega^2 x = \frac{F_0}{m}\sin(\Omega t) \tag{4}$$

Using the Newmark discretization scheme above, with γ=0.5, gives

$$x^{n+1} - \left(2 - \frac{\omega^2 \Delta t^2}{1 + \beta\omega^2\Delta t^2}\right)x^n + x^{n-1} =$$

$$= \frac{\Delta t^2 \beta F_0}{m(1 + \beta\omega^2\Delta t^2)}\left[\sin(\Omega\Delta t(n+1)) + \frac{1-2\beta}{\beta}\sin(\Omega\Delta tn) + \sin(\Omega\Delta t(n-1))\right] \tag{5}$$

A particular solution to this equation is

$$x^n_{part} = \frac{F_0}{m}\cdot\frac{1 + 2\beta(cos(\Omega\Delta t) - 1)}{\omega^2 - 4(\frac{1}{\Delta t^2} + \beta\omega^2)\sin^2(\frac{\Omega\Delta t}{2})}sin(\Omega\Delta tn) \tag{6}$$

It is seen that in order to to put a spike at the eigenfrequency in the approximate solution, β should be chosen such that the denominator of (6) is zero when Ω=ω. This leads to the condition

$$\beta = \frac{1}{4\sin^2(\frac{\omega\Delta t}{2})} - \frac{1}{\Delta t^2\omega^2} \tag{7}$$

It is interesting to compare the present analysis with a similar analysis by Newmark [4]. Here the properties of the discretization scheme is studied by considering the free vibration of a single degree of freedom system. Thus, the homogenous solutions of eq. (4) and (5) are found and used to study how the period and maximum velocity are distorted by the discretization. Using the equations in Newmark's paper to calculate the value of β which does not distort the period, eq. (7) above is again recovered.

3 Numerical Example

The example is concerned with a beam in plane motion, shown in fig. 1. It was decided to use a single beam element in order not to get too many eigenfrequencies to complicate the analysis. The first undamped eigenfrequency with no springs (infinite gap) is 4.996 rad/s, with springs attached (infinite grip) it is 15.04 rad/s and with one spring attached (approximately corresponding to zero gap) it is 11.65 rad/s. The second eigenfrequency is 49.22 rad/s, 54.24 rad/s and 51.69 rad/s with zero, two and one spring attached, respectively.

Fig. 1 shows to the top right the evolution of the gap for an excitation with constant spectral density between 0 rad/s and 20 rad/s for a few values of the coefficient of friction. To the bottom is the gap evolution for zero friction, but with 2 rad/s wide "holes" in the spectral density of the excitation. To the left the excitation is between 0 rad/s and 20 rad/s, except for the "holes", and to the right is the response to a broader band excitation with frequencies between 0 rad/s and 60 rad/s, thus exciting also the higher eigenfrequency. It is seen that the when there is a broader band excitation, the behaviour is different; there is no longer a significant delay in wear when the excitation of the 15 rad/s frequency is removed.

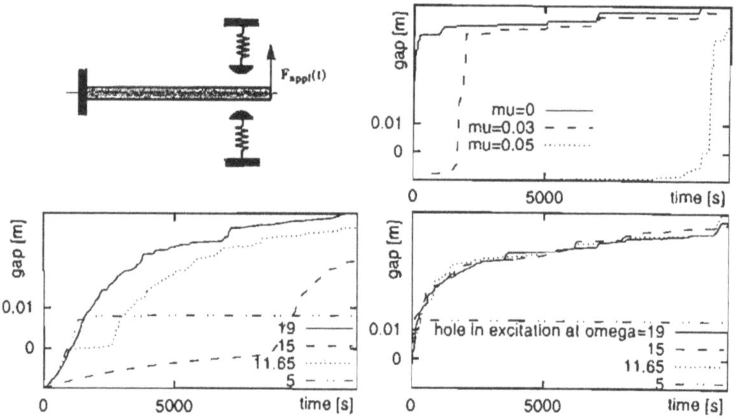

Figure 1. The geometry of the example and some evolutions of the gap.

4 References

[1] Shaw, S.W. & Holmes, J.P., (1983) A Periodically Forced Piecewise Linear Oscillator, *Journal of Sound and Vibration,* Vol. 90, pp. 129–155.

[2] Johansson, L., 1997, Beam Motion With Unilateral Contact Constraints and Wear of Contact Sites, *ASME Journal of Pressure Vessel Technology,* Vol. 119, pp. 105–110.

[3] Soong, T.T. & Grigoriu, M., 1993, *Random Vibration of Mechanical and Structural Systems*, Prentice–Hall.

[4] Newmark, N.M., 1959, A Method of Computation for Structural Dynamics, *Journal of the Engineering Mechanics Division*, Proc. ASCE 85, EM3, pp. 67–94.

Control of Impacting Elastic Elements

Cristian Diaconescu[1]
Dan B. Marghitu[2]
A. Guran[3]

ABSTRACT The paper presents the control of a repetitive impacting elastic link in rotational motion. A fuzzy logic controller is designed to suppress the vibration of the flexible beam resulting after the impact with an external rigid body.

1 Introduction

In recent years, fuzzy logic controllers demonstrated their superiority in many cases over conventional control schemes. This is mainly due to their capacity to handle our inexact knowledge about a real world system. In fact, most of the dynamic processes have nonlinearities that are difficult to be exactly modeled in mathematical forms. The first succesfull industrial application was Sugeno's fuzzy car, which has the capacity of learning from examples [5]. Boghiu et al. [2] studied the flap motion control of a rotating flexible beam. An overview of intelligent control with emphasis on fuzzy logic control design techniques was presented. A controller, was constructed to suppress the vibrations of an elasic beam resulting after the impact with an external rigid body. A similar strategy was used by Boghiu and Marghitu [1] to control the motion of an elastic link.

Marghitu et al. [4] developed a fuzzy logic control scheme to investigate the vibration suppression of a parametrically excited impacted beam. A three mode approximationof the beam was considered. The fuzzy logic controller demonstrated his applicability to control the processes with non-linearity. Kubica and Wang [3] applied a fuzzy logic strategy to control a rigid body and the first flexural mode of vibration in a single link robotic arm. Two fuzzy logic controllers were constructed to govern the rigid mo-

[1]Department of Mechanical Engineering Auburn University, Auburn, AL 36849, USA
[2]Department of Mechanical Engineering, Auburn University, Auburn, AL 36849, USA
[3]Canadian Institute for Structronics, 275 Slater Street, Ottawa, Ontario, Canada

tion and to attenuate the vibrations which result from the rigid body motion. The results showed an improvement over conventional multivariable control techniques.

In this article the control of a repetitive impacting elastic link in rotational motion is considered. A fuzzy logic controller was designed to suppress the vibrations resulting after the impact with an external rigid body. The momentum balance method and an empirical coefficient of restitution was used in the collision of the two bodies.

2 System Model

The system is composed by a slender flexible beam cantilevered onto a rigid massless base, with negligible dimensions. The base is attached to a rigid link of length L_o. The base can perform small rotational deflections with angle $\phi(t)$ about an axis perpendicular on the plan of motion. A spring (of constant k) and a damper (of constant c) are connected to the base and the rigid link in order to avoid large rotation of ϕ. The whole system rotates with a constant angular velocity Ω in the horizontal plane.

Lagrange's method was used to develop the nonlinear equations of motion. To describe the elastic motion, a three modes approximation was used. The impact equations of motion were determined neglecting the friction during impact. A basic assumption was that the configuration of the bodies was held constant in the analysis of the collision process, with no significant change in mass and moments of inertia.

A fuzzy logic controller is designed such that the elastic vibration of the beam are eliminated. The inputs of the fuzzy logic controller are the total deflection of the elastic link denoted by y_t and its time derivative $\dot{y}_t$. The output of the controller is the torque u applied to the base of the elastic link.

Figure 1(a) shows the fuzzy sets for the total beam deflection denoted by μ_{y_t}. There are 7 grades, that correspond to *negative huge* (NH), *negative big* (NB), *negative small* (NS), *zero* (ZE), *positive small* (PS), *positive big* (PB), and *positive huge* (PH) values of the total beam deflection y_t. The support for y_t is between -1 and 1 m, but can be extenguish to $\pm\infty$. For the time derivative $\dot{y}_t$, the fuzzy sets are represented in Figure 1(b). It has only 3 grades, which correspond to *negative* (N), *zero* (Z), and *positive* (P) values of $\dot{y}_t$. The support vector is taken between -2 and 2 m/s, but $\dot{y}_t$ can have any value, by extension.

The fuzzy sets of the controller output μ_u is shown in Figure 1(c). It consists of equidistant triangle grades. The control force fuzzy set has 5 components, that correspond to *negative big* (NB), *negative small* (NS), *zero* (ZE), *positive small* (PS), *positive big* (PB) values of the controller output. The support vector of the control torque was between -4 and 4 Nm.

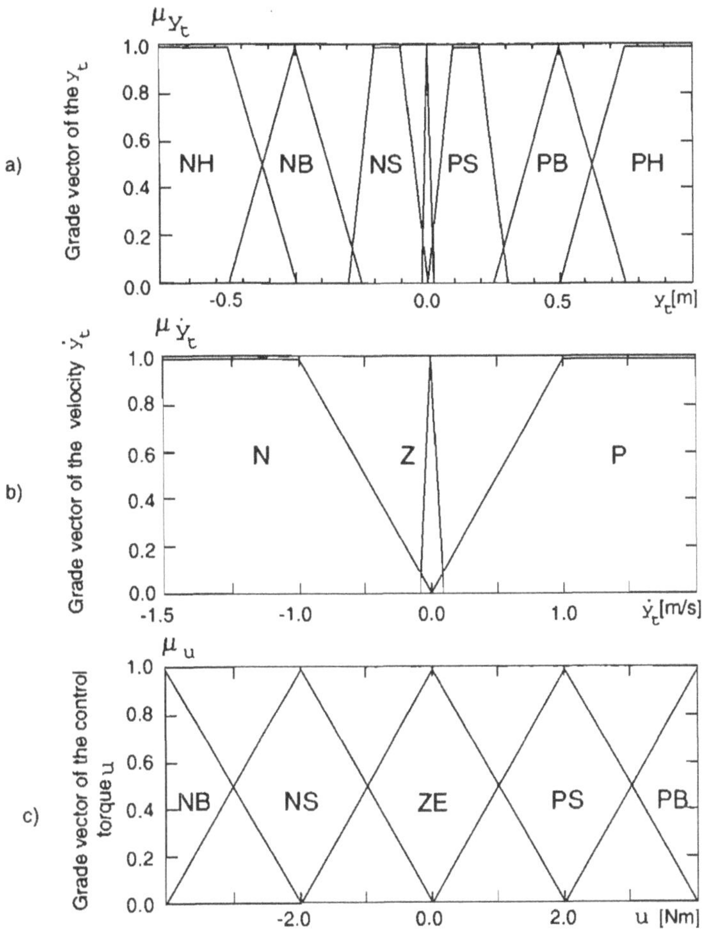

FIGURE 1. Fuzzy sets of the coontroller inputs and output

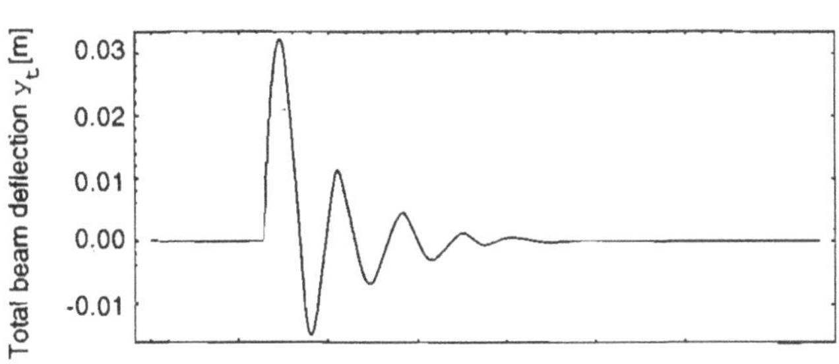

FIGURE 2. Behavior of the controlled system

The controller output is generated using "If-Then" rules, based on a consequent table. Each time, according to the values of y_t and $\dot{y}_t$, one or more "If-Then" rules may be defined. The computation of the control torque u is also based on the intersection and union of fuzzy sets, as well as on the centroid method that returns the "center of mass" of a fuzzy set.

3 Simulations and Results

The simulation of controlled system for an elastic link of length L=1.5 m was performed. The rigid link has length L_o=0.05 m and the angular velocity of rotation was considered Ω=4 rad/s. Figure 2 shows the total beam deflection y_t of the undamped controlled system. The oscilations vanish after less than 1 s.

4 Conclusions

The control of an impacting rotating elastic beam with a rigid link was studied. The controller was designed using fuzzy logic strategy. Several simulations were performed showing the behavior of the controlled system.

5 References

[1] Boghiu, D. and Marghitu, D., The control of an impacting flexible link using fuzzy logic strategy, accepted for publication in the *Journal of Vibration and Control.*

[2] Boghiu, D., Marghitu, D. and Guran A., Fuzzy logic control system design of an impacting flexible link, B. M. Ayyub, A. Guran and A. Haldar (ed.) Uncertainty Modeling in Vibration and Fuzzy Analysis of Structural Systems. *Series on Stability, Vibration and Control of Systems, Series B*, 10, 319-345, 1997.

[3] Kubica, E. and Wang, D., A fuzzy-control strategy for a flexible single link robot, *American Control Conference Proceedings*, 236-241, 1993.

[4] Marghitu, D. B., Diaconescu, C. and Boghiu, D., Fuzzy logic control of parametrically excited impacting flexible system, accepted for publication in the *Journal of Applied Mechanics.*

[5] Sugeno, M. and Murakami, M., Fuzzy control of a model car, *Fuzzy Set System*, 16, 103-113, 1985.

SESSION II – (A)

Nonlinear Dynamics and Vibrations

Chairmen: Prof. M. Nakai, Kyoto University, JAPAN
Prof. T. Kapitaniak, Technical University of Lodz, POLAND

The Dynamics of Resonance Capture

R. Rand[1]

ABSTRACT The concept of resonance in dynamics may be described as follows: Two distinct dynamical components (such as modes or drivers) are said to be in resonance if their associated frequencies are commensurate, that is, if their ratio is a rational number. The most important case is certainly 1:1 resonance, although many important examples involve 2:1 resonance, as well as other ratios involving small integers. The importance of resonance is that it permits energy to travel efficiently between the components that are in resonance. The most famous example is of course the destructive effects caused by forcing an elastic system at a frequency close to its natural frequency. This lecture is concerned with another resonance phenomenon called capture into resonance.

Capture into resonance may be illustrated rather than defined by the following three examples:

1. Imagine a slightly unbalanced wheel attached to an elastic support. If a constant torque is applied to the wheel, then we would expect that the wheel would begin to spin faster and faster. A surprising thing happens, however, when the wheel's angular speed gets close to the natural frequency of the elastic support. The wheel may cease to spin up, and the energy of the applied torque may instead produce large motions of the support. The wheel is said to be captured into resonance. [17],[24],[23],[25],[18].

2. Imagine a pendulum which is rotating (going over the top) in a direction opposite to an applied torque. As time goes by, the torque reduces the angular velocity of the pendulum and eventually reverses the direction of the motion. We would expect that the pendulum would eventually begin to rotate over the top in the direction of the applied torque. This is indeed the case if the torque were a constant. However, if the magnitude of the torque smoothly changes from one value to another (maintaining the same direction), the pendulum may cease to spin up, and may instead end up undergoing small oscillations about a non-vertical equilibrium position (in which the moment of the applied torque and that of gravity cancel each

[1]Department of Theoretical and Applied Mechanics, Cornell University, Ithaca, NY 14853

other.) [22].

3. It is well known that as the Moon travels in its orbit about the Earth, it always keeps one face pointing towards the Earth. How might this situation have arisen? In order that the Moon keep one side facing the Earth, it must complete one revolution about its axis in one orbital period. That is, this motion involves a 1:1 resonance between the motion of the Moon as a particle in orbit, and its motion as a rotating solid body. Imagine that at some time in its history the Moon did not have one side always facing the Earth. A model of the Moon's motion shows that tidal friction could provide a torque which gradually changed the Moon's rotation rate until it was captured into its present 1:1 resonance. [9],[21].

Many other examples exist [8],[6], including applications to dual-spin spacecraft [17],[16],[14],[29],[18],[19],[15],[12].

In all these examples the mathematical model reveals that there is a slowly moving separatrix which surrounds a center. *Captured motions correspond to those orbits which cross the separatrix and end up circulating around the center.* Some motions do not enter the separated region and are not captured. The question of the fate of a particular motion is found to be determined by its initial conditions.

A great deal of work has been done on the mathematical theory of the asymptotics of separatrix crossing. [28],[13],[26],[10],[27],[3],[5],[20],[1],[4],[2],[7],[11].

References

[1] D.L. Bosley and J. Kevorkian, "Sustained Resonance in Very Slowly Varying Oscillatory Hamiltonian Systems", *SIAM J. Appl. Math.* 51:439-471 (1991).

[2] D.L. Bosley and J. Kevorkian, "Adiabatic Invariance and Transient Resonance in Very Slowly Varying Oscillatory Hamiltonian Systems", *SIAM J. Appl. Math.* 52:494-527 (1992).

[3] F.J. Bourland and R. Haberman, "Separatrix Crossing: Time-Invariant Potentials with Dissipation", *SIAM J. Appl. Math.* 50:1716-1744 (1990).

[4] F.J. Bourland and R. Haberman, "Capture and the Connection Formulas for the Transition Across a Separatrix", in *Asymptotic Analysis and the Numerical Solution of Partial Differential Equations,* H.G. Kaper and M. Garbey, eds., Dekker (1991), pp. 17-30.

[5] F.J. Bourland, R. Haberman and W.L. Kath, "Averaging Methods for the Phase Shift of Arbitrarily Perturbed Strongly Nonlinear Oscillators with an Application to Capture", *SIAM J. Appl. Math.* 51:1150-1167 (1991).

[6] J.Bridge and R.Rand, "Chaos and Symbol Sequences in Systems with a Periodically-Disappearing Figure-Eight Separatrix"' in *Bifurcation Phenomena and Chaos in Thermal Convection,* eds. H.H.Bau, L.A.Bertram, S.A.Korpela, pp.47-55, Amer.Soc.Mech.Eng., HTD-Vol.214, (1992).

[7] T.J. Burns and C.K.R.T. Jones, *"A Mechanism for Capture into Resonance",* National Institute of Standards and Technology paper no. NISTIR 5104, U.S.Dept.Commerce, Technology Administration, NIST, Gaithersburg, MD 20899, (1993).

[8] V.T.Coppola, R.H.Rand, "Chaos in a System with a Periodically Disappearing Separatrix", *Nonlinear Dynamics* 1: 401-420 (1990).

[9] B.Gladman, D.D.Quinn, P.Nicholson, and R.Rand, "Synchronous Locking of Tidally Evolving Satellites", *Icarus* 122:166-192 (1996).

[10] R. Haberman, "Energy Bounds for the Slow Capture by a Center in Sustained Resonance", *SIAM J. Appl. Math.* 43:244-256 (1983).

[11] R. Haberman and E.K. Ho, "Boundary of the Basin of Attraction for Weakly Damped Primary Resonance", *J. Appl. Mech.* 62:941-946 (1995).

[12] R. Haberman, R. Rand and T. Yuster, "Resonant Capture and Separatrix Crossing in Dual-Spin Spacecraft", submitted for publication.

[13] J. Henrard, "Capture in Resonance: An Extension of the Use of Adiabatic Invariants", *Celestial Mechanics* 27:3-22 (1982).

[14] C.D. Hall, "Resonance Capture in Axial Gyrostats", *J.Astronauticsl Sciences* 43:127-138 (1995).

[15] C.D. Hall, "Momentum Transfer in Torque-Free Gyrostats", Chapter 3 in *Nonlinear Dynamics: The Richard Rand 50th Anniversary Volume,* ed. A.Guran, World Scientific Pub.Co., pp.60-88 (1997).

[16] C.D. Hall and R.H. Rand, "Spinup Dynamics of Axial Dual-Spin Spacecraft", *J.Guidance, Control, and Dynamics* 17:30-37 (1994).

[17] R.J. Kinsey, D.L.Mingori and R.H.Rand, "Spinup Through Resonance of Rotating Unbalanced Systems with Limited Torque", *Proceedings of the 1990 AIAA/AAS Astrodyamics Conference,* Part 2, AIAA paper no. AIAA-90-2966, AIAA, Washington, DC, 1990, pp.805-813.

[18] R.J.Kinsey, D.L.Mingori and R.H.Rand, "Limited Torque Spinup of an Unbalanced Rotor on an Elastic Support", *Proceedings of the 1995 American Control Conference,* Seattle, Washington, June 21-23, 1995 Vol.6, pp.4368-4373.

[19] R.J. Kinsey, D.L.Mingori and R.H.Rand, "Nonlinear Control of a Dual-Spin Spacecraft During Despin Through Precession Phase Lock", *J.Guidance, Control, and Dynamics* 19:60-67 (1996).

[20] A.I. Neishtadt, "Probability Phenomena Due to Separatrix Crossing", *Chaos* 1:42-48 (1991).

[21] D.Quinn, B.Gladman, P.Nicholson, and R.Rand, "Relaxation Oscillations in Tidally Evolving Satellites", *Celestial Mechanics and Dynamical Astronomy* 67:111-130 (1997).

[22] D.D.Quinn and R.H.Rand, "A Perturbation Approach to Resonant Capture", Chapter 6 in *Smart Structures, Nonlinear Dynamics and Control,* eds. A.Guran and D.J.Inman, pp.226-246, Prentice Hall (1995).

[23] D.Quinn, R.Rand, and J.Bridge, "The Dynamics of Resonant Capture", *Nonlinear Dynamics* 8:1-20 (1995).

[24] R.H. Rand, R.J.Kinsey, and D.L.Mingori, "Dynamics of Spinup through Resonance", *Int.J. Nonlinear Mechanics,* 27:489-502 (1992).

[25] R.H.Rand and D.D.Quinn, "Resonant Capture in a System of Two Coupled Homoclinic Oscillators", *J.Vibration and Control* 1:41-56 (1995).

[26] C. Robinson, "Sustained Resonace for a Nonlinear System with Slowly Varying Coefficients", *SIAM J. Math. Anal.* 14:847-860 (1983).

[27] J.L. Tennyson, J.R. Cary, D.F. Escande, "Change of the Adiabatic Invariant Due to Separatrix Crossing", *Physical Review Letters* 56:2117-2120 (1986).

[28] A.V. Timofeev, "On the Constancy of an Adiabatic Invariant When the Nature of the Motion Changes", *Zh. Eksp. Teor. Fiz.,* 75:1303-1308 (1978) (in Russian), *Soviet Physics JETP* 48:656-659 (1978).

[29] R. Tsui and C.D. Hall, "Resonance Capture in Unbalanced Dual-Spin Spacecraft", *J.Guidance, Control, and Dynamics* 18:1329-1335 (1995).

Stick-Slip Oscillations - Quo Vadis

H. Dankowicz & A.B. Nordmark[1]

ABSTRACT Methods of dynamical systems analysis are developed to explain the origin and bifurcations of stick-slip oscillations in a low-dimensional model of macroscopic friction. It is found that a discontinuity in the derivative of the vector field results in a sequence of grazing bifurcations with which a quantitative and qualitative agreement is achieved to impressive accuracy. It is claimed that the obtained results are generic for systems with this type of discontinuity and thus for a large class of friction models.

1 Introduction

In recent decades, a large number of friction models have been suggested to capture dynamically observed frictional interactions. With varying complexity, the models have attempted to include a dependence on relative displacements and displacement velocities through the introduction of time delays or evolution of internal state variables. With suitable choices for model parameters, the predicted dynamics have been quite successful in achieving at least qualitative agreement with experimental observations.

Where the physical interpretations and motivations of some of these models is somewhat questionable, their qualitative behavior supports further investigations. This is based on the assumption that the observed behavior is characteristic of the general form of the models, albeit not of the details. From this perspective, there are potential benefits to be reaped from studying models whose immediate derivation from first principles is out of reach.

In the present paper, a recently proposed friction model capable of capturing a large class of frictional properties is studied. The model treats the surface interactions between two bodies in contact through the inclusion of internal state variables describing an equivalent local deformation of asperities at the interface and a corresponding separation between the

[1]Department of Mechanics, Royal Institute of Technology, S-100 44 Stockholm, Sweden

surfaces. Particular emphasis is put on the stick-slip type oscillations observed for the model and their bifurcations under changes in experimental parameters.

To analyze the observed bifurcations, methods of dynamical systems analysis are implemented. In particular, a new local analysis tool is introduced based on the concept of grazing bifurcations. The latter are typical features of impact oscillators. Their application to the friction model is motivated by the presence of a discontinuity in the first derivative. It is shown that the stick-slip oscillations originate in such a grazing bifurcation with the discontinuity and proceed to bifurcate largely as a consequence of the local dynamics about the discontinuity.

Although the predicted bifurcation behavior is specific to the present model, we believe that it has implications to a general class of dynamic friction models characterized by the presence of discontinuities in some derivative. This class includes many of the suggested models. Furthermore, the analysis tool is model-independent and should find a great variety of applications in problems with impact.

2 The Friction Model

In the past decades, friction models have been developed by a large number of researchers (see Armstrong-Hélouvry *et al* [1]). While more or less of *ad hoc* origin, the models can largely be divided into those which try to explain friction and those which try to reproduce its dynamics. The former includes a variety of micro-mechanical and physical deliberations of interactions between rough, possibly lubricated surfaces. In the latter category are models of mathematical origin possessing sufficient complexity to, at least qualitatively, capture observed dynamics.

Dankowicz [2, 3, 4] discusses a model which combines some of the above features. In the typical stick-slip-type experiment of a mass pulled over a rough surface by a string whose other end moves with constant velocity, a typical bifurcation diagram is shown in Figure 1. The model has a discontinuity in the vector field's derivative at $y = 1$. This is clearly manifested in the bifurcations. Note that orbits for which $y > 1$ during some part of the motion appear as stick-slip oscillations in physical variables.

3 Discontinuity mappings

For orbits which stay away from the discontinuity, standard techniques suffice to calculate stability and dynamics of nearby orbits. However, for orbits with a tangency with the discontinuity, so-called grazing trajectories,

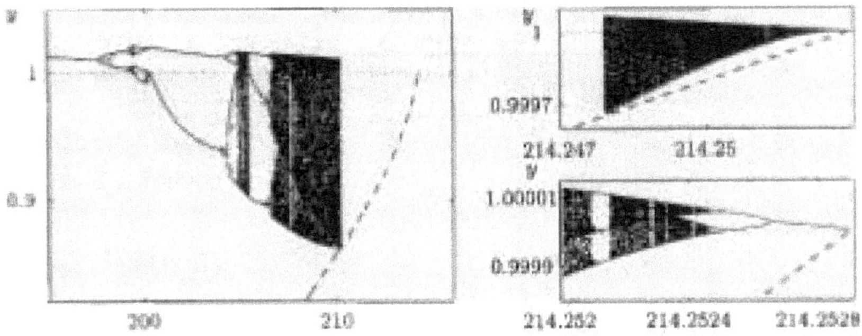

FIGURE 1. "Successive enlargements of a typical bifurcation diagram. $y = 1$ corresponds to the zero velocity relative to the ground. The dashed line is a line of unstable limit cycles. The parameter on the horizontal axis is the stiffness in units of N/m."

such analysis fails. Instead, it is possible to derive a discontinuity mapping which, when combined with the mapping over the full period, provides a complete determination of the dynamics of nearby orbits in state and parameter space (cf. Nordmark [5, 6]). Figure 2 shows the prediction of such a mapping compared to numerical simulations. Here, $v = 0$ corresponds to the discontinuity in suitably rescaled variables.

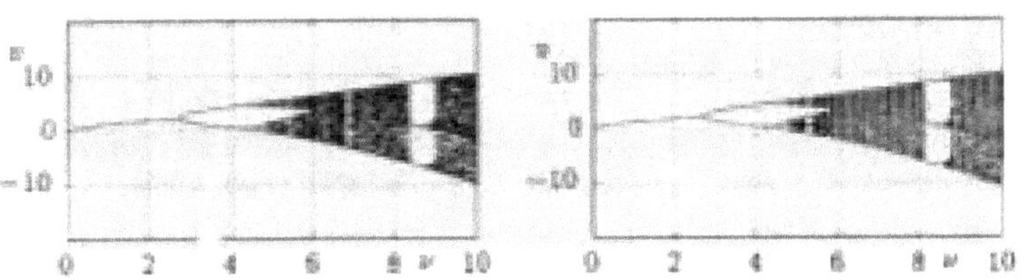

FIGURE 2. "Comparison between numerical simulation (left panel, cf. Figure 1) and discontinuity mapping approach (right panel)."

For large deviations in parameter space from the original grazing trajectory, the discontinuity mapping predicts a bifurcation behavior characteristic of impact oscillators near grazing bifurcations. For example, one expects to find successive windows in the bifurcation diagram between chaotic attractors for which periodic orbits exist with the period changing by one between each window (Nordmark [6]). Figure 3 shows a numerical simulation of the friction model for slightly different parameter values. This clearly exhibits the period-adding phenomenon alluded to.

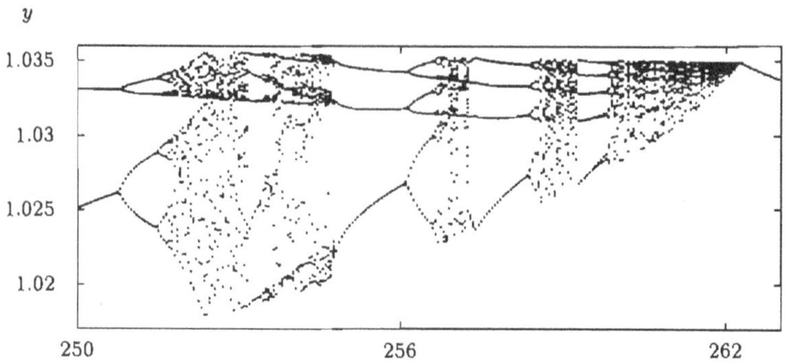

FIGURE 3. **"The period-adding phenomenon typical of impact oscillators appears for relatively larger deviations from the grazing bifurcation. The horizontal axis refers to the stiffness in units N/m."**

4 Acknowledgements

Funding from the Swedish Research Council for Engineering Sciences and the Swedish Foundation for Strategic Research is gratefully acknowledged.

5 References

[1] B. Armstrong-Hélouvry, P. Dupont, and C. Canudas de Wit. A survey of models, analysis tools and compensations methods for the control of machines with friction. *Automatica*, 30(7):1083–1138, 1994.

[2] H. Dankowicz. Dynamical friction modeling. In M. H. Aliabadi and A. Samartin, editors, *Computational Methods in Contact Mechanics III*, pages 227–236. Computational Mechanics Publications, 1997.

[3] H. Dankowicz. On the modeling of dynamical friction phenomena. in submission, 1998.

[4] H. Dankowicz and A. B. Nordmark. On the origin and bifurcations of stick-slip oscillations. in submission, 1998.

[5] A. B. Nordmark. Non-periodic motion caused by grazing incidence in an impact oscillator. *Journal of Sound and Vibration*, 145(2):279–297, 1991.

[6] A. B. Nordmark. Universal limit mapping in grazing bifurcations. *Physical Review E*, 55(1):266–270, 1997.

Nonconvergence in Assumed-Mode Projections of a Distributed Sliding System

C. M. Jung[1]
B. F. Feeny[1]

1 Introduction

In this paper, we investigate a one dimensional system having a distributed frictional contact. We consider the system with constant friction coefficient and focus on the dynamic stability with system parameters. The exact and approximated solutions are derived and there exist contradictory results between them. This work is motivated by studies of higher dimensional systems without analytical solutions, and can be used as a cautionary example for applying assumed mode projections.

2 One Dimensional System

Consider the system shown in Figure 1. The linear elastic media which is between the moving belt and a frictionless linear bearing is used as a one-dimensional continuous model with distributed sliding contact. The friction coefficient between the material and belt is a constant. The equation of the axial motion is

$$A(x)\frac{\partial \sigma_x(x,t)}{\partial x} + f(x,t) = \rho \ddot{u} \tag{1}$$

where $A(x)$ is the cross sectional area of elastic media, ρ is the mass density of elastic material, $\sigma_x(x,t)$ is the stress over the cross section, $u(x,t)$ is the axial displacement and $f(x,t)$ is the friction force. Since $\sigma_x(x,t) = E\,\epsilon_x(x,t)$, where E is the Young's modulus, and considering Poisson effects, the friction force is given by

$$f(x,t) = -\mu\sigma_y(x,t) = -\mu(\sigma_0 + \nu\sigma_x(x,t)) \tag{2}$$

where μ is the friction coefficient, σ_0 is the pre-loaded normal stress per unit length which always less than zero for maintaining a contact condition.

[1]Department of Mechanical Engineering, Michigan State University

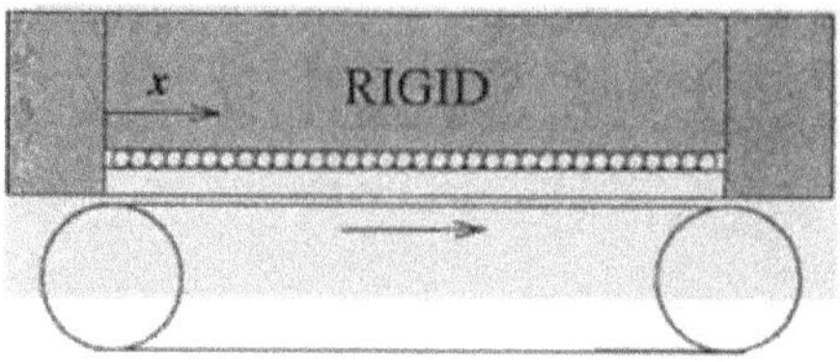

FIGURE 1. One dimensional model with distributed contact

We consider the linear elasticity problem. The non-dimensional equation of motion of the displacement $u(x,t)$ for the effect of distributed contact by considering linear strain, $\epsilon_x(x,t) = \frac{\partial u(x,t)}{\partial x}$, is

$$\frac{\partial^2 u}{\partial^2 x^{*2}} - \alpha \frac{\partial u}{\partial x^*} + \beta = \frac{\partial^2 u}{\partial^2 t^{*2}} \tag{3}$$

where $x \in [0,1]$, $t \in [0,\infty)$. The dimensionless parameters used in (3) are $\alpha = \frac{\mu\nu L}{A}$, $\beta = \frac{-\mu\sigma_0 L^2}{AE}$, $x^* = \frac{x}{L}$, $t^* = \frac{t}{\sqrt{\frac{\rho L^2}{AE}}}$, where x^*, t^* is the dimensionless coordinate and time respectively. The boundary conditions for this system can be written as $u(0,t) = u(1,t) = 0$. This system is non-self-adjoint boundary value problem. (However with a cleverly defined inner product, it is in fact self-adjoint.)

3 Comparison Between Exact and Approximated Solution

The exact solution for this non-self-adjoint linear system shows that the system stability is not dependent on the system parameters. The system is neutrally stable, behaving like an undamped vibration system with natural frequencies of $\Omega_j = \sqrt{(j\pi)^2 + \frac{\alpha^2}{4}}$. The natural frequencies depend on the parameter, α, but stability does not depend on it. The effect of β changes the system's static solution and has no influence on the linear stability. With the addition of modal damping, the eigenvalues will have negative real parts and steady sliding will be asymptotically stable. The effect of generalized internal damping on stability is unknown.

Let us apply the assumed mode method to recast it as ordinary differential equations and compare the results to exact solution. We assumed that its solution can be represented as $u(x,t) = \sum_{j=1}^{\infty} \sqrt{2}\sin(j\pi x)a_j(t)$. The assumed modes, $\sin(j\pi x)$, satisfy the boundary conditions. The resulting ordinary equations are

$$m_{ij}\frac{d^2 a_j}{dt^2} + \sum_{j=1}^{\infty} k_{ij}a_j = f_i, \quad i = 1, 2, \ldots \infty, \tag{4}$$

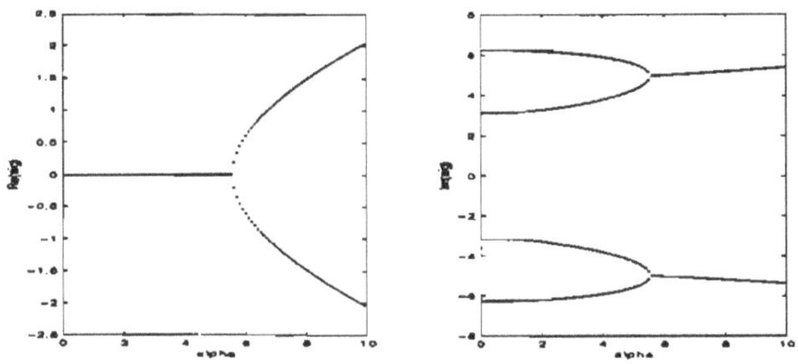

FIGURE 2. Eigenvalues for the one dimensional linear system vs. α by the assumed mode method with two modes included. Here $E = 0.14GN/m^2, \nu = 0.45, \rho = 0.91 \times 10^3 kg/m^3$

where

$$\begin{aligned} m_{ij} &= \delta_{ij}, \quad k_{ij} = k_{Sij} + k_{Aij} \\ k_{Sij} &= (j\pi)^2 \delta_{ij} \\ k_{Aij} &= \begin{cases} \frac{4\alpha ij}{i^2 - j^2} & |i-j| = \text{odd}, \\ 0 & \text{otherwise}, \end{cases} \\ f_i &= \begin{cases} \frac{2\sqrt{2}\beta}{i\pi} & i = \text{odd}, \\ 0 & \text{otherwise}. \end{cases} \end{aligned}$$

If we only focus on the low-dimensional dynamics, we can represent the system with N coupled ordinary differential equations. The real parts of the eigenvalues of this system indicate the predicted stability characteristics. The dependence of eigenvalues on parameters are shown in Figure 2. Instability apparently occurs when the real part of an eigenvalue is positive. This instability mechanism resembles flutter, and has been seen in other friction systems [1, 2, 3, 4, 5].

However, these results contradict the exact solution since it has no instability mechanism. Bolotin [1] had investigated this "paradox" for flow across a membrane. The work showed non-convergent characteristics in the assumed mode projections, and gave a theoretical criterion for convergence based on the linear operator. In summary, conservative systems with second-order operators are not guaranteed to converge in assumed mode approximations.

Nonconvergence can be demonstrated by increasing the number of assumed modes. Figure 3 shows the eigenvalues for 3 to 5 modes. Considering Figure 2 also, the two lowest-frequency modes interact at $\alpha = 5.7$, and $\alpha = 9.0$ for two- and four-mode approximations, but do not interact for three- and five-mode approximations. This shows that the approximated solution by using assumed mode methods for finding the smallest interac-

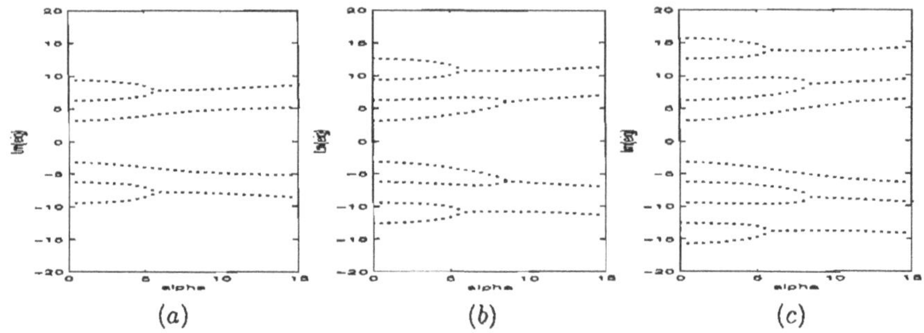

FIGURE 3. **Eigenvalues for the one-dimensional linear system vs. α by the assumed mode method for (a) 3 modes, (b) 4 modes, (c) 5 modes. Here** $E = 0.14GN/m^2, \nu = 0.45, \rho = 0.91 \times 10^3 kg/m^3$

tion value α has divergent properties as we increase the modal coordinates. This hints at faulty results when applying the assumed mode methods to this problem.

4 Conclusion

A one dimensional model with distributed sliding contacts was investigated in order to study the dynamic instabilities. The mathematical model was established and contradictory results between the exact solution and assumed modes method results give us a cautionary example when applying the assumed mode method. In this case, non-convergence of the assumed modes method can be easily detected.

5 References

[1] V.V. Bolotin, Nonconservative Problems of the Theory of Elastic Stability, Pergamon Press, Inc. 1963.

[2] M. Nakai and M. Yokoi, "Generation mechanism of friction noises in dry friction", Japanese Journal of Tribology, **35**(5), 513-522, 1990.

[3] M. Nakai and M. Yokoi, "Band brake squeal", Journal of Vibration and Acoustics, **118**, 1996.

[4] G.G. Adams, "Self-excited oscillations of two elastic half-spaced sliding with a constant coefficient of friction", Journal of Applied Mechanics, **62**, 867-872, 1995.

[5] G.G. Adams, "Self-excited oscillations in sliding with a constant friction coefficient - a simple model". Journal of Tribology, **118**, 819-823, 1996.

Interactions of Harmonic Waves with a Dry Friction Support

M.J. Leamy[1]
J.R. Barber[1]
N.C. Perkins[1]

ABSTRACT This study is motivated by the need to understand the elastodynamic response of belts in frictional contact with pulleys. To this end, a simplified model for belt/pulley contact is used to investigate the dynamic response of a belt subject to a train of harmonic tension waves. Through a non-dimensionalization, a single dimensionless parameter Ω *is identified which governs the dynamic response. A numerical solution is developed and exercised over a wide range of values of* Ω. *An approximate closed form solution is derived assuming the belt stretches quasi-statically, and is shown to yield accurate results for small values of* Ω. *Reported results include the distortion of an initially harmonic tension wave, the energy reflected from the frictional support, and the distance harmonic waves penetrate into the support. The results suggest that the quasi-static stretching assumption may be further utilized as a modeling simplification for belt drives characterized by values of* $\Omega < 1/3$.

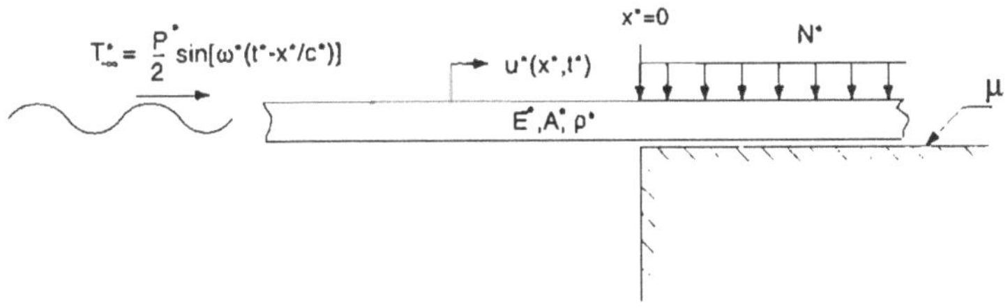

Figure 1: System Model

[1]Department of Mechanical Engineering and Applied Mechanics, University of Michigan, Ann Arbor, MI 48109

1 Introduction

This study is principally motivated by belt drive mechanics which have been studied by many researchers, starting with Leonard Euler[1]. Euler proposed the well known *capstan* formula for the tension distribution in a belt wrapped around a fixed pulley or a capstan. A comprehensive survey of belt drive mechanics after Euler and up to 1981 is provided by Fawcett[2]. Much of the research cited in [2] and work since its appearance has been concerned with either: 1) frictional contact (slip) between a belt and a steadily rotating pulley (neglecting vibratory effects), or 2) vibration of belts neglecting frictional contact.

In this paper, we consider a simplified model of the belt and the belt/pulley interface as a starting point for understanding the coupled response. The model, illustrated in Figure 1, consists of a length of belt resting partly on a frictional interface. We consider incoming *harmonic* tension waves that impinge upon the support and are reflected back with distorted shape. Quantifying this distortion, the reflected wave energy, and the distance the waves penetrate into the support, are objectives of this study.

2 System Model

The displacement $u^*(x^*, t^*)$ of the infinite rod shown in Figure 1 is governed by the *nonlinear* wave equation,

$$\frac{\partial^2 u^*}{\partial x^{*2}} - \frac{\rho^* A^*}{E^* A^*}\frac{\partial^2 u^*}{\partial t^{*2}} - \frac{\mu N^*}{E^* A^*}\text{sgn}\left(\frac{\partial u^*}{\partial t^*}\right) H(x^*) = 0 \tag{1}$$

on the infinite domain ($-\infty < x^* < +\infty$), where E^*, ρ^*, and A^* represent the Young's modulus, volume density, and cross-sectional area, respectively, of the rod. The coefficient of friction, μ, describes the Coulomb surface supporting the rod in the region $x^* \geq 0$ where N^* denotes the normal force per unit length. In (1), sgn denotes the *signum* operator while $H(x^*)$ is the Heaviside step function. The rod is subject to a steady train of harmonic tension waves of amplitude $P^*/2$ radiating to the right from $x^* = -\infty$ and defined by

$$T^*_{-\infty} = \frac{P}{2}\sin\left[\omega^*\left(t^* - \frac{x^*}{c^*}\right)\right] \tag{2}$$

where

$$T^*(x^*, t^*) = E^* A^* \frac{\partial u^*}{\partial x^*} \tag{3}$$

is the tension in the belt, ω^* is the excitation frequency, and $c^* = \sqrt{E^*/\rho^*}$ is the longitudinal wave speed.

Introducing the dimensionless quantities,

$$u = \frac{\mu N^* E^* A^*}{P^{*2}}u^*, \quad x = \frac{\mu N^*}{P^*}x^*, \quad t = \frac{\mu N^* c^*}{P^*}t^*, \tag{4}$$

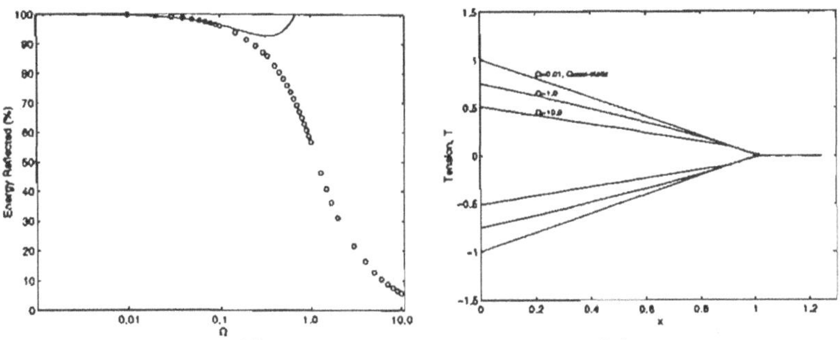

(i) (ii)

Figure 2: (i) Percent energy reflected versus Ω for the quasi-static solution (–) and the numerical solution (o). (ii) Envelope of tension distribution in support region. Plotted are envelopes of dimensionless tension (T) versus dimensionless position (x) for Ω = 0.01, 1.0, and 10.0. The amplitude of the envelope monotonically decreases with increasing Ω.

reduces (1)-(3) to:

$$\frac{\partial^2 u}{\partial x^2} - \frac{\partial^2 u}{\partial t^2} - \operatorname{sgn}\left(\frac{\partial u}{\partial t}\right) H(x) = 0 \tag{5}$$

$$T_{-\infty} = \frac{1}{2}\sin[\Omega(t-x)] \tag{6}$$

$$T(x,t) \equiv \frac{T^*}{P^*} = \frac{\partial u}{\partial x} \tag{7}$$

where

$$\Omega = \frac{\omega^* P^*}{\mu N^* c^*} \tag{8}$$

The dimensionless equations (5)-(7) reveal that the single parameter Ω classifies all problems of the type shown in Figure 1.

3 Results

A numerical solution is used to investigate the behavior of a harmonic disturbance which propagates into a frictional support as compared to that predicted by an approximate, closed form solution derived using a quasi-static stretching assumption. In particular, attention focuses on the Fourier content of the reflected wave, the penetration distance, and the percent energy reflected from the support. Details of the numerical solution method and the closed form approximate method are contained in [3].

The loss of energy due to frictional dissipation is apparent in Figure 3, which shows the energy of the reflected wave as a function of Ω. Less than 5% of the incoming energy is reflected for values of Ω exceeding 10. The quasi-static solution follows the numerical solution closely for small values of Ω, but for $\Omega > \frac{2}{3}$, the quasi-static solution predicts the non-physical result of energy increase.

The gradual distortion of the harmonic wave with Ω is evaluated by examining the Fourier content of the reflected wave form. Figure 4 illustrates the magnitude of the first three non-zero Fourier coefficients (first, third, and fifth harmonics) of the reflected wave form, as well as the phase of the first harmonic, as functions of Ω. This phase defines a lag between the first harmonic of the reflected wave and the reference wave reflected from a fixed end. For small values of Ω in Figure 4, the wave reflected from the frictional support deviates little in either amplitude or phase from the pure harmonic reflected from the fixed end. As Ω increases, the amplitude of the reflected wave decreases, and the relative contributions of the higher harmonics increase. As the value of Ω increases still further, the higher harmonics achieve a maximum magnitude (near $\Omega = 1$) and then decay with increasing Ω. The phase lag of the first harmonic approaches the limiting value of $\pi/2$. As Ω approaches infinity, all coefficients approach zero and the frictional support acts as an anechoic termination.

The coefficients for the quasi-static solution converge to those of the numerical solution as $\Omega \to 0$. The convergence of the coefficients for the higher ($n > 5$) harmonics is ultimately limited by the discretization size. A trend captured by both the numerical and quasi-static solutions at small Ω is the increasing importance of the super-harmonics as Ω increases from zero, and the decreasing importance of the fundamental harmonic. This trend continues in the numerical solution until $\Omega \cong 1.0$, beyond which all harmonics decay. This decay is not captured by the quasi-static solution which ultimately predicts for larger values of Ω that the reflected wave possesses a greater amplitude (and thus energy) than the originating wave (a physical impossibility); refer also to Figure 3.

Observe from Figure 4 that a regime for Ω exists in which the quasi-static solution accurately predicts the reflected waveform and that this regime can be identified by examining the fundamental harmonic alone. The higher harmonics have a negligible effect on the magnitude and phase of the reflected wave. Values for $\Omega < 1/3$ ensure that the quasi-static solution predicts the magnitude of the reflected wave to within 8 percent and the phase to within 2 percent, when compared to the elastodynamic solution. Figure 3 also indicates the fidelity of the quasi-static solution. For $\Omega < 1/3$, both solutions predict near identical reflected wave energy.

Consider next the steady state tension developed in the support region. Figure 3 illustrates the envelope of the tension wave that penetrates into the support region. For small Ω, the support creates near perfect reflections and the magnitude of the tension at the entrance to the frictional support is nearly twice the amplitude of the incoming wave, as shown in the Figure. As Ω increases, the penetration distance (dimensionless) increases slightly and the magnitude of the tension envelope at the entrance approaches $\pm\frac{1}{2}$ as the waves pass through with little reflection. From Figure 3, it is also observed that the tension envelope from the numerical solution approaches that of the quasi-static solution for small Ω, with no observable difference between envelopes at $\Omega = 0.01$. Further comparisons reveals an interesting observation; namely, the dimensionless penetration distance (S) of the quasi-static solution predicts the penetration with errors less than 5% over the large range of Ω considered in Figure 3.

4 Conclusions

A single dimensionless parameter Ω has been identified which governs the dynamic response of an infinite rod subject to harmonic tension waves and interacting with a dry friction surface. Incoming harmonic tension waves partially reflect from the frictional support and both the degree of reflection and the degree of distortion are controlled solely by Ω. This distortion and energy reflection have been quantified as a function of Ω using an approximate, closed-form quasi-static solution and a numerical solution. The results of this study suggest that a quasi-static model for belt/pulley interaction will remain accurate for belt drive systems for which $\Omega < 1/3$. The distance harmonic tension waves penetrate into the frictional support is accurately predicted by the quasi-static solution, even for large Ω. For $\Omega > 10$, the frictional support acts effectively as an anechoic termination.

5 References

[1] Euler, M. Leonard, 1762, "Remarques sur l'effect du frottement dans l'equilibre," *Mém. Acad. Sci.*, pp. 265-278.
[2] Fawcett, J. N., 1981, "Chain and Belt Drives - A Review," *Shock & Vibration Digest*, Vol. 13, No. 5, pp. 5-12.
[3] Leamy, M. J., 1998, "Influence of Dry Friction in the Dynamic Response of Accessory Belt Drive Systems," Ph.D. dissertation, The University of Michigan, Chapter V.

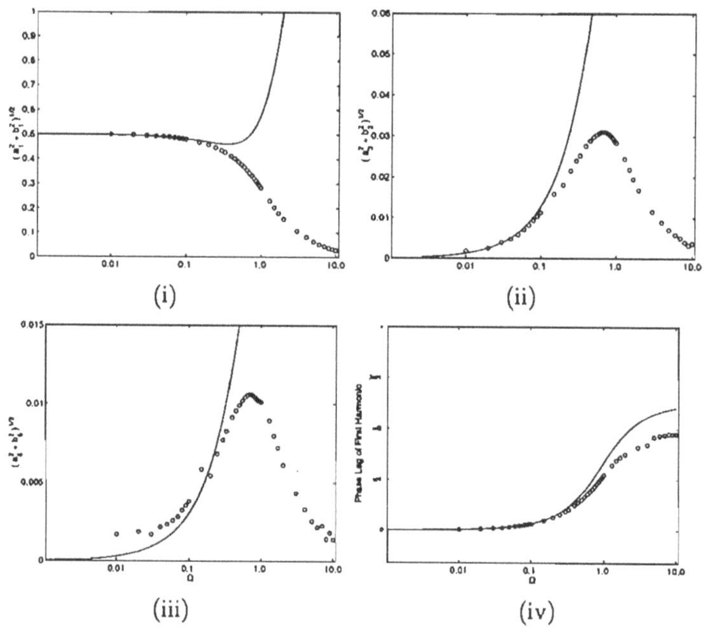

Figure 3: Fourier coefficients vs. Ω. The magnitudes of the first, third, and fifth harmonic are shown in (i)-(iii), respectively, and the phase of the first harmonic is shown in (iv). Numerical solution denoted by o; quasi-static solution denoted by —.

Railway Wheel Squeal and Competition in Species for Population Biology

Mikio Nakai[1]

ABSTRACT *This paper describes a close analogy between railway wheel squeal and competition in species for population biology.*

1 Introduction

For railway wheel squeal, the nonlinear analyses and the experiments using a disk served as a wheel and a rod as a rail have demonstrated that only a squeal having a single mode occurs, if there are no internal resonances among axial natural frequencies of the disk(1). The quadratically nonlinear equations for railway wheel squeal are similar to the nonlinear Lotka-Volterra-Gause equations for two competing species in population biology(2)~(4). A close analogy between railway wheel squeal and competition of species in population is investigated in this paper.

2 Competition in species for population

The Lotka-Volterra-Gause equations have been proposed for two competing species in population biology as follows(2)~(4):

$$\begin{aligned}\frac{du_1}{dt} &= (R_1 - a_{11}u_1 - a_{12}u_2)u_1 = R_1 u_1\left\{1 - \frac{(u_1 + au_2)}{K_1}\right\} \equiv f_1(u_1, u_2)\\ \frac{du_2}{dt} &= (R_2 - a_{21}u_1 - a_{22}u_2)u_2 = R_2 u_2\left\{1 - \frac{(bu_1 + u_2)}{K_2}\right\} \equiv f_2(u_1, u_2)\end{aligned} \tag{1}$$

where u_1, u_2 are the numbers of individuals, R_1, R_2 are the intrinsic growth rates, a_{11}, a_{22}, a_{12} and a_{21} are competition coefficients within and between species and are positive, K_1 and K_2 are carrying capacities, a (b) is a competition coefficient measuring the extent to which the second (first) species affects the growth rate of the first (second).

Figure 1 shows behavior of the solutions of Eq.(1)(2), where solid and dotted lines indicate $f_1 = 0$ and $f_2 = 0$, and ● and ○ indicate stable and unstable solutions. Only u_1 or u_2 survives for the cases (i−1), (ii−1), (i−3)

[1] Department of Precision Engineering, Kyoto University, Kyoto, 606-8501 Japan

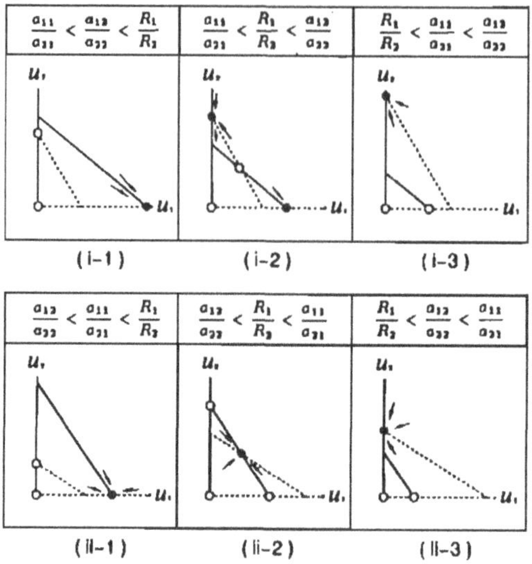

Figure 1: Stability properties of equilibrium points for two competing species (● : stable equilibrium point, ○ : unstable equilibrium point, solid line : $f_1 = 0$, dotted line : $f_2 = 0$)(2)

and (ii–3). Two stable solutions exist for the case (i–2), but they do not coexist. For the case (ii–2), $\frac{a_{12}}{a_{22}} < \frac{R_1}{R_2} < \frac{a_{11}}{a_{21}}$ $(ab < 1)$, both species coexist. These results show that a steady state does not depend on the intrinsic growth rates R_1 and R_2, and that a coexistent solution is stable if $ab < 1$, and unstable if $ab > 1$. This suggests that both species can coexist easily when the competition within species is severer than that between species.

3 Railway Wheel Squeal

A frictional force μP_0 is applied in the axial direction at the point (r_o, θ_o) of the circumference of the disk. By neglecting the vibration of the beam, the equation of motion for transverse displacement $w(r, \theta, t)$ of the disk from the equilibrium (the static condition $v_r = V$) can be expressed as

$$D\nabla^4 w + c\frac{\partial w}{\partial t} + \rho h \frac{\partial^2 w}{\partial t^2} = \{[\mu(v_r) - \mu(V)]P_0\} \frac{1}{r}\delta(r - r_o)\delta(\theta - \theta_o) \quad (2)$$

where δ : Dirac's delta function, P_0 : contact load between the beam and the disk, μ : coefficient of friction which is a function of relative sliding velocity v_r, $\mu(v_r)$, $v_r = V - (\partial w/\partial t)_{r_o,\theta_o}$,$V$: sliding velocity of the beam, D : flexural stiffness of the disk, ρ : density, h : thickness, t : time, c : damping coefficient, r_i : inner radius, r_O : outer radius.
Applying the Galerkin's procedure and the method of averaging, we get the

following equations for an amplitude a_i and a phase ψ_i:

$$\dot{\psi}_i = 0. \quad \psi_i = \text{const.} = \varphi_i^0 \tag{3}$$

$$2\dot{a}_i = a_i(A_{i1} + A_{i2}a_1^2 + \cdots + A_{iN+1}a_N^2) \quad (i = 1, N) \tag{4}$$

where [$\dot{}$] : time derivatives, $A_{i1} = D_i B_i - C_i$, $A_{ii+1} = \frac{\varepsilon_i \varphi_i}{\varepsilon_1 \varphi_1}\left(\frac{\omega_i \varphi_i}{\omega_1 \varphi_1}\right)^2 A_{12}$, $A_{ij} = 2\frac{\varepsilon_j \varphi_i}{\varepsilon_1 \varphi_1}\left(\frac{\omega_{j-1}\varphi_{j-1}}{\omega_1 \varphi_1}\right)^2 A_{12}$ $(j \neq 1, j \neq i+1)$, $A_{12} = -\frac{3}{4}D_1 \mu_2 \varphi_1^3 \omega_1^2$, $B_i = \left(\frac{\varphi_i}{\varphi_1}\right) B_1$, $\varphi_i = \varphi_{ns}$, $\omega_i = \omega_{ns}$, $N_i = N_{ns}$, $N_{ns} = \int_0^{2\pi}\int_{r_i}^{r_0} \varphi_{ns}^2(r,\theta) r dr d\theta$,
$D_i = P_0 \varphi_i / \rho h N_i$, $C_i = c/\rho h$, $\varepsilon_i = D_i \varepsilon_1 / D_1$, $\varepsilon_1 = D_1 B_1 - C_1$,
$\mu(v_r) = \mu_0 \text{sgn}(v_r) + \mu_1 v_r + \mu_2 v_r^3$, $B_1 = -(\mu_1 + 3\mu_2 V^2)\varphi_1$, $\varphi_{ns}(r,\theta)$ is obtained from the boundary conditions under which the disk is clamped at $r = r_i$ and free at $r = r_o$, ω_{ns} : axial natural circular frequencies of the disk, n, s : the numbers of nodal diameters and circles.
Further, by introducing $y_i(t) = \{a_i(t)\}^2 > 0$, Eq.(4) is rewritten as

$$\dot{y}_i = y_i(A_{i1} + A_{i2}y_1 + A_{i3}y_2 + \cdots + A_{iN+1}y_N) \quad (i = 1, N) \tag{5}$$

Explicitly, Eq.(5) is the same differential equation as Eq.(1) for the two competing species. Figure 2 shows solutions in six cases classified by coefficients A_{ij} in Eq.(5) for the two-degree of freedom. This figure reveals that squeals having two or more modes do not occur simultaneously in a steady state and only a squeal having a single mode occurs.

4 Relation between Competition in Species and Railway Wheel Squeal

Intrinsic growth rates R_1, R_2 and carrying capacities $K_1 = \frac{4R_1\rho h N_1}{3P_0\mu_2\varphi_1^4\omega_1^2}$, $K_2 = \frac{4R_2\rho h N_2}{3P_0\mu_2\varphi_2^4\omega_2^2}$ correspond to the potentials of squealing. Figures 3(a),(b) show

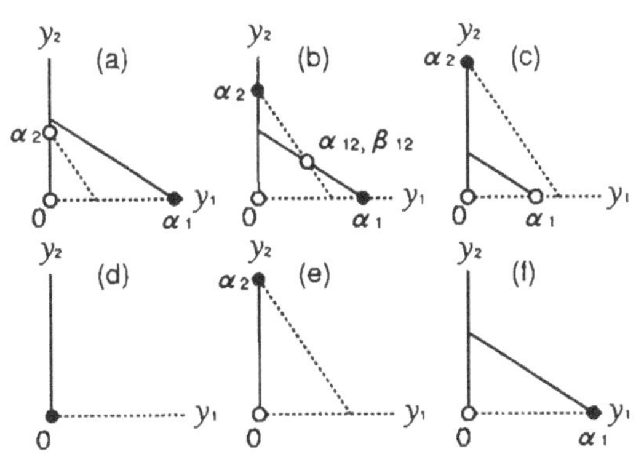

Figure 2: Stability properties of solutions for wheel squeal (● : stable solution, ○ : unstable solution, solid line : $\dot{y}_1 = 0$, dotted line : $\dot{y}_2 = 0$)

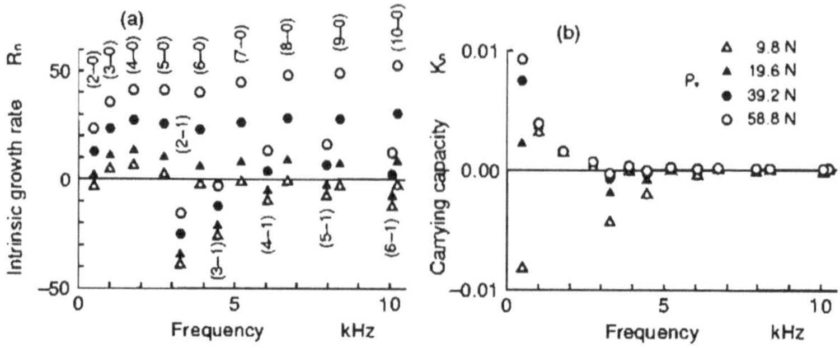

Figure 3: (a): Intrinsic growth rates at different modes, (b) carrying capacities at different modes

calculated R_n and K_n at each mode versus the variation of contact loads using measured damping coefficients of the disk. It is found that a squeal occurs easily as the values of R_n at nodal diameter modes increase with higher load, and that (2-0)$\sim$(5-0) modes with small damping coefficients tend to have the greatest likelihood of squealing. In case (ii$-$2) which two species coexist for population biology, $ab < 1$ must hold. For wheel squeal, $ab = 4$, which cannot satisfy $ab < 1$. It is therefore concluded that a squeal with two modes does not occur. This is because coefficients can change flexibly for species, but they can be rigidly determined by dynamic relation for wheel squeal. That is, a species is "soft", but a wheel is "hard".

5 Conclusions

Coefficients in the Lotka-Volterra-Gause equations for the competition in species have a close analogy to those for wheel squeal. Only a squeal having a single mode occurs. However, unlike the railway wheel squeal, the two competing species exhibit that the asymptotic behavior is convergence to one of the four possible points (both coexisting), depending on the relative magnitudes of the competition coefficients.

References

[1] Nakai, M., *et al*, Railway Wheel Squeal (1st and 3rd Reports), *Bulletin of the JSME*, Vol.25, No.205, 1982, p.1127, Vol.28, No.237, 1985, p.500.

[2] Mimura, M., Competition within and between species for population biology (in Japanese), *Mathematical Sciences*, No.229, 1982, p.59.

[3] Iwasa, Y., An Introduction to Mathematical Biology (in Japanese), HBJ Public. 1994.

[4] May, R.M. and Leonard, W.J., Nonlinear Aspects of Competition between Three Species, SIAM *J. Appl. Math.*, Vol.29, No.2, 1975, p.243.

Relaxing Nonholonomic Constraints

Richard Rand and Deepak Ramani [1]

Introduction

A constraint is a geometric restriction on the motion of a dynamical system. It is enforced by the application of one or more constraint forces. Instead of being prescribed, the constraint forces take on whatever values are necessary to produce the constrained motion.

It is often desirable to be able to relax a given constraint without eliminating it altogether. For example, the rolling-without-slipping of a rigid body on a plane is well known to be a nonholonomic constraint. By relaxing such a constraint, one could study the effect of rolling with limited slipping.

The idea of this work is to provide a one-parameter family of systems which serve to relax a given constraint. This is accomplished by replacing the constraint force by a force directed in the same direction, but with a value which is a fractional multiple $0 < \lambda < 1$ of the value necessary to enforce the given constraint.

We illustrate the idea by applying it to an example based on the motion of a supercavitating underwater projectile [3]. The presence of tail fins on such a body suggests a nonholonomic constraint which requires the velocity of the tail to be directed along the axis of the body, similar to an arrow in flight. This idealized constraint is unattainable, however, and a more realistic situation may be achieved by relaxing this nonholonomic constraint as described above.

Kiceniuk Forces

When a body moves through water at sufficient speed, the fluid pressure may drop locally below a level which sustains the liquid phase, and a low-density gaseous cavity can form. Flows exhibiting cavities enveloping a moving body entirely are called *supercavitating,* and, since the liquid phase does not contact the moving body through most of its length, skin drag is almost negligible. In [1], Kiceniuk experimentally measured lift and drag forces on a supercavitating body. As in [3], we idealize his data by assuming that (i) the net force F acts along the axis of the body, (ii) the net moment

[1]Department of Theoretical and Applied Mechanics, Cornell University, Ithaca, NY 14853

by the fluid on the body is zero, and (iii) the magnitude of F is given by the expression:

$$F = \frac{1}{2}\,\rho\,A\,k\,s^2\,\cos\alpha \tag{1}$$

where $s =$ speed,
α is the angle between the body's velocity vector and its symmetry axis,
$\rho =$ density of water,
$A =$ cross-sectional area at the body's tip,
$k =$ a nondimensional constant.

Equations of Motion

A supercavitating projectile with fins may be modeled by assuming that the fins do not permit motion perpendicular to the projectile's axis at the point where the fins are attached. This nonholonomic constraint has been called *Caplygin's sleigh* [2]. The equations of plane motion are (see Fig.1):

$$F_c = m\,(\dot{v} + \omega\,u + a\,\dot{\omega}) \tag{2}$$

$$-F = m\,(\dot{u} - \omega\,v - a\,\omega^2) \tag{3}$$

$$I\,\dot{\omega} = -F_c\,a \tag{4}$$

where u=projectile velocity in axial direction,
v=projectile velocity at fin attachment point in direction normal to axis,
ω=angular velocity of projectile,
F_c=constraint force,
F=Kiceniuk force of eq.(1),
a=distance from center of mass to fin attachment point,
m=mass of projectile,
I=moment of inertia about center of mass.

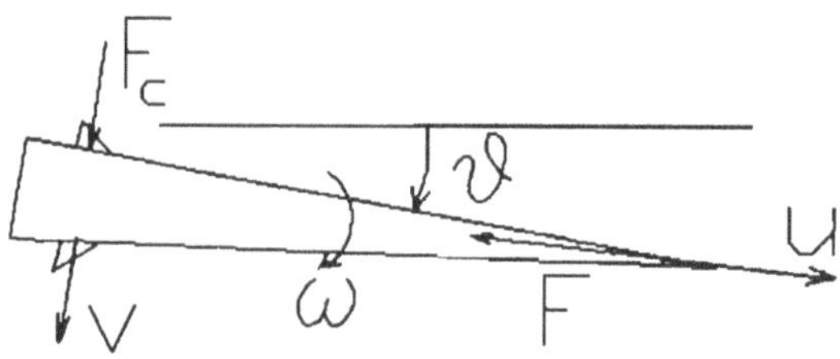

Fig.1. A supercavitating projectile with fins.

If the fins were perfectly effective in administering the constraint $v = 0$, the associated constraint force F_c could be obtained by setting $v = 0$ in eq.(2) and substituting $\dot{\omega}$ from eq.(4):

$$F_c = \frac{mI}{I + ma^2}\,u\,\omega \tag{5}$$

However, the fins are expected to be rather ineffective since in the super-cavitated state they interact only with the water vapor inside the cavity rather than with the water itself. As explained in the introduction, we will model the relaxed constraint by replacing F_c by λ times the value given by eq.(5):

$$F_c = \lambda \frac{mI}{I + ma^2} u\,\omega, \qquad 0 < \lambda < 1 \tag{6}$$

The expression (1) for the Kiceniuk force F can be rewritten in terms of the state variables u, v, ω:

$$F = \beta\, m\, u\, \sqrt{u^2 + (v + a\omega)^2} \tag{7}$$

where $\beta = \frac{\rho A k}{2m}$. When expressions (6) and (7) are substituted into eqs.(2)-(4), we obtain:

$$\dot{v} = (\lambda - 1)\, u\, \omega \tag{8}$$

$$\dot{u} = a\,\omega^2 + \omega\, v - \beta\, u\, \sqrt{u^2 + (v + a\omega)^2} \tag{9}$$

$$\dot{\omega} = -\frac{\lambda J}{a}\, u\, \omega \tag{10}$$

where $J = \frac{ma^2}{I+ma^2}$. In addition to modeling the relaxed constraint force by eq.(6), our approach is to also require that the initial conditions be consistent with the constraint. In this case we require that the projectile have no sideways velocity v at the fin attachment point:

$$v(0) = 0 \tag{11}$$

Eqs.(8)-(11) are the governing equations for the system with relaxed constraint. They include the two special cases $\lambda = 1$ (the original unrelaxed constraint) and $\lambda = 0$ (the unconstrained problem).

Stability of Motion

Eqs.(8)-(11) admit the exact solution:

$$u(t) = \frac{U_0}{1 + \beta U_0 t}, \qquad v(t) = 0, \qquad \omega(t) = 0. \tag{12}$$

We investigate the stability of this motion due to a small initial angular velocity ω_0:

$$u(0) = U_0, \qquad v(0) = 0, \qquad \omega(0) = \omega_0 \tag{13}$$

Linearizing eqs.(8)-(10) about the solution (12) for the initial conditions (13) leaves $u(t)$ unchanged and yields the following expressions for the disturbed motions $v(t)$ and $\omega(t)$:

$$v(t) = \frac{a(1-\lambda)}{J\lambda}\left[(1 + \beta U_0 t)^{-p} - 1\right] \omega_0 \tag{14}$$

$$\omega(t) = \omega_0\ (1 + \beta U_0 t)^{-p} \tag{15}$$

where $p = \frac{J\lambda}{a\beta}$. In the special case $\lambda = 0$, these become:

$$v(t) = -\frac{1}{\beta} \ln(1 + \beta U_0 t)\, \omega_0, \qquad \omega(t) = \omega_0 \tag{16}$$

The displacements x, y of the attachment point on the fin relative to a nonrotating frame, and the rotation angle θ may be obtained from:

$$\dot{x} = u \cos\theta - v \sin\theta, \qquad \dot{y} = -u \sin\theta - v \cos\theta, \qquad \dot{\theta} = \omega. \tag{17}$$

We numerically integrated eqs.(17) for $\omega_0 = 0.5\ rad/sec$, using the parameters given in [3]:

$$U_0 = 1500\ m/sec, \quad a = 0.06\ m, \quad \beta = 0.00942\ m^{-1}, \quad J = 0.8823. \tag{18}$$

The resulting trajectories for one second of flight are shown in Fig.2 for λ = 0, 0.001, 0.01 and 1. Note that although the perfectly constrained motion ($\lambda = 1$) completely eliminates the instability, even a relatively small value of λ has a large stabilizing effect.

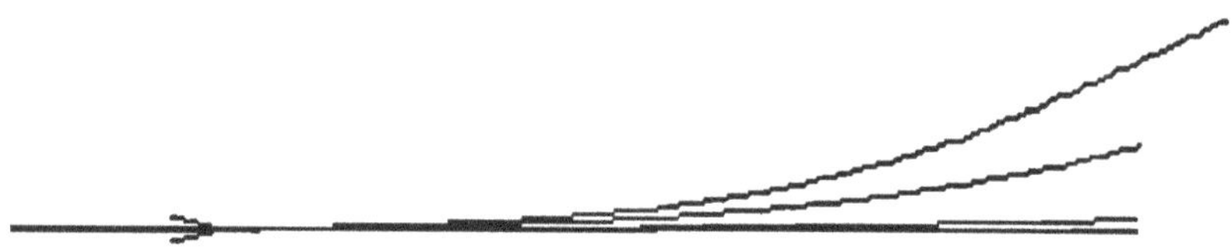

Fig.2. Trajectories after 1 second of flight for (top to bottom) $\lambda = 0,\ 0.001,\ 0.01,\ 1$. For $\lambda = 1$, the horizontal distance traveled is 288 m.

References

[1] Kiceniuk, T., An Experimental Study of the Hydrodynamical Forces Acting on a Family of Cavity-Producing Conical Bodies of Revolution Inclined to the Flow, *Cal.Inst.Tech. Hydrodynamics Lab. Report No.E-12.17*, 1954.

[2] Neimark, J.I. and Fufaev, N.A., *Dynamics of Nonholonomic Systems*, American Math.Soc., 518 pp., 1972.

[3] Rand, R., Pratap, R., Ramani, D., Cipolla, J., Kirschner, I., Impact Dynamics of a Supercavitating Underwater Projectile, *Proceedings of the 1997 ASME Design Engineering Technical Conferences*, 16th Biennial Conference on Mechanical Vibration and Noise, Sacramento, CA,Sept. 14-17, 1997, paper no.DETC97/VIB-3929 (CD-ROM)

Dynamics of the "Wild Mouse"

Th. Rossmann[1]
A. Stiegelmeyr[1]
F. Pfeiffer[1]

ABSTRACT The "Wild Mouse" represents a roller coaster system which can be characterised mainly by frequent and abrupt changes of the vehicle's course. Thereby and due to the slackness between wheels and rails sequences of impacts with friction stress the undercarriage. To model the contact dynamics, methods of nonlinear and linear complementarity (NCP/LCP) are applied. To show the numerical efficiency the simulation is compared with a model based on classical force elements.

1 Introduction

The investigation of the dynamical behaviour of the roller coaster "Wild Mouse" can be divided in two subproblems. On the one hand the overall motion has been optimized in such a way that the visitor experiences a maximum of thrills and spills. On the other hand detailed computations of the undercarriage loads in consideration of the unilateral contact mechanics between wheels and rails have been carried out. In this paper we concentrate on the second part. We use this example in order to show the applicability and validity of the contact models to practical problems.

Particularly in consideration of the wheel wear the contact problem wheel/rail becomes non-trivial from the geometrical point of view. Figure 1 sketches the roller coaster with some enlarged details. The vehicle consists of four wheel packages with six wheels, each. Thus altogether 24 possibly active unilateral contacts have to be handled.

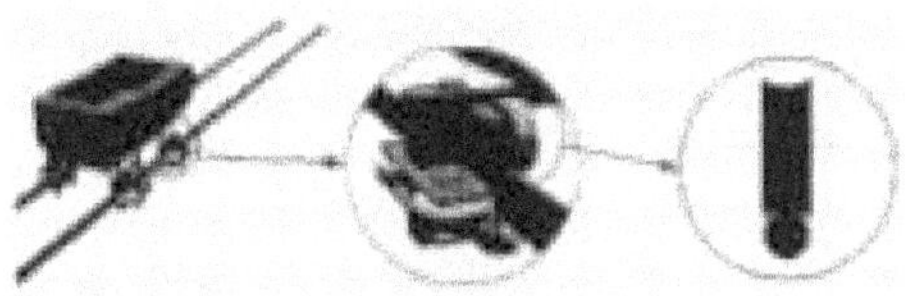

Figure 1 Wheel/Rail Contact

2 Contact Kinematics

To provide information about the status of the potential contacts between wheels and rails some kinematical equations have to be formulated. The relative velocities between two bodies in normal $\dot{g}_{N_i}$ and two orthogonal tangential directions $\dot{g}_{T_i}$ for the $i-th$ contact write:

$$\dot{g}_{N_i} = w_{N_i}^T \dot{q} + \tilde{w}_{N_i} \quad \in \mathbb{R}^1; \quad \dot{g}_{T_i} = w_{T_i}^T \dot{q} + \tilde{w}_{T_i} \quad \in \mathbb{R}^2 \qquad (1)$$

[1]Lehrstuhl B für Mechanik, Technische Universität München, Germany

Collecting the velocities of (1) of all contacts the differentiation with respect to time results in the relative accelerations.

$$\begin{pmatrix} \ddot{g}_N \\ \ddot{g}_T \end{pmatrix} = \begin{pmatrix} W_N^T \\ W_T^T \end{pmatrix} \ddot{q} + \begin{pmatrix} \bar{w}_N \\ \bar{w}_T \end{pmatrix} \quad \text{where: } \begin{pmatrix} W_N \\ W_T \end{pmatrix} = \begin{pmatrix} \cdots & w_{N,i} & \cdots \\ \cdots & w_{T,i} & \cdots \end{pmatrix} \tag{2}$$

3 Equations of Motion

Basically, the vehicle consists of six components. These are the body, front and transaxle, and the four wheel packages. All components are modelled as rigid bodies and are coupled among each other by appropriate force elements. Due to their low mass the plastic wheels do not have any degree of freedom. Their physical properties are modelled by different resistance coefficients in axial and circumferential direction, representing frictional and rolling resistance, respectively. Applying the method of Newton-Euler, for example, to systems with unilateral constraints [2] one obtains:

$$M(q,t)\ddot{q} - h(q,\dot{q},t) - (W_N + H_R)\lambda_N - W_T\lambda_T = 0 \tag{3}$$

M is the mass matrix and the vector $h(q,\dot{q},t)$ contains the gyroscopical and active forces. The additional constraint forces are taken into account by the vectors λ_N and λ_T. More precisely, $\lambda_{N,i}$ are all active normal forces while the elements of λ_T are tangential forces that belong to possibly sticking unilateral contacts. Finally, the matrices W_N and W_T project λ_N and λ_T into the space of the generalized coordinates $q \in \mathbb{R}^f$. The friction forces of sliding contacts are expressed with the aid of Coulomb's friction law by the appropriate normal forces. The respective mapping is realized by H_R (see also [2], [3]).

4 Formulation of the Contact Problem

In order to integrate (3) the contact forces are required. For these purposes contact laws are introduced. In detail, the derivation of capable laws is given for example in [1] or [3]. Basically, the contact law in normal direction can be described by use of the following complementarity condition for the i-th contact:

$$\ddot{g}_{N,i} \geq 0; \quad \lambda_{N,i} \geq 0; \quad \ddot{g}_{N,i}\lambda_{N,i} = 0 \tag{4}$$

In tangential direction the variational inequality

$$\ddot{g}_{T,i}^T(\lambda_{T,i}^* - \lambda_{T,i}) \geq 0; \quad |\lambda_{T,i}| \leq \mu_{0,i}\lambda_{N,i}; \quad \forall\, |\lambda_{T,i}^*| \leq \mu_{0,i}\lambda_{N,i} \tag{5}$$

is an appropriate representation for the spatial friction cone (see fig. 2). For later use we introduce the friction saturation $\lambda_{T0,i}$ describing the "distance" of a sticking force from its limit, where it starts sliding.

$$\lambda_{T0,i} = \mu_{0,i}\lambda_{N,i} - |\lambda_{T,i}| \geq 0 \tag{6}$$

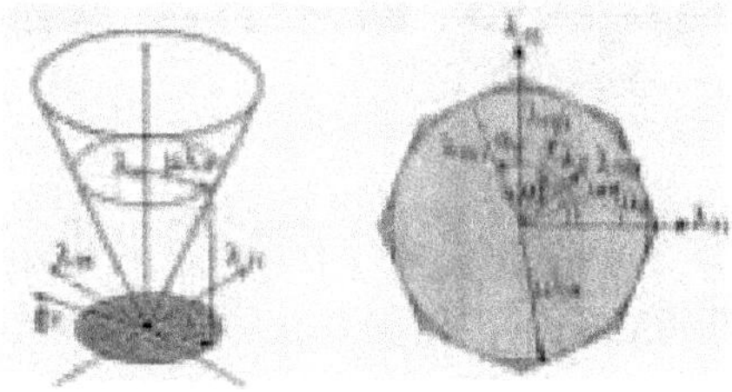

Figure 2: Friction Cone and Linearized Friction Cone

In order to express the tangential contact law in a linear complementary form the friction cone is approximated by an l-sided pyramid (see fig. 2). The decomposition of the tangential forces orthogonal to the l directions of the linearized cone results in a new $\lambda_{T0,i}$

$$\begin{aligned} \lambda_{T0,i} &= \begin{pmatrix} \mu_{0,i1} \\ \vdots \\ \mu_{0,il} \end{pmatrix} \lambda_{N,i} - \begin{pmatrix} \cos\alpha_{i,1} & \sin\alpha_{i,1} \\ \vdots & \vdots \\ \cos\alpha_{i,l} & \sin\alpha_{i,l} \end{pmatrix} \lambda_{T,i} \\ &= G_{N,i}^T \lambda_{N,i} - G_{T,i}^T \lambda_{T,i} \geq 0 \quad \text{with} \quad \lambda_{T0,i} \in \mathbb{R}^l \end{aligned} \tag{7}$$

With this the variational inequality (5) reduces to

$$-\dot{g}_{T,i} = G_{T,i}\kappa_i; \quad \lambda_{T0,i} \geq 0, \quad \kappa_i \geq 0, \quad \kappa_i^T \lambda_{T0,i} = 0 \tag{8}$$

For further steps it is necessary to decompose κ_i and $\lambda_{T0,i}$.

$$\begin{aligned} \lambda_{T0,i} &= \begin{pmatrix} \lambda_{T01,i} \\ \lambda_{T02,i} \end{pmatrix} \quad \text{with} \quad \lambda_{T01,i} \in \mathbb{R}^2 \quad \lambda_{T02,i} \in \mathbb{R}^{l-2} \\ \kappa_{T0,i} &= \begin{pmatrix} \kappa_{1,i} \\ \kappa_{2,i} \end{pmatrix} \quad \text{with} \quad \kappa_{1,i} \in \mathbb{R}^2 \quad \kappa_{2,i} \in \mathbb{R}^{l-2} \\ & \lambda_{T01,i} \geq 0, \quad \kappa_{1,i} \geq 0, \quad \kappa_{1,i}^T \lambda_{T01,i} = 0 \\ \text{and} \quad & \lambda_{T02,i} \geq 0, \quad \kappa_{2,i} \geq 0; \quad \kappa_{2,i}^T \lambda_{T02,i} = 0 \end{aligned} \tag{9}$$

In order to formulate an LCP in standard form the forces $\lambda_{T,i}$ and $\lambda_{T02,i}$ have to be replaced by exploiting (7)

$$\begin{aligned} \lambda_{T,i} &= G_{T1,i}^{-T} G_{N1,i}^T \lambda_{N,i} - G_{T1,i}^{-T} \lambda_{T01,i} \\ \lambda_{T02,i} &= (G_{N2,i}^T - G_{T2,i}^T G_{T1,i}^{-T} G_{N1,i}^T) \lambda_{N,i} - G_{T2,i}^T G_{T1,i}^{-T} \lambda_{T01,i} \end{aligned} \tag{10}$$

Taking together the quantities of all possibly sticking contacts and substituting the matrix products in (8) and (10) by

$$\begin{aligned} \lambda_T &= \bar{G}_{N1}^T \lambda_N - \bar{G}_{T1}^T \lambda_{T01}, \quad \lambda_{T02} = \bar{G}_{N2}^T \lambda_N - \bar{G}_{T2}^T \lambda_{T01}, \\ \kappa_1 &= G_{T1}\dot{g}_T - G_{T2}\kappa_2, \end{aligned} \tag{11}$$

the expressions (2), (3), (4), (9), (10) and (11) finally result in the following LCP (in standard form) to detect the contact forces

$$\begin{pmatrix} \ddot{g}_N \\ \kappa_1 \\ \lambda_{T02} \end{pmatrix} = \begin{pmatrix} W_N^T M^{-1} W^* & -W_N^T M^{-1} W_T \bar{G}_T^{-T} & 0 \\ -G_T W_T^T M^{-1} W^* & G_T W_T^T M^{-1} W_T \bar{G}_T^{-T} & G_T \\ 2\bar{G}_N^T & -\bar{G}_2^T & 0 \end{pmatrix} \begin{pmatrix} \lambda_N \\ \lambda_{T01} \\ \kappa_2 \end{pmatrix}$$

$$+ \begin{pmatrix} W_N^T M^{-1} h \\ G_T - W_T^T M^{-1} h \\ 0 \end{pmatrix} \tag{12}$$

$$\begin{pmatrix} \ddot{g}_N \\ \kappa_1 \\ \lambda_{T02} \end{pmatrix} \geq 0 \quad \begin{pmatrix} \lambda_N \\ \lambda_{T01} \\ \kappa_2 \end{pmatrix} \geq 0 \quad \begin{pmatrix} \ddot{g}_N \\ \kappa_1 \\ \lambda_{T02} \end{pmatrix}^T \begin{pmatrix} \lambda_N \\ \lambda_{T01} \\ \kappa_2 \end{pmatrix} = 0$$

With the abbreviation $W^* = W_N + H_R + W_T G_N^T$.

Further approaches are necessary if two originally separated bodies come into contact. In general this entails a transfer of impulses across the active contacts which can be computed again by solving a linear complementarity problem. In detail, the derivation of appropriate contact laws is given for example in [2] and will not be shown in this paper.

5 Simulation Results

One goal of this investigation consists in comparing the above contact formulation with the classical method using force elements. Basically the application of these contact models to systems with unilateral constraints reduces the actual simulation time with regard to this example to about one third. A further reduction of simulation time is not possible because of the fact that the determination of contact points including wear effects becomes dominant. For other examples we found time factors of one tenth to one hundredth. Since the quantitative results of both models are identical, within the limits of the parameter accuracy, the complementarity formulation is the better choice. It was also possible to reproduce the various mechanisms of wheel wear which finally leads to typical wheel profiles. This also confirms the efficiency of the presented contact model.

6 References

[1] A. Klarbring, *Mathematical Programming and Augmented Lagrangian Methods for Frictional Contact Problems*, Proceedings Contact Mechanics International Symposium, Lausanne, 1992

[2] F. Pfeiffer, Ch. Glocker, *Multibody Dynamics Unilateral Contacts*, Wiley Series in Nonlinear Science, John Wiley & Sons, Inc., New York, 1996

[3] Th. Rossmann, F. Pfeiffer, Ch. Glocker, *Efficient Algorithms for Nonsmooth Dynamics*, Proc. of ASME, Texas, Nov. 1997, Dallas

Controlling Chaos in a Piece-Wise Linear Oscillator

H. Yabuno[1]
T. Okada[2]
N. Aoshima[3]

ABSTRACT In this paper, the chaotic motion in a piece-wise linear system under a harmonic excitation source is converted to a periodic motion by employing the delayed feedback method proposed by Pyragas. From the practical view point, an actuator attached to the original system is used for the control, without directly actuating the harmonic excitation source. Comparing time histories between the cases with and without control, it is experimentally confirmed that the chaos is suppressed.

1 Introduction

Many of the mechanical components with clearance can be modeled by systems with piece-wise linear damping and restoring forces. Recent studies have shown that the system can exhibit chaotic motion under an external harmonic source(for example [1] and [2]). In this paper, it is experimentally shown that the chaotic motion can be converted to a periodic one by employing the delayed feedback method [3]. The application of this method for the chaos in the mechanical systems has been performed by superimposing the control signal on the external harmonic source. In this paper, the control is carried out by attaching the actuator for the control to the original system.

2 Analytical Model

We consider a simple piece-wise linear system under an external harmonic force, which has different stiffness constants and damping coefficient for

[1]Institute of Applied Physics University of Tsukuba, Tsukuba, 305-8573, JAPAN
[2]Mitsubishi Heavy Industry, Wada-saki, 1-1-1, Hyogo ward 652, JAPAN
[3]Institute of Applied Physics University of Tsukuba, Tsukuba, 305-8573, JAPAN

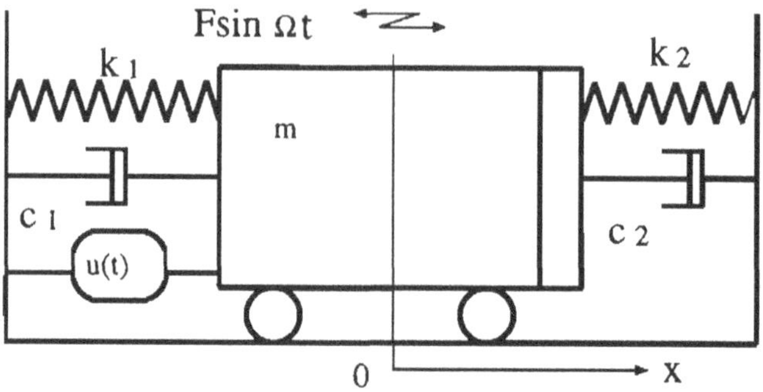

FIGURE 1. Analytical model.

positive and negative deflections, as shown in Figure 1. Putting the origin O of a static coordinate X the position of at the mass in the static equilibrium state, the equation of motion can be written as follows:

$$m\ddot{x} + D(\dot{x}) + E(x) = F(t) + u(t), \tag{1}$$

where the dot expresses the derivative with respect to time, $F(t) = a \sin \Omega t$ is the external harmonic force, and u(t) is the control force by an actuator attached to the original system. The $E(x)$ is the stiffness in the system expressed as follows:

$$E(x) = \begin{cases} (k_1 + k_2)x, & x \geq 0 \\ k_1 x, & x < 0. \end{cases} \tag{2}$$

The $D(\dot{x})$ is the damping effect expressed as follows:

$$D(x) = \begin{cases} (c_1 + c_2)\dot{x}, & x \geq 0 \\ c_1 \dot{x}, & x < 0. \end{cases} \tag{3}$$

We use the following parameter values corresponding to the subsequent experiment: m = 1.74kg, k_1 = 497N/m, k_2 = 9795N/m, c_1 = 2.697kg/s, c_2 = 4.99kg/s, a = 3.0N. By putting $u = 0$, and numerically simulating Eq. (1), we can obtain a Poincare map without control. The bifurcation diagram whose Poincare section satisfies $\Omega t \equiv 0(mod 2\pi)$ and control parameter is Ω is shown in Figure 2(a). The Lyapunov exponent is also obtained as in Figure 2(b). It is theoretically predicted that the system exhibits chaotic motions in the vicinity of $\Omega/2\pi$ = 2Hz, because there are a large number of plots and the Lyapunov exponent is positive. In the next section, we experimentally carry out converting the chaos to a periodic motion.

3 Controlling Chaos

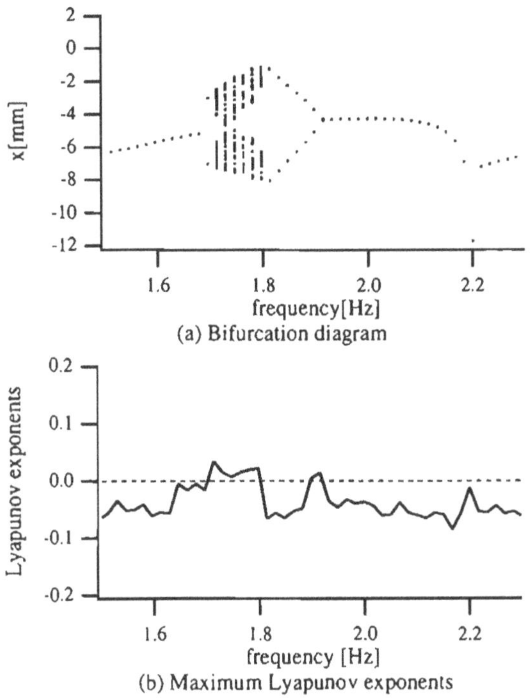

FIGURE 2. Poincare map.

A linear motor is used as the external harmonic source, $F(t)$, and the control force, $u(t)$, is applied by the other linear motor. We convert the chaos in the case of $\Omega/(2\pi) = 1.9$Hz to the periodic motion by employing the delayed feedback method [3]. We introduce the control force expressed as follows:

$$u(t) = -k\{x(t) - x(t - 2\pi/\Omega)\}, \tag{4}$$

where constant feedback gain, $k = 100$ is selected by trial and error. Figure 3 illustrates an experimental result under the control. The control starts at $t = 20$s. Then the actuating signal which is proportional to the control force is applied according to Eq. (4). It is found that the control signal becomes very small when the chaos is converted to a periodic motion and after this time it keeps very small.

4 Conclusions

In this paper, it is experimentally shown that the chaos in a piece-wise

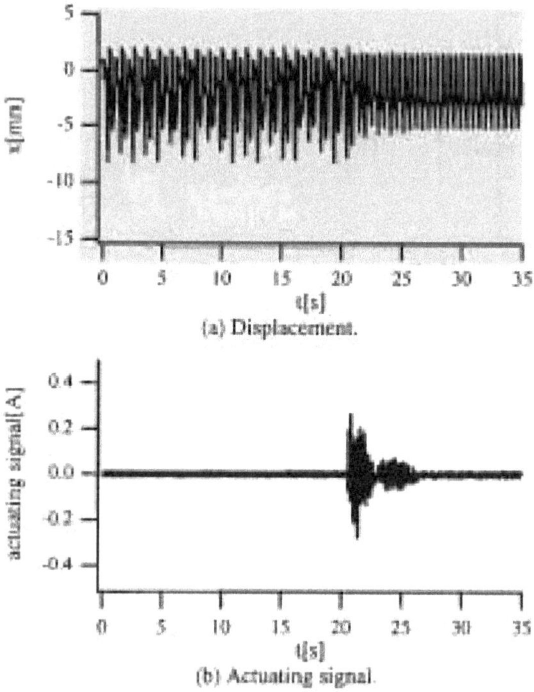

FIGURE 3. Conversion to a periodic motion. ($\Omega/2\pi$ = 1.9Hz, a = 3.0N, k = 100)

linear system under an external harmonic excitation is converted to the periodic motion by the delayed feedback method. In the future work, the suppressions of the chaos and of the local bifurcation in the wide range of the control parameter, i.e., Ω will be experimentally discussed.

5 References

[1] Tung, P. C., and Show, S. W., The Dynamics of an Impact Print Hammer, *ASME, Journal of Vibration, Acoustics, Stress, and Reliability in Design*, Vol 110, 193-200, 1988.

[2] Chancellor, R. S., Alexander, R. M., and Noah, S. T., Detecting Parameter Changes Using Experimental Nonlinear Dynamics and Chaos, *ASME, Journal of Vibration and, Acoustics*, Vol 118, 375-383, 1996.

[3] Pyragas, K., Continuous Control of Chaos by self-controlling Feedback, *Physics Letters, Series A* 170, 421-428, 1992.

A History of Friction in Dynamics

B. F. Feeny[1]
A. Guran[2]
N. Hinrichs[3]
K. Popp[3]

1. Introduction

Friction had been a topic of technological attention long before the dawn of science, and is still a hot topic in scientific and engineering research today. As a nonlinearity, it poses challenges to the dynamical systems researcher. For one, friction is very difficult to model, since the underlying mechanism is not entirely understood. Coulomb had mentioned more than two hundred years ago:
"To prove to the specialist how desirable it would be to know more about the law of the resistance of rubbing surfaces would be useless - to tell somebody about this who does not know anything about friction is for other, but also known reasons, a fruitless business." Thus, friction as a nonlinearity is the current focus of a great number of research activities throughout the world.

The manifestations of friction took the attention of the ancients as they developed critical technologies. It took incredible creativity and ingenuity for the first people of a community to recognize properties of friction, and then implement them in the design of a new machine. As time progressed, and people yearned to understand machines and nature, the awareness of friction eventually made its way into scientific thought. Friction laws were formulated, and the ideas of dissipation and motion were conceived. Later, researchers began to question the mechanism of friction. This continues to the present day. Friction laws have also been included in system equations

[1]Department of Mechanical Engineering, Michigan State University, 2555 Engineering Building, East Lansing, MI 48824

[2]Electrical Engineering - System Department, University of Southern California, Los Angeles, CA 90089-2560

[3]Institute of Mechanics, University of Hannover, 30167 Hannover, Germany

of motion, which have then been used to study the nonlinear dynamics.

2. Friction in Technology

Stick and slip have been used in machines and devices in clever ways, which enriched human societies by improving technologies that facilitated survival, provided a competitive edge, and enhancing pastimes which defined culture. Ancient evidence of such technologies occurs in the archeological records, murals, tombs and pictographs, and more recently on paper. Among developments which used friction were

- fire making
- cord drive and belt drive
- hafting of stone tools
- friction grip
- ropes and cordage
- brakes
- clutches
- toys and sports
- bowed instruments
- crayons and pencils
- electric machines

In other cases, friction was a nuisance which made work extra difficult and caused wear. The desire to reduce the frictional work load and wear led to other creative developments, such as

- sledges
- the wheel
- roads and rails
- skis and skates
- sports
- lubrication
- bearings

- domestication of animals
- levitation

3. Friction in Mechanics

There has always been a desire to describe the behavior of nature and machines. The first step in developing a macroscopic model is to obtain the equations of motion. The key ideas in analytical mechanics were first formulated as words and ideas, and later as equations. Early works were expresses with an implied awareness of frictional resistance. Applying analytical mechanics to systems with friction and intermittent contacts is a research topic that continues today.

Da Vinci is the first known to develop a friction law that could be inserted into equations of motion. It took time for his simple law to take hold, as it was rediscovered by Amontons, who received criticism, and tested experimentally by Coulomb, who won a prize for his work in mechanics. More recent studies on surfaces and structures have aimed at measuring, predicting, and describing friction with as many details as needed. This is an ongoing effort.

4. Friction and Dynamical Response

Dry and lubricated friction are extremely nonlinear, and models are discontinuous. Discontinous differential equations are not guaranteed to have unique solutions. Thus, friction has motivated works regarding existence and uniqueness of solutions. Due to the nonlinearity, when solutions do exists, there is the potential for all sorts of nonlinear behavior, which may be dissipative, periodic, or nonperiodic.

The study of the dynamics of systems with friction has two important applications: friction damping and frictional excitation. Some of the developments in frictional damping are

- free responses and identification
- forced responses
- paradoxical destabilization of friction damping
- tidal friction

Perhaps Galileo made the first careful descriptions of frictional excitation, regarding the fingered goblet and the chattering of a scraping tool. Other systems include

- bowed instruments
- squeak and squeal
- sonorous sands
- earthquake models

5. References

Details of this discussion can be found in the reference below, and in all of the citations therein.

1. Feeny, B. F., Guran, A., Hinrichs, N., and Popp, K., 1998, "A historical review on friction and stick-slip phenomena," *Applied Mechanics Reviews,* **51** (5) 321-341.

SESSION II–(B)

Nonlinear Dynamics and Experiments

Nonlinear Phenomena in Rotor Systems Due to Clearance, Friction and Impact

Y. Ishida[1]

ABSTRACT In turbomachinery, the nonlinearity appears in restoring forces and damping forces due to various causes, such as, clearances in bearings, frictions and impacts. In this lecture, various kinds of nonlinear phenomena in rotor systems are introduced.

1 Introduction

As the operating speed of rotating machinery increased and the rotors became lighter, various kinds of nonlinear phenomena were encountered. The study on nonlinear oscillations in rotor systems started in the 1950's. T.Yamamoto, A.Tondl and F.Ehrich are the main contributors in the earlier stage and numerous papers have been published thereafter. Various kinds of nonlinear phenomena observed in our laboratory are introduced in this lecture. Important sources of nonlinear oscillations are illustrated in Figure 1.

2 Nonlinear Resonances in Symmetrical Rotor Systems.

A rotor system composed of an elastic shaft with a circular cross section and a

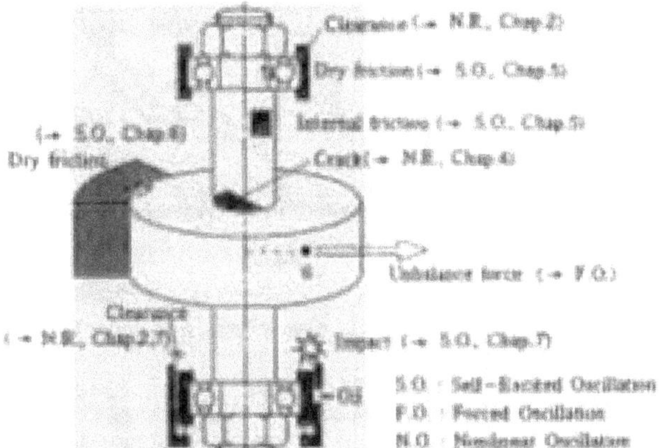

Figure 1. Various sources of oscillations

1 Department of Electronic-Mechanical Engineering, Nagoya University, Nagoya, Aichi, 464-8603, Japan

disk is called a symmetrical rotor system. If the shaft is supported by single row deep groove ball bearings which have small clearances around the balls, the shaft has spring characteristics shown in Figure 2. Symmetrical nonlinear charac- teristics appear when the system is well-assembled and unsymmetrical ones appear when it has misalignment. An example of resonance curves is shown in Figure 3. The resonance curves at the major critical speed become a hard spring type and various kinds of subharmonic and combination resonances appear.

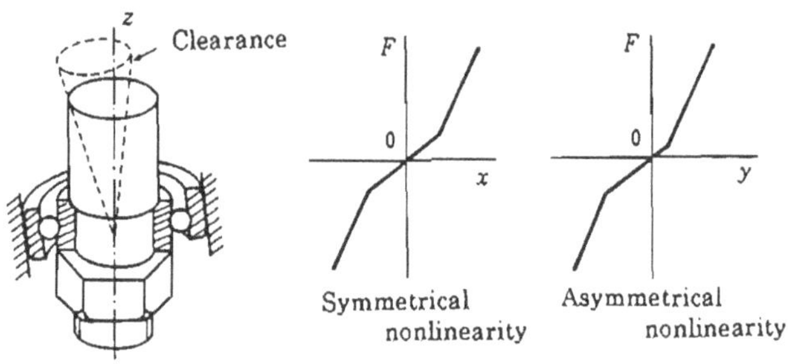

Figure 2. Clearance in a ball bearing and spring characteristics

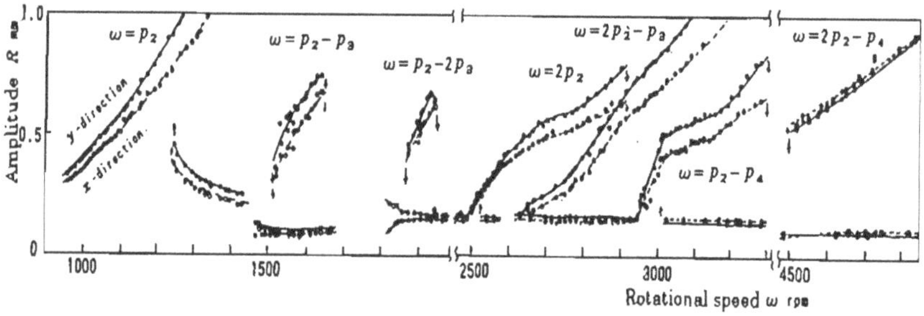

Figure 3. An experimental result of a rotor supported by ball bearings

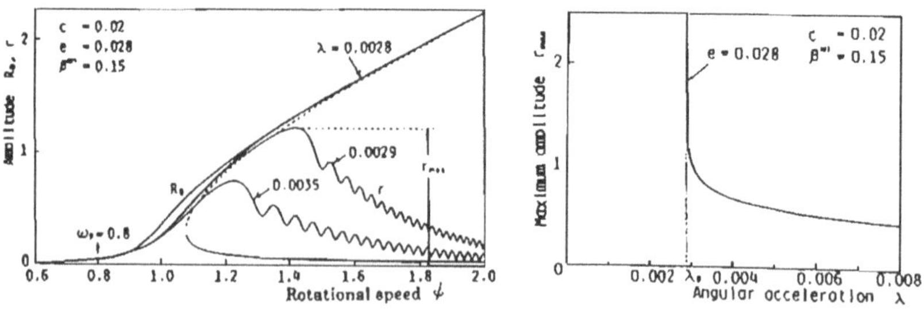

Figure 4. Nonstationary oscillation during passage through a critical speed

3 Nonstationary Oscillations

When the rated speed is above the major critical speed, the rotor must pass through the critical speed with a small amplitude for safety. When a rotor has symmetrical nonlinear characteristics, the resonance curves become a hard spring type. In such a case, when acceleration λ is comparatively small, the amplitude increases along the resonance curve and the rotor cannot pass the critical speed. Figure 4 shows amplitude variation curves for various angular accelerations and a relationship between the maximum amplitude and the acceleration. The rotor cannot pass under the critical value λ_0.

4 Resonances Due to Cracks

A crack due to fatigue is one of the most serious causes of rotating machinery accidents. In order to develop a monitoring system to detect a crack, it is important to investigate resonances caused by a crack. Due to open-closed mechanizm of a crack, the rotor becomes to have nonlinear and parametric characteristics and, as a result, resonances shown in Figure 5 appears.

5 Self-Excited Oscillations Due to Internal Frictions

Due to frictions in a shaft material or dry friction between a shaft and an

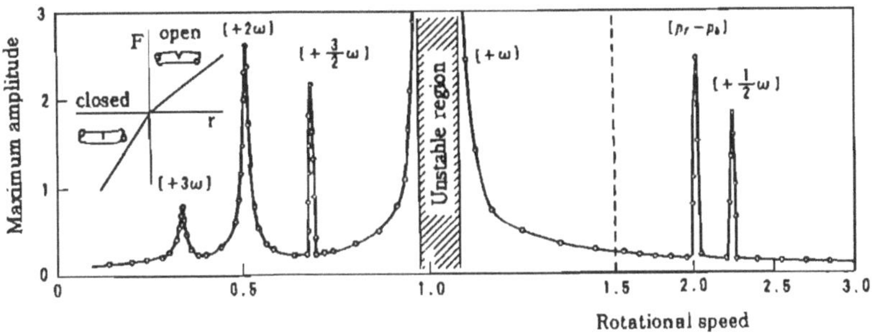

Figure5. Resonances due to a transverse crack

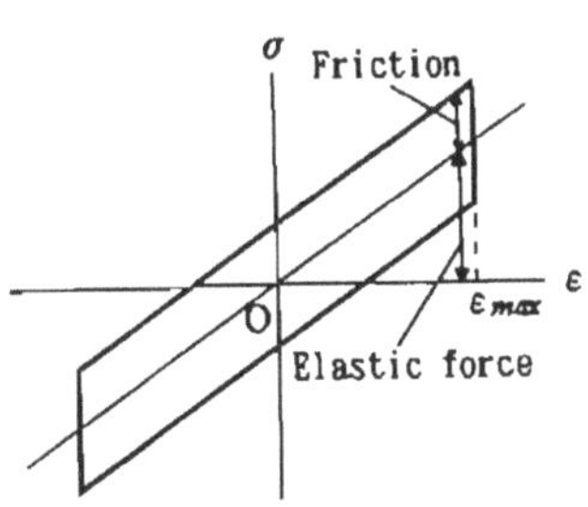

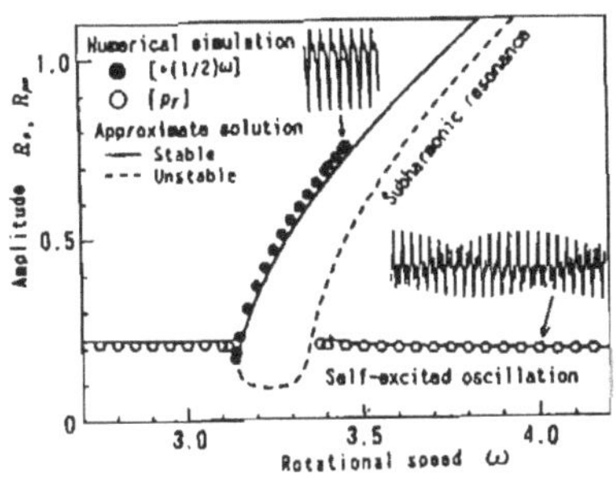

Figure 6 Self-excited oscillations and entrainment

element attached to it, the stress-strain relationship forms a hysteresis loop shown in Figure 6. During a whirling motion, energy corresponding to this loop area enters the shaft and a self-excited oscillation of a forward whirling mode appears above the major critical speed. Figure 6 also shows self-excited oscillations and the entrained phenomena at a subharmonic resonance.

6 Self-Excited Oscillations Due to Contacts with a Casing

When a shaft touches the casing, dry friction works between the shaft and the casing, and the self-excited oscillations of a backward whirling mode occur.

7 Self-Excited Oscillations Due to Impacts

In a squeeze film damper bearing, impacts occur between a bearing holder and a casing when oil is not supplied enough and self-excited oscillations of a forward whirling mode occur. As these two elements do not rotate, rubbing does not occur. Figure 7 shows an experimental result with self-excited oscillations and nonlinear resonances. Adopting the bouncing mechanism used by Li and Paidoussis[1] to explain vibrations of a rotating disk flying and bouncing in a cylinder, we can obtain the same response curve numerically.

8 Conclusions

In this lecture, representative nonlinear phenomena observed in our research were introduced. Concerning their detailed references, refer the review article[2] by the author. Besides the phenomena mensioned above, chaotic vibrations reported by F.Ehrich[3] are noticeable.

9 References

[1]G.X.Li and M.P.Paidoussis,Nonlinear Dynamics.5, p.53, 1994, [2]Y. Ishida, JSME International J. Ser.C, 37-2, p.1, 1994, [3]F.Ehrich,Trans. ASME,113, p.50,1991.

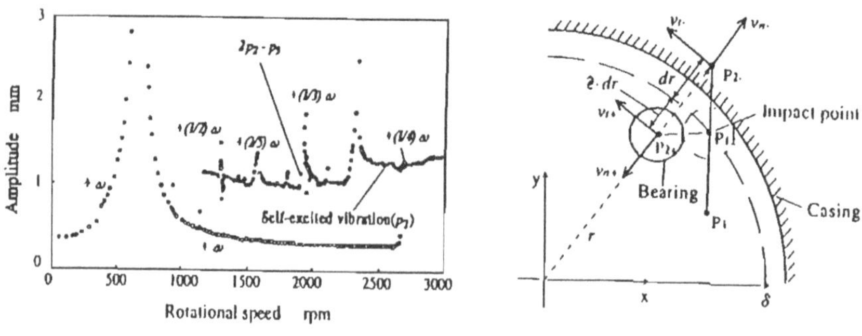

Figure 7 Self-excited oscillations and nonlinear resonances due to impacts

Non-smooth, Non-reversible Dry Friction Oscillator: Design and Measurements

M. Wiercigroch[1]
W.T.V. Sin[2]
F.K.Z. Liew[3]

ABSTRACT A Coulomb oscillator with a variable normal force was designed to carry out a wide range of experimental dynamic analysis, especially the study of the non-reversibility of dry friction characteristics. The design choice was made based on the criteria of accuracy of representing the mathematical model and flexibility in terms of parameter changes such as natural frequency of the system, coefficient of friction and normal force. The system consists of a block mass attached to the two coil springs and the dry frictional damper in which the friction force is varied by a pneumatic actuator. The experimental set-up, calibration procedure for the clamping device, and typical results are discussed.

1 Introduction

As noted by Den Hartog almost seventy years ago (1931), damping mechanisms in practical engineering vibration problems are of great variety. One of the most notorious types is dry friction. Despite the fact that dry friction forces have been known for centuries, solving a dynamic system having dry friction elements still causes a major problem, especially, where the relative motion of subsystems (components) of the considered system undergo a qualitative change of excitation type from force to kinematic and vice-versa. This can be easily explained for a system undergoing stick-slip phenomenon. During the slip portion, at any moment of time, the system components are acting on each other, and this can be classified as the force

[1]Department of Engineering University of Aberdeen, King's College, Aberdeen AB24 3UE, UK
[2]Siemens Components Pte Ltd, 168 Kallang Way, Singapore
[3]Dactt Engineering Pte Ltd, 540 Sims Avenue Centre, #03-06, Singapore

type of excitation. When the stick phase occurs, the previously separately moving subsystems (components) are now joined, and the motion is governed via a kinematic principle, i.e. velocity of the sticking surfaces are the same. In practice, this transition from the stick to slip phase is also accompanied by a change of friction coefficient; from static to dynamic. The main factor responsible for this phenonenon is 'time of stick' (Rabinowicz, 1958), which allows bonds to be developed between two interacting surfaces. Topologically, the system during stick-slip is changing (decreasing) its dynamical dimension.

The transition from static to kinematic friction is associated with another important phenomenon; non-reversibility of the friction force-velocity characteristics. This will be commented on in more detail in the latter part of the paper. This characteristics is conspicuous while viewing dynamics of the considered system via a time history of the friction force or directly looking at the friction force - relative velocity of the sliding surfaces characteristics. Moreover, the non-reversibility of the friction characteristics become crucial for dynamical systems exhibiting chaotic motion, where any a slight difference in initial conditions, parameter values or forcing can produce qualitelively different responces. Therefore, the main objective of this work was to design an experimental rig capable of providing accurate dynamic data which would allow a better understanding and modeling of this phenomenon.

Dry friction problems, especially stick-slip motion have been investigated extensively for years, and there is a body of valueable references, for example Den Den Hartog (1931), Rabinowicz (1958), Shaw (1986), Stelter (1992), Feeny (1992) and Mottershead and Chan (1995), to name a few. However, only few works have reported non-reversiblity of dry friction characteristics. Among them, the study by Bell and Burdekin (1970) gives a clear evidence of different slopes of friction forces for the acceleration and deceleration phases. This was investigated for an important industrial application, i.e., machine tool feed drives where stick-slip vibration is a major problem. This work gives a thorough experimental account and comparison between unlubricated and lubricated cases for the dynamic friction force - relative velocity characteristics. Bo and Pavelescu (1982) carried out rigorous analytical studies on a similar system using the Krylov-Bogolubov method to analyse stick-slip vibrations. Also, in their work non-reversibility of the friction force is not mentioned directly.

The term non-reversibility of the dry friction characteristics was introduced by Powell and Wiercigroch (1992), where an analytical approach was undertaken to describe the dynamics of a based excited oscillator with an electrorheological damper which actied like a Coulomb damper. Non-reversibility of the dry friction characteristics can be easily explained using the friction force - relative velocity graph, as depicted in Figure 1b. If one considers a simple oscillator with a Coulomb damper (Figure 1a), the force across the damper drops from the static to dynamic value during the acceleration phase; transition from point 1 to 2. If the systems undergoes steady state oscillations, the deceleration phase occurs and the relative velocity changes its sign; transition from point 3 to 4. This can be modelled conveniently by a pair of the following exponential functions (Wiercigroch, 1993).

2 Dry Friction Oscillator

The experimental oscillator is comprised of a pneumatic viscous damper and a mass consisting of two parts which oscillates along two guide posts. The first part of the mass weighs 1.294 kg, and comprises the main aluminium block with two mid-bushes mounted the the two ends. Inside these mid-bushes are housed two teflon bushes providing a low coefficient of friction for sliding along the guide posts. Four springs retainers used to secure the two compression springs to the mass are included in the mass of 1.294 kg. The springs have a combined stiffness of 4780 N/m and the other ends are held to the lower bush on the base, again by four spring retainers. The second part of the mass weighing 0.254 kg consists of the vertical plate, the vertical plate mount and a force transducer mounted between the main mass and the vertical plate mount. Slots are machined onto the two clamping faces of the vertical plate in order to remove debris generated during the experiments. For measuring the relative displacement of the mass, the LVDT rod is connected to the top surface of the main mass while the coil portion of LVDT is mounted to the top plate. B&K accelerometers are used to measure the acceleration time histories of the base and the mass, respectively.

3 Experimental Setup

The experimental rig was mounted onto the Ling Dynamic Systems shaker. The base excitation, y, provided by the shaker is controlled by a Wavetek Arbitrary Waveform Generator. A B&K accelerometer used to measure the base acceleration, was mounted on the shaker table. The signal from this accelerometer was passed through a B&K charge amplifier, before being captured by a MicroLink Data Storage System. An oscilloscope was also used to monitor this signal.

As mentioned earlier the relative displacement of the mass, z, is measured by the LVDT. The signal from the LVDT transducer was passed trough a low pass filter to remove noises of high frequency (above 300 Hz) and the initial offset value was brought to zero using the offset compensating facility that was incorporated in the filter. The output from the filter was than split with one cable connected to channel 1 of the Microlink to register the relative displacement of the mass. The signal from the other cable was passed to a differentiator to obtain the relative velocity of the mass. The output from the differentiator was also split into two with one cable connected to channel 2 of the Microlink while the other was connected to a second differentiator. The output from the second differentiator was fed into the channel 3 of the Microlink to register the relative acceleration of the mass. The force between the oscillating mass and the tongue (combined dry friction and inertia forces) was measured using B&K force transducer and passed through a B&K charge amplifier, before being captured on channel 4 of the Microlink. Discrete time series of the data were obtained from the Microlink using the Wavecap software installed on a PC. Two oscilloscopes and a Spectrum Analyser were used to monitor the phase plane, frequency spectrum and the force versus velocity characteristics during the experiment.

4 Discussion

Another convenient means of representing the friction force characteristics is friction force - relative displacement relationship. This can provide useful information regarding the friction force variation and relating it to motion of the mass, especially to study stick-slip phases and number of stops per cycle of excitation. For example, if the investigated oscillator uses a simple sign function to model the friction force, the friction force-relative displacement characteristics would be a plain rectangular, assuming that inertia force generated by base excitation is greater than the resultant resistance force across the dry friction and viscous dampers, and the spring. The arrows indicate the progression of time. Going to more realistic dry friction characteristics, i.e., those having static and dynamic coefficients, would change the right angles of the rectangular to four symmetric sharp corners mirroring the transition from static to dynamic. The discussed curves portray steady state oscillations; where all initial transients have been truncated. Otherwise, the curve will start with the static friction force value at zero relative velocity. Finally, if one considers non-reversible dry friction characteristics, there must be a different path for the acceleration and deceleration phases.

The discussion outlined above is supported by the experimental data obtained from the rig described in the previous sections of this paper. Before comparing the theoretical and experimental force-velocity and force displacement characteristics, it is worth looking closely at the time histories of the experimental results. They are prepared for most typical frictional pair, i.e. steel-steel, where the base acceleration is 25.6 ms-2, natural and forcing frequencies are 9 and 18 Hz respectively, and the clamping force is 35 N. The stick portions causing two stops per cycle are clearly visible as flat maxima in the dispacement and friction force diagrams, and steps at zero values in the velocity time history. It must be pointed out that the measured friction force contains the inertia compent of the tongue. However, since for the choosen set of parameters the generated inertia force is relatively small, this effect can be neglected. This is confirmed by looking at the force-velocity and force displacement characteristics, where two different paths for the acceleration and deceleration phases are easily distinguishable. Especially, if one examines the force-displacement characteristics, where there is a distinctive diference for the transition from stick to slip and vice versa.

5 References

[1] Bell, R. and Burdekin, M., 1970, A study of the stick-slip motion of machine tool feed drives, *Proc. Instn Mech Engrs* 184, 543-557.

[2] Bo, L.C. and Pavelescu, D., 1982, The friction-speed relation and its influence on the critical velocity of stick-slip motion, *Wear* 82, 277-289.

[3] Den Hartog, J.P., 1931, Forced vibrations with combined Coulomb and viscous friction, *Trans. ASME* APM-53-9, 107-115.

[4] Feeny, B., 1992, A non-smooth Coulomb friction oscillator, *Physica D* 59, 25-38.

[5] Mottershead, J.E. and Chan, S.N., 1995, Flutter instability of circular plate on a floating central collar, *Trans. ASME; J. Vib. Acoust.* 117, 161-163.

[6] Powell, J.A. and Wiercigroch, M., 1992, Influence of non-reversible Coulomb characteristics on the response of a harmonically excited linear oscillator, *Machine Vibration* 1(2), 94-104.

[7] Rabinowicz, E., 1958, The intrinsic variables effecting the stick-slip process, *Proceedings of the Physical Society* 71, 668-673.

[8] Shaw, S.W., 1986, On the dynamic responce of a system with dry friction, *Journal of Sound and Vibration* 108, 305-325.

[9] Stelter, P., 1992, Nonlinear vibrations of structures induced by dry friction, *Nonlinear Dynamics* 3, 329-345.

[10] Wiercigroch, M., 1993, Comments on the study of a harmonically excited linear oscillator with a Coulomb damper, *Journal of Sound and Vibration* 167(3), 560-563.

[11] Wiercigroch, M., 1994, A Note on the switch function for the stick-slip phenomenon, *Journal of Sound and Vibration* 175(5), 700-704.

Recent Signal Processing Techniques for the Detection of Localized Faults in Gearboxes

D. Boulahbal[1], F. Golnaraghi[1] and F. Ismail[1]

ABSTRACT This paper presents two new techniques for the detection of localized damage in gear trains. These are the Continuous Wavelet Transform (CWT) and the Short-Time Beta Kurtosis (STBK). In contrast with the classical spectral analysis, the proposed techniques extract signal features in the time domain and are able not only to detect the presence of a local defect but also pinpoint its exact location along the circumference of the gear under study. The mathematical foundations of the techniques are presented along with the results of applying them to three different test cases: a mathematically perfect gear, an off-the-shelf realistic gear, and lastly one with a cracked tooth. Another novelty is that the techniques are applied in conjunction to both the amplitude and phase modulations of the vibration synchronous average signals. Using both amplitude and phase yields a more positive assessment of both the gear condition and the type of fault at hand.

Introduction

Up until few years ago, and because of the availability of low-cost FFT analyzers, the most widespread diagnostic technique for gears was spectral analysis. Numerous studies [1, 2] have shown that, in the presence of localized defects, the Synchronous Time Averaging (STA) technique and its extensions offer superior results. In the present paper we examine two such methods, these are the CWT and the STBK. The CWT conciliates the strengths of both time and frequency domain techniques, as it gives the evolution in *time* of a signal's *frequency* content. This property makes the CWT a preferred choice for extracting transients caused by localized faults. The STBK extends on work pioneered by the third author and relies on statistical modeling, which makes it robust in the presence of noise, and weakly sensitive to changes in system operating conditions.

Recent applications of the CWT have focused on the use of only the amplitude map. In this study the phase wavelet map is also used since it was found to display distinctive features in the presence of a cracked tooth. The reciprocal of the STBK exhibits sharp "spikes" when applied to both the amplitude and phase modulation of the gear vibration STA. When the STBKs of both the amplitude and phase modulation derivative are used together, a more positive assessment of the type of gear fault is obtained.

[1] Department of Mechanical Engineering, University of Waterloo, Waterloo, Ontario, Canada, N2L 3G1

The Continuous Wavelet Transform

If we let $x(t)$ be the gear vibration signal obtained from the STA technique, then its CWT is defined as a circular convolution integral.

$$\mathcal{W}_x(t,f) = \frac{1}{\sqrt{fT}} \int_{-T/2}^{T/2} x(\tau) \cdot w\left(\frac{t-\tau}{fT}\right) \cdot d\tau \tag{1}$$

where T is the period of rotation of the gear under study and $w(t)$ is the mother wavelet. In this study we use the Morlet wavelet.

$$w(t) = \exp\left\{-\frac{t^2}{2b^2}\right\} \cdot \exp\{i2\pi f t\} \tag{2}$$

The parameter $b = \pi\sqrt{2/\ln 2}$ controls the effective support ($\approx 6b$) of the wavelet. Since $w(t)$ is complex-valued, the wavelet transform $\mathcal{W}_x(t,f)$ is also complex-valued. Thus, it gives the amplitude and phase modulation of the signal $x(t)$ over the entire time-frequency plane.

The Short-Time Beta Kurtosis

The kurtosis (KU) is often used as an indicator of the "peakedness" for an entire time trace. Here however, we use a running version of the KU that extends over one gear meshing period only, and slides along the time-axis. Hence, the qualifier "Short-Time". This approach enables us to compare local features of the vibration signal as opposed to providing a global indicator. The Beta distribution is defined as

$$B(\alpha,\beta) = \int_0^1 x^{\alpha-1} \cdot (1-x)^{\beta-1} \cdot dx \tag{3}$$

The above definition requires that the signal $x(t)$ be normalized between 0 and 1. The parameters α and β control the shape of the beta distribution and they are extracted from the mean m and variance σ^2 of $x(t)$

$$m = \frac{\alpha}{\alpha+\beta} \qquad \textit{and} \qquad \sigma^2 = \frac{\alpha\beta}{(\alpha+\beta)^2(\alpha+\beta+1)} \tag{4}$$

The beta kurtosis is then calculated from

$$KU = \frac{3(\alpha+\beta+1)(2\alpha^2 - 2\alpha\beta + \alpha^2\beta + \alpha\beta^2 + 2\beta^2)}{\alpha\beta(\alpha+\beta+2)(\alpha+\beta+3)} \tag{5}$$

We have experimented with both the ordinary and Beta kurtoses. The latter is robust because it *fits* a Beta distribution to the observed data, then computes the KU from the parameters of the distribution, and not from the raw data as is the case for the ordinary KU. This "statistical modeling" step is the reason behind the STBK's robustness.

Experimental Results and Discussion

The CWT and STBK techniques are applied to vibration signals collected from the instrumented test rig described in [2]. Three different test cases were selected: (a) mathematically average gear, (b) actual average precision gear, and (c) gear with a 20% deep transverse crack cut at the root of one tooth. The results for these three cases are shown below.

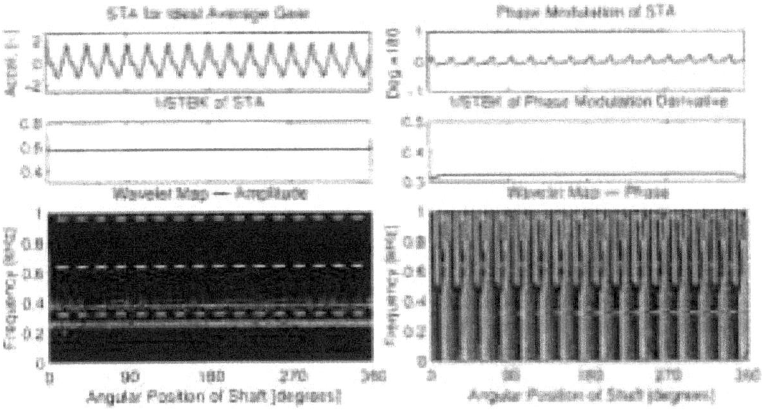

FIGURE 1. Results for Mathematically Average Gear

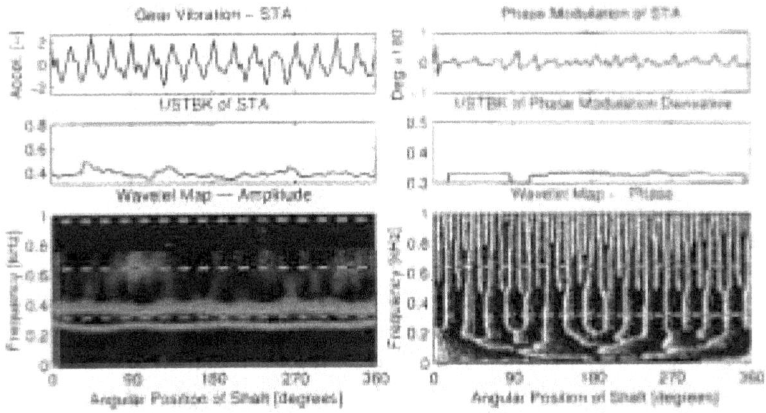

FIGURE 2. Results for Actual, average Precision Gear

Because the vibration signal in Figure 1 consists of only the gear meshing frequency (GMF) and its harmonics, the wavelet maps and the STBK traces are very regular. The maps and traces of Figure 2 show little modulations that are realistic for the gear train, and representative of various geometric errors in the gear pair. For the gear with a cracked tooth, Figure 3, characteristic signatures are visible in all graphs. The humps in the STBK trace for the phase derivative leads that in of the STA by about 1/16 of shaft revolution. The gear under study has 16 teeth and this is consistent with the vibration signature for a crack [3] i.e., when a cracked tooth comes in mesh, there is a small change in speed, followed by a stronger impact when the next tooth enters the mesh. The wavelet maps also exhibit features delayed by 1/16 of the gear's revolution. While the CWT offers a map with great detail, the traces from the STBK are rather convincing. The conclusion we shall retain from this study is that both amplitude and phase information need to be used in conjunction, and that the STBK "picks-up" the gear fault reliably.

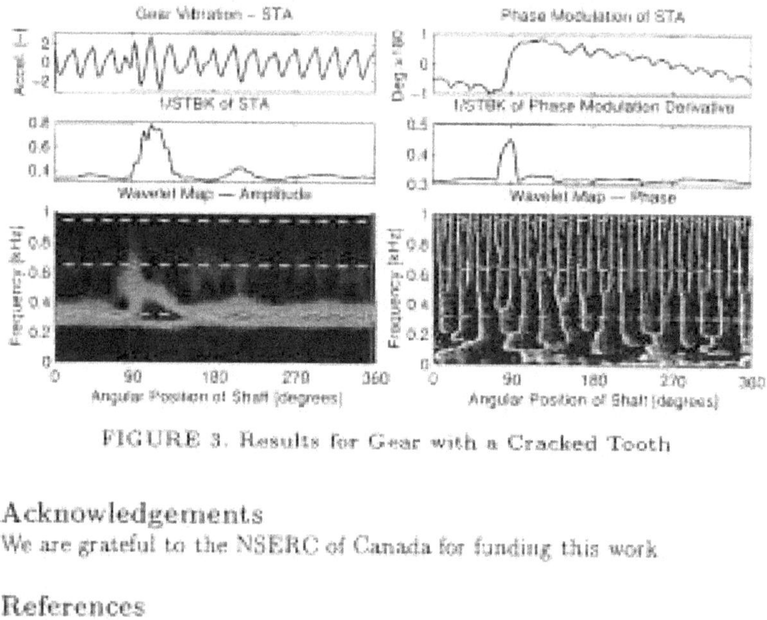

FIGURE 3. Results for Gear with a Cracked Tooth

Acknowledgements

We are grateful to the NSERC of Canada for funding this work.

References

[1] McFadden, P.D., Detecting fatigue cracks in gears by amplitude and phase demodulation of the meshing vibration. *ASME Journal of Vibration, Acoustics, Stress and Reliability in Design*, 108:165-170, April 1986.

[2] Boulahbal, D., Golnaraghi, M.F. and Ismail, F., Detection of fatigue cracks in gears with the continuous wavelet transform. *AGMA Technical paper, presented at the AGMA FTM'97*, SanDiego, November 1997.

[3] F. Ismail, F., Martin, H.R. and Omar, F., A statistical index for monitoring tooth cracks in a gearbox. *ASME Biennial Conference on Vibration and Noise*, DE-Vol. 84-1 1413-1418, Boston, 1995.

Dynamics of the Impact Force Generator

B. Blazejczyk-Okolewska[1]
K. Czolczynski[1]
T. Kapitaniak[1]

ABSTRACT During the last years the interest of scientists in multibody mechanical systems in which the phenomenon of impact occurs has been still growing. In the present paper a principle of operation of the impact force generator being an element of the rotor of the heat exchanger has been presented. Step disturbances of the rotational velocity of the rotor caused by the generator are aimed at intensification of the heat exchange process.

1 Introduction

In numerous industrial machines the impact of their movable parts is either the basic principle of their operation or the effect which improves their operating efficiency. The classic examples of such machines or devices are: a pneumatic hammer, impact dampers, or heat exchangers [1]. One of the factors that contribute to intensification of the heat exchange process are disturbances in the rotational velocity of the rotor of the heat exchanger [2].

One of type of disturbances which can result in intensification of the heat exchange are step disturbances of the rotational velocity. The simplest way to generate such disturbances is to employ the phenomenon of impact which causes, according to the Newton's hypothesis, step variations of the velocity of the bodies impacting on each other.

2 Physical Model of the Generator

The object of considerations is a system composed of two main parts (Figure 1): a rotor equipped with a fender, driven by an electric engine, a hammer in the form of a cylinder with a semicircular end, mounted on the end of a cantilever beam. During the operation of the generator, the rotor fender impacts on the hammer, which causes vibrations of the hammer on one hand and the desired step variations of the rotational velocity of the rotor on the other hand. As the geometry of the generator is rather sophisticated, two various kinds of impacts occur during its operation: I - the fender collides with the cylindrical part of the hammer, II - the

[1] Division of Dynamics, Technical University of Lodz, Stefanowskiego 15, 90-924 Lodz, Poland

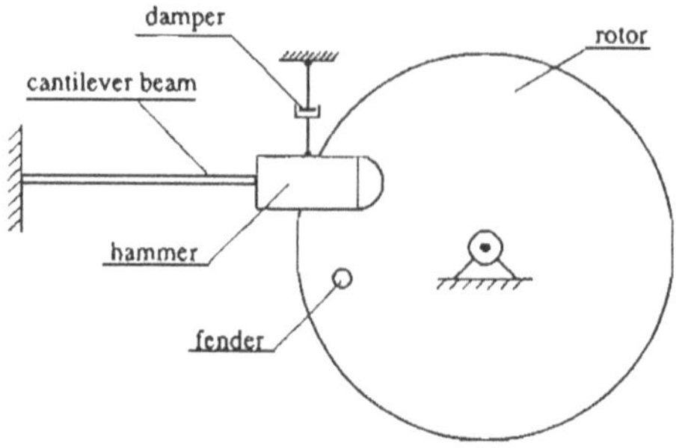

Figure 1. Schematic diagram of generator.

fender collides with the spherical part of the hammer. Equations of motion of the generator, and equations of impact have been presented in details in [3].

3 Influence of the Cantilever Beam Length on the System Operation

In order to evaluate the generator usefulness, the most important information is the average value of its angular velocity ω_a and the value of the velocity before the impact ω_{bi} and after the impact ω_{ai}.
Figure 2 presents a map of impacts for many generators which differ in the length of the cantilever beam l_s. The values of the angular velocity of the rotor before (ω_{bi}) and after (ω_{ai}) impacts and, additionally, the curves showing the values of the fractions of the basic frequency of free vibrations of the hammer: $1/2\alpha_1$, $1/3\alpha_1$, $1/4\alpha_1$ and $1/5\alpha_1$ have been shown on this map. In Here, a close relation between the variations of both the velocities ω_{ai} and ω_{bi} and the function of α_1 can be easily observed. For instance: when l_s=0.07 m, the rotor velocity ω_{bi} (close to its average angular velocity ω_a) is equal to approximately 1/3 of the value of the basic

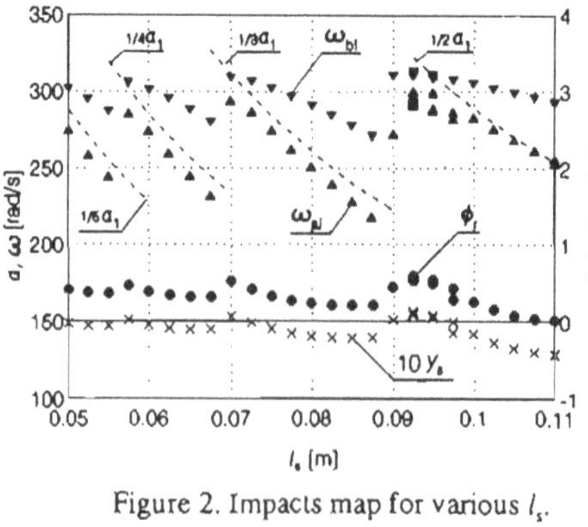

Figure 2. Impacts map for various l_s.

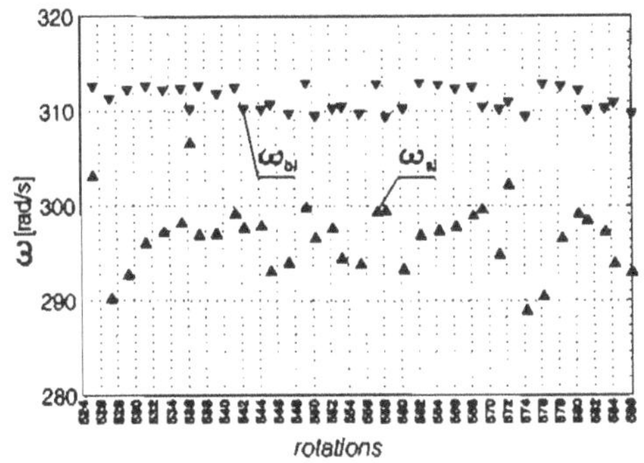

Figure 3. Global map of impacts; l_s=0.0925 m, Δ=ln(1.5), k=0.9.

frequency of free vibrations of the hammer α_1. It means that the impact forcing of these vibrations has a subharmonic frequency.

An increase in the cantilever beam length causes of course a decrease in the value of $1/3\alpha_1$. Despite it, however, the hammer "wants" the impact forcing of its free vibrations to have a subharmonic character: the impacts are stronger and stronger, which, as a consequence, causes the diminishing of both ω_{bi} and ω_{ai}. This way of affecting the angular velocity by the hammer has been called the self synchronisation of the system. When l_s exceeds the value 0.0875 m, the conditions for easy generation of the hammer vibrations with the next subharmonic frequency $1/2\alpha_1$, arise and the situation repeats.

4 Regular and Irregular Motion

In the majority of cases presented in Figure 2, distinct, sharp contours of the markers can be seen. It means that the system motion is fully regular in these cases. In the vicinity of the first subharmonic resonance (l_s = 0.0925 m), blurred columns of the markers can be seen, which indicates that the system motion has become irregular.

Figure 3 depicts a global map of impacts, on which it can be easily observed that the values of ω_{bi} and ω_{ai} during individual impacts are different, and, moreover, that impacts do not occur at all during some rotations of the rotor (e.g. rotation No. 526, 528, 530). In order to investigate the character of the system motion with l_s=0.0925 m in detail, maps of impacts showing the impacts occurring in the subsequent time intervals of the generator motion have also been drawn (Figure 4). As can be seen, despite the fact that the time passes, the motion does not become stable, and the motion assumes a chaotic character. Such a character of the motion is caused here by the occurrence of the impacts of the I and II kind and by the fact that the hammer and the fender often miss each other without impact.

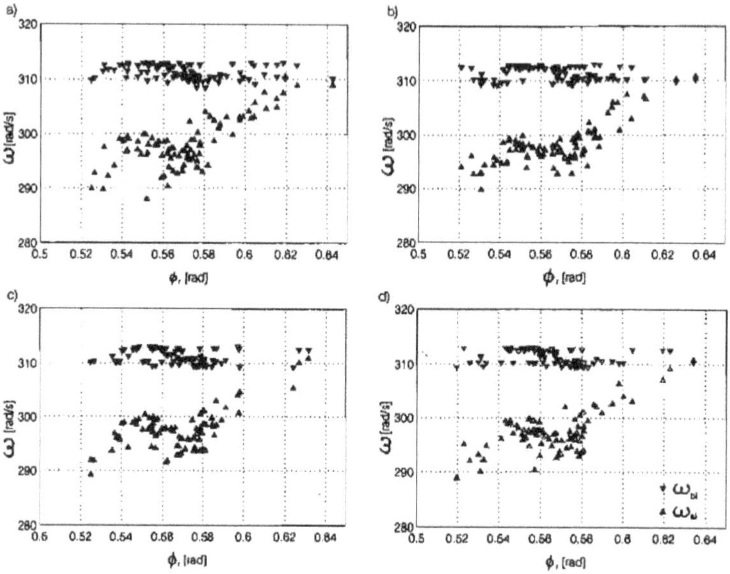

Figure 4. Local map of impacts; l_s=0.0925 m, Δ=ln(1.5), k=0.9; a) t=0-3 s, b) t=3-6 s, c) t=6-9 s, a) t=9-12 s.

5 Conclusions

The object of the numerical investigations presented here is a mechanical impact force generator. During the investigations it was found that in the majority of cases the system exhibited a regular motion. By a proper choice of the value of the basic frequency of free vibrations of the hammer it was possible to obtain desired changes both in the average rotational velocity of the generator rotor, and in the intensity of impacts. When the hammer is subject to the forcing close to the synchronic one, the motion of the system with the features of a chaotic motion can even occur.

6 References

[1] Wawszczak, W., Wysokoobrotowy wymiennik ciepla, *Zeszyty Naukowe PL*, **705**, 207, Lodz 1994 (in Polish).

[2] Wawszczak, W. and Jagiełło, B., Generation and analysis of torsional vibration, *Mechanics and Mechanical Engineering* **1** (1), 43-60, 1997.

[3] Blazejczyk-Okolewska, B., Czolczynski, K., Jach, A., Dynamics of the impact force generator, *Mechanics and Mechanical Engineering* **1** (2), 103-109, 1997.

Dimension Estimation of an Impacting Rotor/Stator System

C. Craig[1]
R. D. Neilson[1]
J. Penman[1]

ABSTRACT The feasibility of monitoring the correlation dimension for use in machine health monitoring of systems exhibiting a clearance is considered by analyzing a four degree of freedom non-linear rotor/stator system. A comparison of measured results with theoretical predictions of the correlation dimension computed from the pseudo phase space is presented for the rotor/stator system. Investigations of both the numerical and experimental models indicate the existence of a trend where the value of the correlation dimension decreases as the clearance between the rotor and stator is increased, suggesting that these techniques may provide a viable method of monitoring the condition of systems exhibiting clearance.

1 Introduction

To date, little attention has been paid to the occurrence of chaotic or 'randomlike' vibrations in plant or machinery. These vibration signatures may, however, be more constructively analysed by utilizing chaos techniques to detect and ultimately predict the occurrence of damaging defects. Although the field of condition monitoring already embodies an extensive, yet diverse range of techniques currently used to predict and diagnose faults, it continually welcomes the introduction of new techniques, as, if many different techniques are used in conjunction with each other, unambiguous diagnoses may result.

The correlation dimension is one quantitative chaotic measure which may be used to characterise so called 'strange' phase space attractors. It is not generally feasible, however, to acquire data for all the active states which contibute to the dynamics of the entire system, therfore, phase space at-

[1]Department of Engineering, University of Aberdeen, Scotland, U.K.

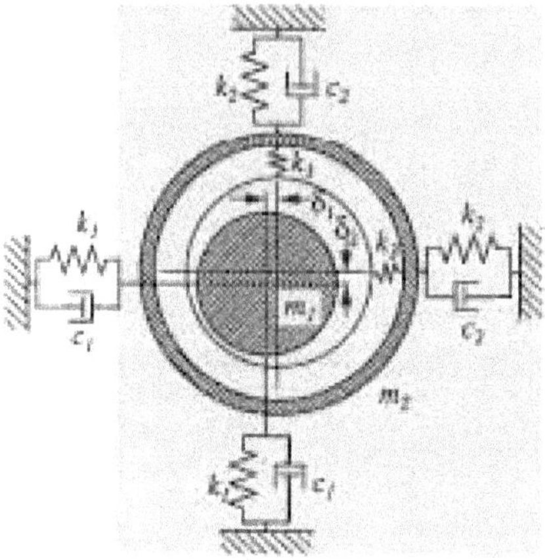

FIGURE 1. Dynamic Model of the Rig.

tractors are typically reconstructed using a technique known as the method of delays [1]. This is more useful in condition monitoring, as ordinarily only a limited number of active states of the system is known *a priori*. From the pseudo phase space attractor a value of the correlation dimension, comparable to that from the equivalent full phase space attractor, can be computed.

Fault diagnosis of rolling element bearings has been achieved by monitoring the correlation dimension both theoretically and experimentally in studies by Logan and Mathew [2] [3]. Craig *et al* on the other hand have studied the relationship between clearance in a system and the correlation dimension for a simple two degree of freedom system [4]. To be beneficial and viable for application in condition monitoring this relationship would have to be detectable from a measurement taken on the outer casing. This paper extends the work to a four degree of freedom non-linear rotor/stator system.

2 Dynamical Model Description

Illustrated in figure 1 is the dynamic model used to extend the investigation of the relationship between clearance and the correlation dimension by using a four degree of freedom model to establish whether the relationship holds from measurements taken at the outer casing. This type of analysis has previoulsy been carried out on a similar two degree of freedom system in which the snubber ring was modelled as being massless. This is, however, in contrast to the present study of a four degree of freedom model as the snubber ring now has measurable mass. Details of this model can be found in reference [4].

The equations of motion which descibe the system illustrated in figure 1 are presented below.

$$\ddot{x}_1 + 2\nu_1\omega_1\dot{x}_1 + \omega_1^2 x_1 + H(R-\delta_r)\omega_2^2(x_1 - (x_2+\delta_1) - \delta_r(\frac{x_1 - x_2 - \delta_1}{R})) = \rho\Omega^2 \cos \Omega t \quad (1)$$

$$\ddot{y}_1 + 2\nu_1\omega_1\dot{y}_1 + \omega_1^2 y_1 + H(R-\delta_r)\omega_2^2(y_1 - (y_2+\delta_2) - \delta_r(\frac{y_1 - y_2 - \delta_2}{R})) = \rho\Omega^2 \sin \Omega t \quad (2)$$

$$\ddot{x}_2 + 2\nu_2\omega_3\dot{x}_2 + \omega_3^2 x_2 + H(R-\delta_r)\omega_4^2(x_2 - (x_1-\delta_1) + \delta_r(\frac{x_1 - x_2 - \delta_1}{R})) = 0 \quad (3)$$

$$\ddot{y}_2 + 2\nu_2\omega_3\dot{y}_2 + \omega_3^2 y_2 + H(R-\delta_r)\omega_4^2(y_2 - (y_1-\delta_2) + \delta_r(\frac{y_1 - y_2 - \delta_2}{R})) = 0 \quad (4)$$

where

$$\omega_1 = \sqrt{\frac{k_1}{m_1}} \quad ; \quad \omega_2 = \sqrt{\frac{k_3}{m_1}} \quad ; \quad \omega_3 = \sqrt{\frac{k_2}{m_2}} \quad ; \quad \omega_4 = \sqrt{\frac{k_3}{m_2}}$$

$$\nu_1 = \frac{c_1}{2m_1\omega_1} \quad ; \quad \nu_2 = \frac{c_2}{2m_2\omega_3} \quad \text{and} \quad R = \sqrt{(x_1 - x_2 - \delta_1)^2 + (y_1 - y_2 - \delta_2)^2}$$

H() is a heaviside function with $H(R-\delta_r) = 0$ if $\delta_r > R$ and $H(R-\delta_r) = 1$ if $\delta_r \leq R$, δ_r being the clearance between the rotor and stator. The parameter values used to simulate the numerical model are; $\Omega = 2340rpm$, $m_1 = 8.9kg$, $m_2 = 0.6kg$, $k_1 = 79kN/m$, $k_2 = 2354kN/m$, $k_3 = 235400kN/m$, $\nu_1 = 0.0850$, $\nu_2 = 0.0148$, $\rho = 60.6 \times 10^{-6}$, with $\delta_1 = \delta_r = 0.0003m, 0.0005m, 0.00075m$ for the three different gap clearances analysed and $\delta_2 = 0$.

3 Experimental Model Description

Figure 2 illustrates a shematic of the experimental model used as a practical comparison to the dynamic model described in section 2. This is the same experimental rig used in a similar anlysis of a two degree of freedom rotor/stator system, and details of the test rig and data acquisition system can be found in reference [4].

To obtain a direct comparison between the numerical and experimental models, data from the experimental model (from both the rotor and stator) was captured for simulated clearances of 0.3mm, 0.5mm and 0.75mm at a shaft speed of 2340rpm. Sufficient time was allowed to elapse at the aformentioned shaft speed in order to obtain steady state responses before any data was accumulated. The data acquisition system allowed up to 32,000 point time histories to be captured. Both numerical and experimental analyses were performed using 20,000 data points. The test rig was

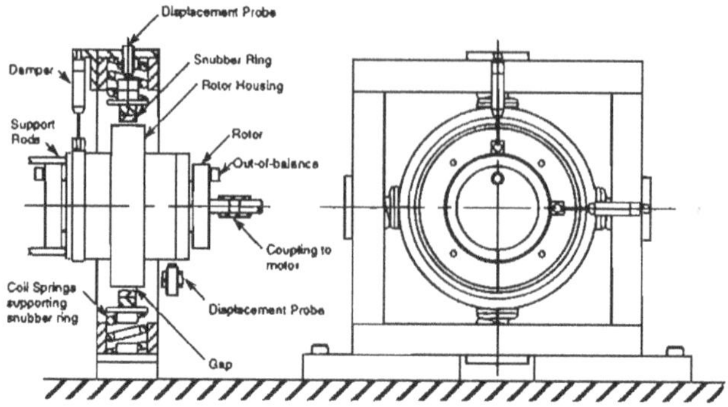

FIGURE 2. Experimental Model Schematic.

Clearance: x-Direction	*Est. Corr. Dim. Rotor*	*Stator*	*Clearance: y-Direction*	*Est. Corr. Dim. Rotor*	*Stator*
0.3mm	2.7730	4.9489	0.3mm	2.7569	4.9316
0.5mm	2.5577	4.4036	0.5mm	2.5530	4.4102
0.75mm	1.7851	3.5127	0.75mm	1.7648	3.5098

TABLE 1. Numerical correlation dimension estimates computed from the pseudo phase space for the four degree of freedom non-linear rotor/stator system.

originally assembled with a 0.5mm clearance between the rotor and stator, therfore, as an alternative to disturbing the original set-up, the 0.3mm and 0.75mm clearances where simulated by altering the level of the out-of balance mass. It has been shown, that from scaling considerations, dynamically similar responses can be elicited at similar values of ρ/δ_r if all other system parameters are held fixed [5].

4 Results

Due to the system's dynamics the method of delays had to be slightly modified to obtain topoligically equivalent reconstructed attractors. This was achieved with the inclusion of values from an additonal observable into the the delay vectors used to specify the pseudo phase space. Detailed descriptions of the algorithms used to reconstruct the systems phase space atttractor's and compute its correlation dimension can be found in reference [4]. Tables 1 and 2 present both the numerical and experimental results of the four degree of freedom non-linear rotor/stator system respectively.

Clearance: x-Direction	*Est. Corr. Dim.* *Rotor*	*Stator*	*Clearance: y-Direction*	*Est. Corr. Dim.* *Rotor*	*Stator*
0.3mm	3.5803	7.5583	0.3mm	3.5846	7.5105
0.5mm	3.3437	6.6541	0.5mm	3.3539	6.6214
0.75mm	3.0804	5.5119	0.75mm	3.0536	5.5067

TABLE 2. Experimental correlation dimension estimates computed from the pseudo phase space for the four degree of freedom non-linear rotor/stator system.

Examining the results from both the numerical and experimental models it is clear that a decernible decrease occurs in the value obtained for the correlation dimension with increase in radial clearance. This, therefore, indicates the existence of a monitorable trend.

5 Discussion and Conclusions

On comparing the results presented in tables 1 and 2 it is evident that there is good agreement between the numerical and experimental models in terms of trend in dimensional value as clearance is varied. The values of correlation dimension obtained for the experimental model are, however, somewhat higher than those of the numerical model. This could be due to the influence of noise inherent in the experimental signals. It is known that noise within a signal has the effect of increasing the value of the correlation dimension, however satisfactory estimates of dimension can still be achieved within reasonables limits of noise to signal ratio. This is more apparent in the results from data acquired at the stator, with the influence of noise being more noticeable in these smaller scale signals.

Despite the inflated values of dimension obtained from the experimental model, both the numerical and experimental models indicate a monitorable trend from measurements taken on either the rotor or outer casing for the rotor/stator system as clearance is varied. This suggests that the condition monitoring of systems exhibiting a clearance by trend monitoring the correlation dimension is suitable and implementable as part of condition monitoring schemes.

6 References

[1] F.C. Moon. *Chaotic Vibrations: An Introduction for Applied Scientist and Engineers*. John Wiley and Sons, Inc., 1987.

[2] D Logan and J Mathew. Using the correlation dimension for vibration fault diagnosis of rolling element bearings - 1. basic concepts. *Mechanical Systems and Signal Processing*, 10(3):241–250, 1996.

[3] D.B. Logan and J Mathew. Using the correlation dimension for vibration fault diagnosis of rolling element bearings - 2. selection of experimental parameters. *Mechanical Systems and Signal Processing*, 10(3):251–264, 1996.

[4] C. Craig, R.D. Neilson, and J. Penman. A study of the use of the correlation dimension to monitor a nonlinear rotor/stator system. *Published in the Proceedings of the ISROMAC-7 Conference, Honolulu, 22-26 February 1998.*

[5] D. Gonsalves, R.D. Neilson, and A.D.S. Barr. A study of the response of a discontinuously nonlinear rotor system. *Journal of Nonlinear Dynamics*, 7:451–470, 1995.

Stability of Sliding with Coulomb Friction Law

Hanbum Cho and J. R. Barber [1]

This paper considers the behavior of an elastically-supported mass with three translational degrees-of-freedom which can make contact with a rigid Coulomb friction support. Even when the quasi-static analysis guarantees uniqueness of solutions, the dynamic solution is shown to deviate from the quasi-static solution and typical dynamic responses include oscillatory slip, limit cycle, and stick-slip. A perturbation analysis is performed on the quasi-static solution and it predicts instability for certain slip directions. Simulation results for general loading scenario are presented to demonstrate the responses passing through unstable slip regions.

1 Introduction

In two-dimensional contact problems, the direction of frictional slip must lie in the plane and hence the contact problem is piecewise linear, since all the states are governed by linear equations. By contrast, in three-dimensional systems, the direction of slip is a vector: the Coulomb friction requirement that the frictional force oppose the instantaneous direction of slip results in a fully non-linear governing equation for slip states. This can introduce considerably more richness into the behavior of the system. In the present paper, we shall develop a dynamic simulation for the three-dimensional equivalent of Klarbring's model [1, 2] and present qualitatively different kinds of friction-induced instabilities.

2 Three-Dimensional Friction Model

A mass M with three translational degrees-of-freedom u_1, u_2, u_3 is constrained to the domain $u_3 \geq 0$ by a rigid plane surface whose coefficient of friction is f. The mass is then connected to a generalized linear elastic support with stiffness and damping matrices $\boldsymbol{K}$, $\boldsymbol{B}$ respectively and is subjected to an externally applied force $\boldsymbol{F}$. If the mass makes contact with the plane ($u_3 = 0$), there will also be a reaction force $\boldsymbol{R}$. The equation of

[1] Department of Mechanical Engineering and Applied Mechanics, University of Michigan, Ann Arbor, MI 48109-2125, USA

motion of the system can be written

$$\boldsymbol{M}\ddot{\boldsymbol{u}} + \boldsymbol{B}\dot{\boldsymbol{u}} + \boldsymbol{K}\boldsymbol{u} = \boldsymbol{F} + \boldsymbol{R} . \tag{1}$$

We assume unilateral contact conditions between the mass and the plane, including Coulomb friction with constant coefficient of friction f. Under these assumptions, the system can at any given time be in any one of the three states *separation*, *stick*, and *slip*. Each of these states is governed by one or more equations and inequalities.

3 Stick-Slip Motion

The dynamic equations of motion and the state equations were integrated in time similar to that described by Cho and Barber [3]. If violations are detected in the inequalities that govern the current state, appropriate state change follows. For the purpose of illustration, stiffness and damping matrices are chosen to be

$$\boldsymbol{K} = \begin{bmatrix} 1 & 1 & -1 \\ 1 & 2 & -2 \\ -1 & -2 & 3 \end{bmatrix} \quad ; \quad \boldsymbol{B} = \boldsymbol{0} \, , \tag{2}$$

and the loading scenario used in this case involves the application of a unit normal load $F_3 = 1$ followed by a monotonically increasing tangential load. For certain sliding directions, the dynamic trajectory significantly deviates from the corresponding quasi-static trajectory and the mass spends significant periods in a state of stick, interspersed with short rapid slip periods.

Figure 1 shows the variation of arc velocity, $\dot{s}$, with time and the corresponding trajectory for such a case, with $\theta = \tan^{-1}(F_2/F_1) = 130^o$, $f = 0.9$. It is noted that the response at the beginning of each slip event resembles the precursor waves preceding earthquake events. The simulation results also show that stick-slip tends to occur at high normal contact force and slow sliding speed. Adams [4] reported a similar trend in instabilities associated with the destabilization of frictional interface waves.

4 General Loading Trajectories

For a loading scenario in which $\dot{\boldsymbol{F}}$ varies slowly, we anticipate that unstable perturbations on the quasi-static response will grow whenever the instantaneous loading direction is predicted to be unstable. However, the extent of such disturbances may be limited if the period of time spent in the unstable range is restricted.

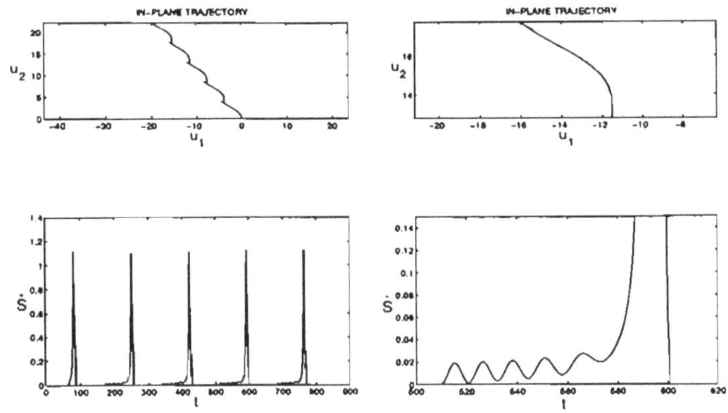

(*a*) Trajectory and velocity (*b*) Precursors

FIGURE 1. Stick-slip motion

Figure 2 shows a typical case in which forces are applied so as to cause the mass to execute a uniform circular motion ($u_1 = 10\cos\theta, u_2 = 10\sin\theta, \theta = 0.001t, f = 0.9$) at a slow constant speed $V = 0.01$ and a constant quasi-static normal reaction force $N = 0.2$. The mass follows the quasi-static trajectory closely except in the sectors AB, CD, EF, where the perturbation analysis indicates instability. In the sector CD, stick-slip motion is generated and there is visible deviation from the intended circular trajectory, as shown in Fig. 2(a). In AB, and EF, oscillatory motion develops: The deviation from the circular trajectory is too small to be seen in the figure, but the instantaneous arc velocity $\dot{s}$ deviates significantly from the intended uniform quasi-static value V, as shown in Fig 2(b).

5 Conclusions

This investigation has shown that elastic systems involving three-dimensional Coulomb friction support can exhibit considerably more complex behavior than corresponding two-dimensional systems. Even for small coefficient of friction, some sliding directions are generally unstable, providing only that there is some coupling between displacements in the sliding plane and spring forces normal to that plane.

A numerical solution for the dynamic behavior of the system shows that various kinds of unstable response are possible, including limit cycle and stick-slip motion. All these non-steady motions involve non-rectilinear motion, even in cases where the time derivative of the applied load is constant in direction. It should be emphasized that stick-slip motion in this sys-

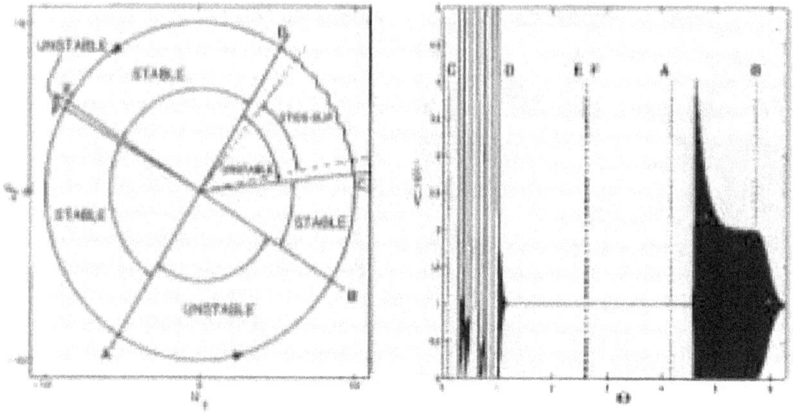

(a) (b)

FIGURE 2. Stability on circular path

tem can be produced with a constant coefficient of friction, in contrast to most previous models of this phenomenon which involve a coefficient that varies with speed or a distinction between static and dynamic coefficient [1]. These results are of concern for the control of positioning mechanisms involving Coulomb friction support.

[1] Klarbring, A., Contact problems with friction, Doctoral Dissertation, Linköping University, Sweden, 1984.

[2] Klarbring, A., Examples of non-uniqueness and non-existence of solutions to quasi-static contact problems with friction. Ingenieur-Archiv, 60, 529-541, 1990.

[3] Cho, Hanbum and Barber, J.R., Dynamic behavior and stability of simple frictional systems. *Mathematical and Computer Modeling*, 1998, in press.

[4] Adams, G.G., Self-excited oscillations of two elastic half-spaces sliding with a constant coefficient of friction. *ASME Journal of Applied Mechanics*, 62, 867-872, 1995.

Tracking Parameter Drift in a Vibro-Impact System

Anindya Chatterjee[1]
Joseph P. Cusumano[2]
David Chelidze[3]

ABSTRACT A vibro-impact system with a drifting parameter is studied experimentally. Feature extraction is difficult as typical "features" are non-smooth. Optimal features are found via solutions of eigenvalue problems.

1 System Description

In systems with slowly evolving damage, "slow" variables act like drifting parameters in a "fast" system. Parameter drift in such systems is tracked by observing the fast variables. In applications, one might only have access to the system's steady state behavior. Then, only steady state characteristics of the system are used for tracking. However, steady state characteristics of some systems can be nonsmooth functions of parameters. Non-model-based feature extraction then becomes difficult. Tracking methods used include reconstructing unstable periodic orbits [1], or full model construction [2]. Non-model-based feature extraction is difficult for such systems. A good tracking metric may not be known *a priori*, and may have to be constructed from several nonsmooth features of the kind mentioned above.

Here, we study a vibro-impact system. Such systems are strongly nonlinear, often chaotic over wide parameter ranges, have higher-dimensional dynamics due to the impacts, and are usually "noisier". We show that the steady-state characteristics of this system are nonsmooth, and develop a non-model-based tracker that captures a hidden drift process.

The system (Fig. 0.1, top) was a beam with stiffeners, and with magnets near its end. One magnet was strengthened electrically by a battery. The system was mounted on a shaker, and forced at about 5.6 Hz. The beam intermittently impacted a hard stop. Forcing amplitude was set to obtain chaotic motions. The battery drained itself over more than 6 hours, weak-

[1,2,3]Engineering Sc. & Mechanics, Penn State University, University Park, PA 16802

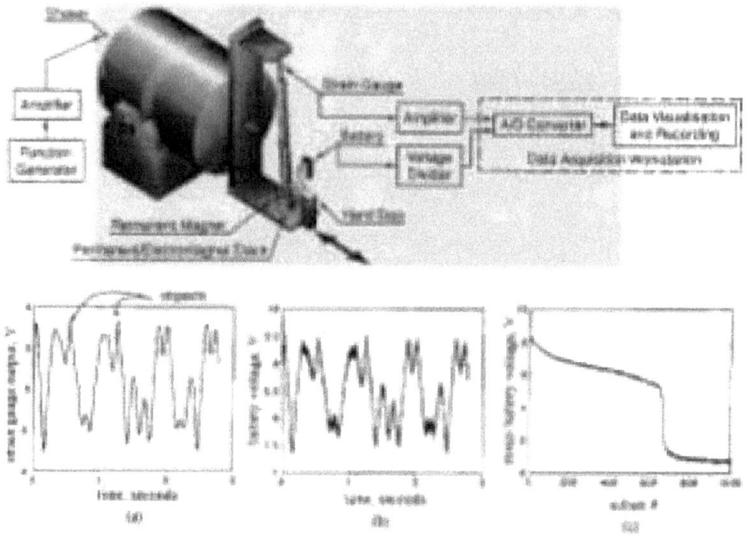

FIGURE 0.1. Top: Experimental System. Bottom (a),(b),(c): Data.

ening the electromagnet. Strain gauge output and battery voltage were sampled at 180 Hz, digitized (12 bit A/D) and stored.

Figure 0.1(a) shows typical strain gauge output. Fig. 0.1(b) shows battery voltage over the same time. Fig. 0.1(c) shows averaged battery voltage (1,000 data subsets of 4,320 points each) for the full experiment.

2 Tracking Metrics

Let $x(k)$ be the discretely-sampled strain gauge output. We divided the total data into 160 records (27,000 points each). For each record, we computed the autocorrelation of x. Specifically, we averaged $x(k) \cdot x(k+d)$, for $d = 1, 2, 3, \cdots, 32$. This gave 32 vectors (160×1); from these vectors we subtracted their means, and scaled them to unit norm. We arranged these in a 160×32 matrix, which we call Y. Thus, element Y_{ij} represents the mean value over record i, of $x(k) \cdot x(k+j)$, mean-subtracted and scaled.

Figure 0.2(a) shows the singular values of matrix Y. Three singular values dominate. About six lie above a "knee" point near 10^{-3}. Singular values below the knee represent noise. This is consistent with the 12 bit A/D data conversion, since 10^{-3} is about 2^{-12} times the largest singular value.

Figures 0.2(b-f) show plots of some columns of Y. The graphs are not smooth, hence unsuitable for tracking parameter drift. We mention that the 24th column, Fig. 0.2(e), best matches the battery voltage. Thus, the

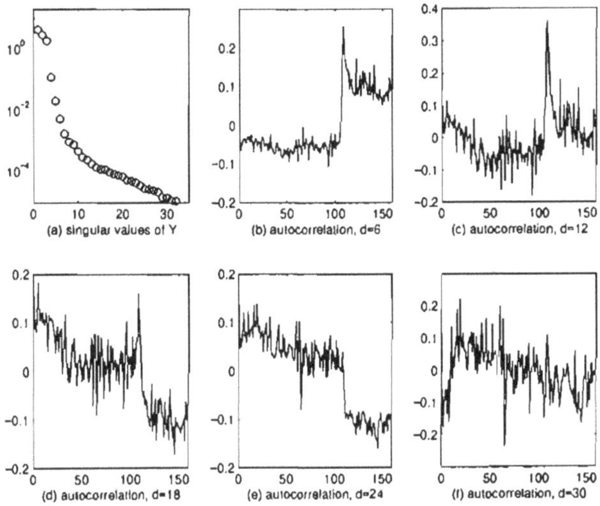

FIGURE 0.2. (a) singular values of Y, (b) through (f): some columns of Y vs. record number.

autocorrelation by itself provides a poor tracking metric for the battery voltage, even at the optimal delay of 24. Note that the optimal delay can only be chosen because the battery voltage is separately monitored. For hidden drift processes, the optimum cannot be found in this way.

We consider linear combinations of the columns of Y, given by $v = Yc$. We expect good tracking metrics to be smooth and monotonic. Thus, we take the 159×160 "derivative matrix" W (for $k = 1, 2, \cdots, 159$, in row k, element k is 1, element $k+1$ is -1, and the rest are zero).

We minimize, with respect to c above, the ratio

$$g(v) := \frac{\|Wv\|^2}{\|v\|^2} \tag{0.1}$$

The rationale is that if v is jagged or nonsmooth, then $\|Wv\|$ is high. Meanwhile, since v has zero mean, $\|v\|$ is a measure of its total variation. In minimizing g, we maximize smoothness and overall variation.

For an extremum of g, c must satisfy $Ac = \lambda Bc$, where $A := (WY)^T(WY)$ and $B := Y^TY$. The miniimum of g is the smallest generalized eigenvalue of (A, B). The corresponding eigenvector gives the optimal c and hence, v.

Figure 0.3(a) shows the eigenvalues of $Ac = \lambda Bc$. The smallest one is significantly smaller than the next. The "optimal" metric is plotted in Fig. 0.3(b) with circles; the measured battery voltage (mean-subtracted and normalized) is plotted with a heavy solid line. Figure 0.3(c) shows the same quantities plotted against each other, for a "calibration curve".

Note that c is "optimal" only in maximizing g, based on the idea that in

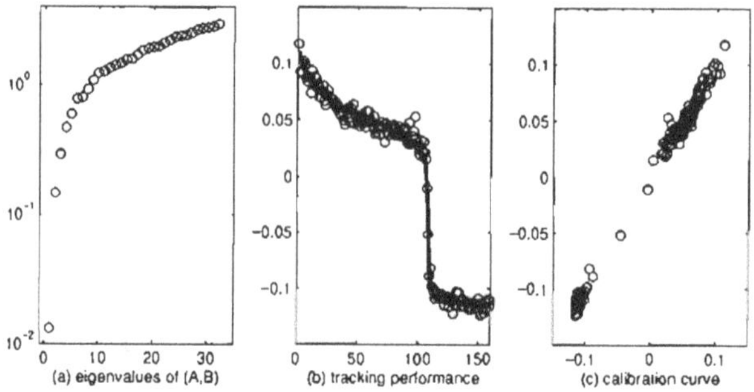

FIGURE 0.3. (a) Eigenvalues of (A, B). (b) Thick line: battery voltage; Circles: tracking metric (c) Calibration curve.

a tracking metric, steady and smooth changes are desirable. The specific form of g is somewhat arbitrary. Could we use other forms? Our studies suggest that other similarly-based criteria produce similar good results.

Recall that the closest match between the battery voltage and the columns of Y is for the 24th column (Fig. 0.2(e)). However, the fit for that column is about 4.5 times poorer than the "optimal" fit of Fig. 0.3(b). We emphasize that beyond setting the mean to zero and normalizing to unit norm for comparison, *no* curve fitting of any kind was performed in determining the tracking metric. Note that the shape of the battery drain curve is captured very well by the tracking metric, though the battery voltage was not used in the calculation. The hidden process was detected successfully. We also obtained similar results with a smooth (non-impacting) chaotic system. Thus, the method is promising.

Theoretical reasons for the success of this method are currently under study. Future work will need to address the effects of parameters like the degree of nonsmoothness of the statistics (columns of Y), the length of these columns (here, 160), the number of these columns (here, 32) and the number of linearly independent columns (here, about 6, given measurement noise). These questions are not yet fully resolved.

3 References

[1] Chancellor, R. S., Alexander, R. M. and Noah, T. S., Detecting parameter changes using experimental nonlinear dynamics and chaos. *Journal of Vibration and Acoustics*, 118:375–383, 1996.

[2] Cusumano, J. P., Chelidze, D., and Chatterjee, A., Experimental application of a method for hidden parameter tracking in a slowly changing, chaotic system. *Emerging Technologies for Machinery Health Monitoring and Prognosis*, R. S. Cowan, ed., ASME vol. TRIB-7, 45–54. 1997.

Nonlinear Resonances and Self-Excited Oscillations of a Rotor Due to Radial Clearance and Impact in Bearings (Experiments and Numerical Simulations)

Y. Ishida[1]
M. Inagaki[2]
R. Ejima[1]

ABSTRACT Oscillations in a rotor system with a radial bearing clearance between the outer ring of the bearing and the casing are investigated by experiments and numerical simulations. Self-excited oscillations of a forward whirling mode occur due to impact . In addition, various kinds of nonlinear resonances, such as subharmonic, super-subharmonic and combination resonances occur. Entrainment phenomena are observed in their resonance ranges.

1 Introduction

In turbomachinery with radial clearances in bearings or casings, various kinds of nonlinear resonances occur due to nonlinear characteristics of the rotor restoring force[1],[2]. It is also known that when a rotor contacts with a stator, such as seals and casings, self-exited oscillations of a backward whirling mode occur due to dry friction[3]. Recently, Li[4] indicated numerically that self-excited oscillations of a forward whirling mode occurred due to rubbings and impacts when a rotating disk flying in a cylinder collided with its inner surface.

In this paper, we investigate oscillation phenomena in a rotor model with a squeeze film damper bearing where the outer ring of the bearing collides with the casing. Nonlinear resonances, self-excited oscillations of a forward whirling mode and entrainment phenomena are found and clarified experimentally and numerically.

2 Experiments

2.1 Experimental setup

Figure 1 shows the experimental setup. A vertical flexible rotor with a disk is supported by two self-aligning double-row ball bearings. The lower bearing has a radial clearance of about 0.1mm between the holder of the outer ring and the casing.

---

1 Department of Electronic-Mechanical Engineering, Nagoya University, Nagoya, Aichi, 464-8603, Japan
2 Structural Dynamics labs., TOYOTA Central R & D Labs., Inc., Nagakute, Aichi, 480-1192, Japan

This rotor system has natural frequencies of a forward and a backward whirling modes, p_1, p_2 and p_3, p_4, respectively, which change as a function of the rotational speed ω.

2.2 *Experimental results*

Figure 2 shows response curves. In addition to a harmonic oscillation with frequency ω, various kinds of oscillations occurred with large amplitudes in the high speed range above 1150rpm. They are classified into two groups. One is nonlinear resonances which occur around specific rotational speeds and the other is self-excited oscillations which occur in a wide rotational speed range. When nonlinear resonances occur, a self-excited oscillations disappears(entraninment phenomena). Figure 3 shows typical examples of these oscillations. At 1200rpm, a self-excited oscillation of a forward whirling mode with frequency p_2 occurred besides a harmonic component with frequency ω. At 1930rpm, combination resonance occurred. At 1970 rpm, subharmonic oscillation of order 1/3 occurred.

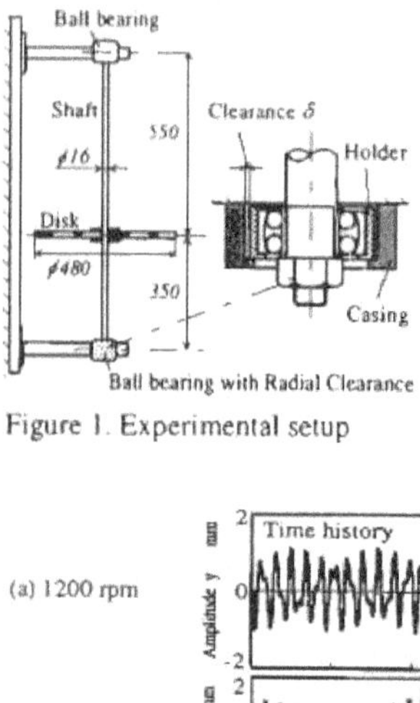

Figure 1. Experimental setup

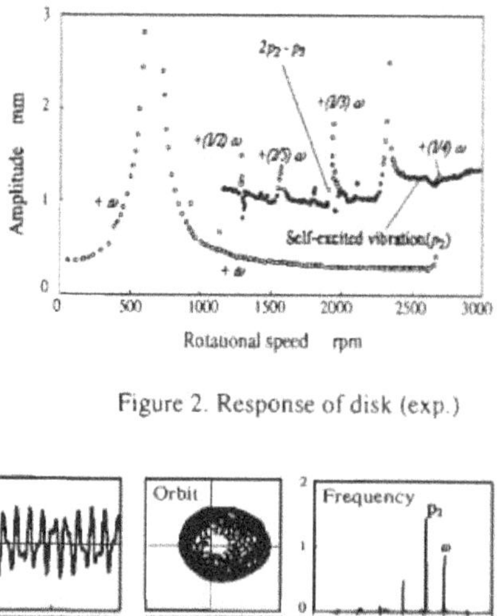

Figure 2. Response of disk (exp.)

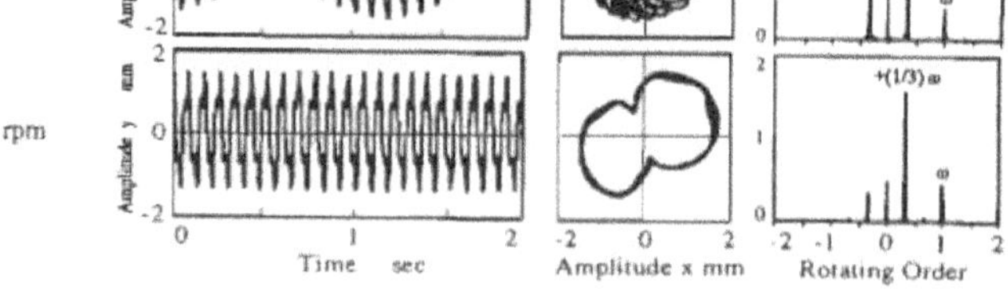

Figure 3. Time history, orbit and frequency of disk (exp.)

3 Numerical Simulations

3.1 Theoretical model and equations of motion

In order to explain oscillations observed in experiments, numerical simulations are performed using a rotor-bearing model considering impacts in clearance between the bearing and the casing(Figures 4 and 5). Dimensionless equations of motion can be expressed as

$$\left.\begin{aligned}
\ddot{x}+i_p\omega\,\dot{y}+c\dot{x}+k(x-x_1)&=e\omega^2\cos\omega t\\
\ddot{y}-i_p\omega\,\dot{x}+c\dot{y}+k(y-y_1)&=e\omega^2\sin\omega t+g\\
\ddot{x}_1+c_1\dot{x}_1+\frac{k}{\mu}(x_1-x)&=0\\
\ddot{y}_1+c_1\dot{y}_1+\frac{k}{\mu}(y_1-y)&=0
\end{aligned}\right\}\qquad(1)$$

where x,y and x_1,y_1 are the displacements of the disk and the bearings in x- and y-directions, μ is mass ratio of the bearing mass m to the disk mass M, c and c_1 are damping coefficients of the disk and bearings, e is imbalance of the disk, ω is the rotational speed and t is time. The terms including i_p express the gyroscopic effect of the disk. The constant force g is induced to express the eccentricity of the bearing in a clearance due to an assembling error. Equations (1) are numerically calculated by Runge-Kutta's time integral method.

3.2 Modeling of impact

The impact between the bearing and the casing are treated following the paper Li and Paidoussis[4]. In Figures 4 and 5, δ is the radial clearance, r is the displacement of the bearing. In a numerical calculation, the bearing orbit is given by a succession of discrete points. Suppose that the bearing moves toward the casing wall from the position P_1. If the position P_{2-} is obtained as the next position in a numerical simulation, then the position is transferred to P_{2+} under the assumption that it bounds at P_{12}. The position of P_{12} is given by

$$\left.\begin{aligned}
dr&=r_- -\delta, & r_+&=\delta-\hat{e}\cdot dr\\
v_{n+}&=-\hat{e}\cdot v_{n-}, & v_{t+}&=v_{t-}
\end{aligned}\right\}\qquad(2)$$

where e is the coefficient of restitution, v_n and v_t are the velocity components in the normal and tangential directions. Dry friction at the impact is ignored.

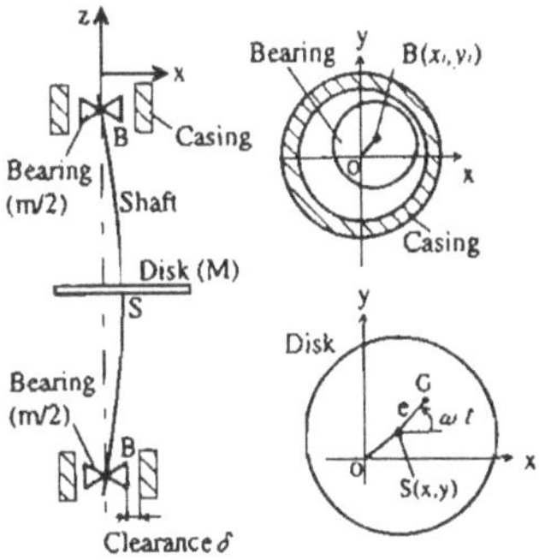

Figure 4. Theoretical model

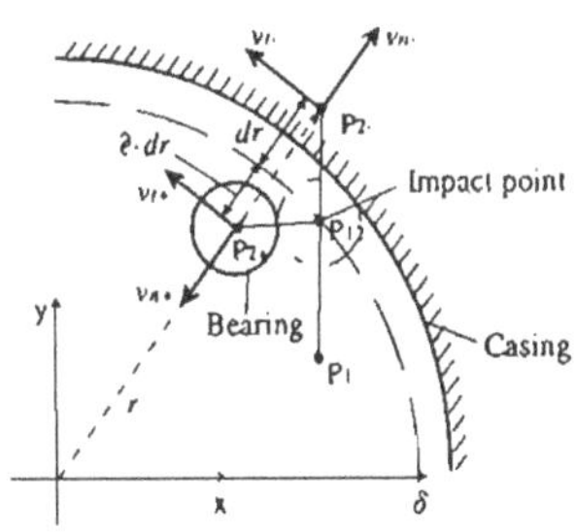

Figure 5. Bearing bounce at impact

3.3 Numerical results

Figure 6 and 7 show numerical results corresponding to the experimental results shown in Figures 2 and 3. As the same kinds of phenomena occur, we know that such a rotor model explains the phenomena well enough.

4 Conclusion

In a rotor-bearing system with radial clearance between the outer ring of the bearing and the casing, self-excited oscillations of a forward whirling mode due to impacts and various kinds of nonlinear resonances occur due to clearance. Entrainment phenomena also occur in these resonance ranges.

5 Reference

[1] Ehrich, F.F., Trans.ASME,J.Vib.Acoust.,110-1, 1988, 9-15

[2] Ehrich, F.F., Handbook of Rotordynamics, Sec.1.7, 1992

[3] Hartog, J.P.D., Mechanical Vibrations,1956

[4] Li, G.X. and Paidoussis, M.P., Nonlinear Dynamics 5, 1994, 53-70

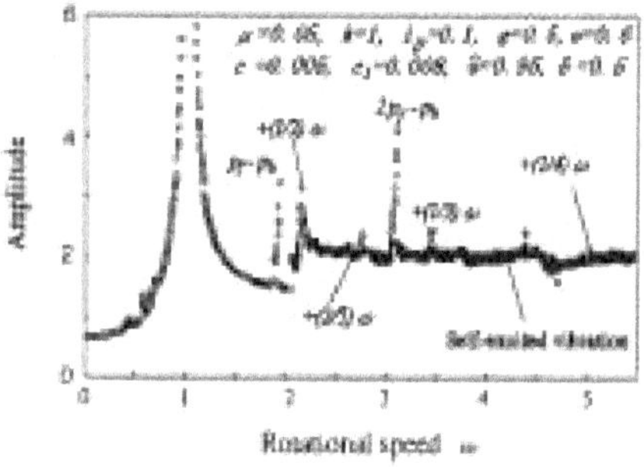

Figure 6. Response of disk (cal.)

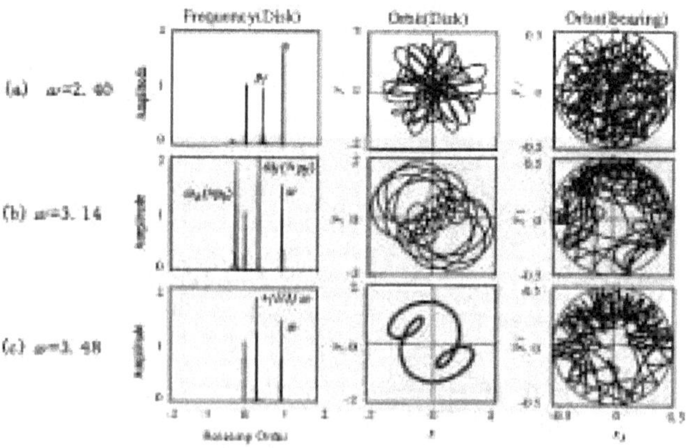

Figure 7. Time history, orbit and frequency of the disk and the bearing (cal.)

Proper Orthogonal Modes of a Beam With Frictional Excitation

R. V. Kappagantu[1]
B. F. Feeny[1]

ABSTRACT We conducted experiments on a cantilever beam subjected to friction from an oscillating surface. We use proper orthogonal decomposition to obtain a basis that has the "optimal" distribution of energy in the system. The system dynamics are rich, and various responses could each result in a different set of proper orthogonal modes. A set of proper orthogonal modes is compared to linear normal modes in the ability to represent the spatial description of the beam dynamics.

1. Experimental Description

A cantilever beam is subjected to frictional excitation. The frictional contact is provided by a solid mass of steel connected to a shaker. The displacements of the beam at various locations can be measured using strain gages. The schematic of experimental setup is shown in Figure 1.

The figure uses two coordinates systems, XYZ and xyz. The cantilever is built by taking a $0.4000 \times 0.0128 \times 0.00086\,\mathrm{m}^3$ mild steel beam ($E=128 \times 10^9$ $\mathrm{N/m}^2$, $\rho=7488$ $\mathrm{kg/m}^3$) and fixing it at one end. At the free end of the cantilever, we attached a small beam ($0.064 \times 0.0134 \times 0.00056$ m^2, $E = 126 \times 10^9$ $\mathrm{N/m}^2$, $\rho = 6777$ $\mathrm{kg/m}^3$) to act as a leaf spring. It is referred to as a loading beam. The bending axes of the two beams are perpendicular to each other, *i.e.* the length of loading beam is along the X axis, the thickness along the Z axis and the width along the Y axis. The loading beam has low transverse (Z) rigidity and high lateral (Y) rigidity. A small hemispherical rubber nub is glued to the free end of the loading beam. This nub makes contact with the oscillator. The loading beam is used to provide a spring-loaded normal load. Easy monitoring of the normal load is achieved by attaching strain gages to this loading beam.

An impulse response test revealed that the first six modes of the cantilever setup have frequencies of approximately 3.75, 18.4, 53.75, 102.5, 172.5 and 265 Hz. Detailed description of the experiment is given in [4].

---

[1]Department of Mechanical Engineering, Michigan State University, East Lansing, MI 48824

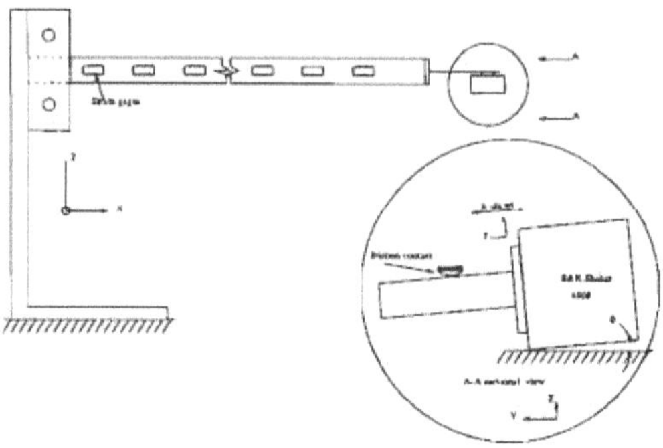

FIGURE 1. Schematic diagram of the experimental setup.

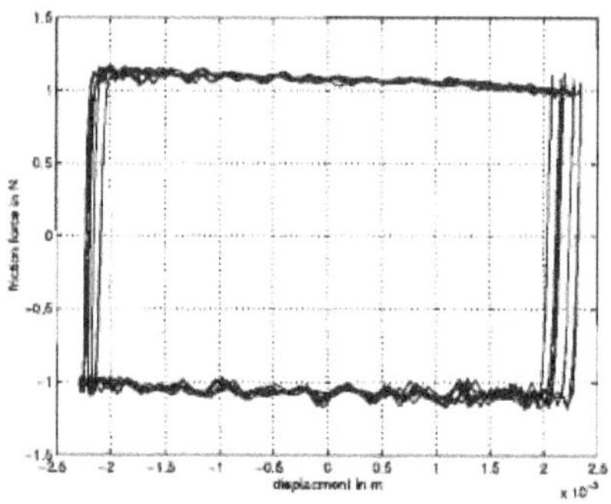

FIGURE 2. Friction-displacement relation.

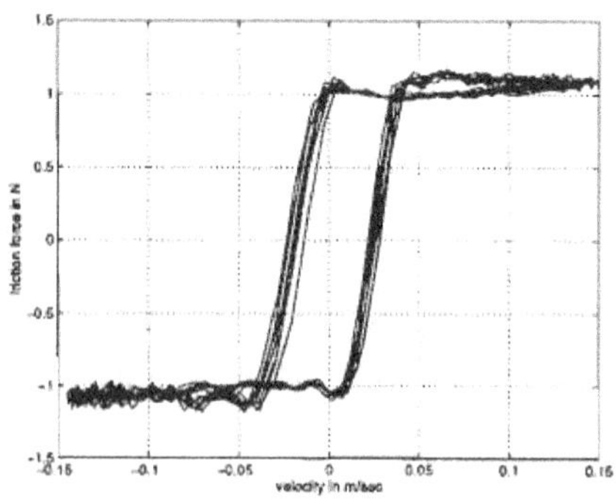

FIGURE 3. Friction-velocity relation at 10 Hz.

The friction was measured. The hysteresis in the friction-velocity plot in Figure 3 suggests that the friction characteristic may be close to a Coulomb model with contact compliance of 20 kN/m. The kinetic friction force limit is close to 1.03 N. Dividing this by the mean normal load of 1.86 N gives us the kinetic friction coefficient of 0.55. The static to kinetic transition if it exists, is generally at high frequencies and gets filtered. As such the static friction limit is difficult to get captured in experiments of this nature and needs a different but simple method. In a separate static experiment, we found $\mu_S \approx 1.38\mu_K$.

From the friction-displacement plot we have $A = 0.0023$ m, $K_C = -20$ KN/m, $F_s = 1.03$ N and $F_k = -1.0$ N. We know the value of $\omega = 20\pi$ rad/s. With some analysis these parameters predict a slipping velocity equal to 0.044 m/s which is consistent with the the friction-velocity plot, in which a microstick-slip transition occurs at approximately $V_s = \pm 0.045$ m/s.

For the displacement measurement, we used six half-bridge strain-gage circuits to obtain the strains at six uniformly distributed locations. The gages were spaced at 0.05 m, with the sixth gage located at 0.05 m from the fixed end. Based on numerical simulations we found that this distribution helps us maximize the strains. The strain $\epsilon(x)$ in a beam subjected to pure bending is related to the transverse displacement $y(x)$ by

$$\epsilon_{xx} = c\frac{\partial^2 y}{\partial x^2} \tag{1}$$

where c is half the width of the beam.

We assume that the displacement of the beam is

$$y(x,t) = \sum_{i=1}^{n} \phi_i(x) u_i(t) = \mathbf{\Phi}(x)\mathbf{u}(t) \tag{2}$$

where $\phi_i(x)$ form a basis satisfying the geometric conditions and where $\mathbf{\Phi}$ and $\mathbf{u}$ are defined appropriately. We use the same technique here which enables us to write ϵ_{xx} in terms of ϕ_i and u_i. Thus

$$\epsilon_{xx}(x,t) = c\sum_{i=1}^{n} \psi_i(x) u_i(t) = c\mathbf{\Psi}(x)\mathbf{u}(t) \tag{3}$$

where $\psi_i(x) = \partial^2\phi_i(x)/\partial x^2$ and $\mathbf{\Psi}$ is defined appropriately. Now by measuring the strain ϵ at n different points on the beam we get n such equations which can be solved for $u_i(t)$ at each t. These in turn can be used to determine $y(x,t)$ using equation 2. For the basis$\{\phi_i\}$ we used the linear natural modes and we used $n = 6$.

We checked the displacement measurements by comparing the strain-gage measured displacements at a point on the beam with a laser-measured displacement.

2. Dynamics

We conducted frequency sweeps to find regimes of rich dynamics, and present the phase-portraits using the displacement and "velocity" of the contact point of the beam, at a few selected frequencies (11, 12 and 41.1 Hz). We measured the displacement using the method described earlier, and used a simple forward-difference scheme to differentiate these displacements to determine the velocity of the beam at the contact point. We have not accounted for any phase deviation that may result due to this process. We present the resulting phase-portraits and the corresponding power-spectra in Figure 4.

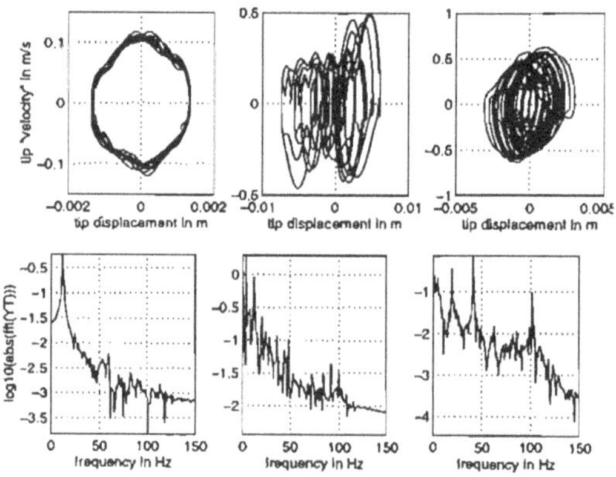

FIGURE 4. Dynamics of the contact point of the beam. The top row indicates the phase-portraits of the contact point at frequencies 11, 12 and 41.1 Hz. The bottom row indicates corresponding power-spectra.

The dynamics at 11 Hz were found to be periodic and is presented in this study as a reference for noise level. The dynamics at 12.0 and 41.1 Hz are non-periodic and are further investigated in parallel studies.

3. Proper Orthogonal Modes

In this section we describe the spatial content of the beam dynamics in terms of proper orthogonal modes and proper orthogonal values (see, e.g., Berkooz *et al.*, 1993; Cusumano *et al.*, 1993; and Feeny and Kappagantu, 1998). The motivation is to use the dominant modes in a chosen dynamical regime for building a reduced-order model for the beam in ongoing studies.

We collected the six strain-signal time series at the operating frequencies 12.0 and 41.1 Hz. Using the method described in the previous sections we have converted the six strain signals into displacement signals at locations 1 – 6. After subtracting the mean displacement from each of these displacement time series, we arranged them to form a matrix $\mathbf{X}$ of six columns. We computed the covariance matrix $\mathbf{R} = \mathbf{X}^T\mathbf{X}$. This matrix is a positive definite matrix and the eigenvectors of this 6×6 matrix, yields the proper orthogonal modes. Using the proper values we identified the dominant modes in the system. As we have the values only at six different discrete locations we need to curve-fit the discrete modes to get continuous modal functions. Though the modes are orthonormal, by doing a curve-fit we lose the orthogonality, as the definition of orthonormality is changed as we move from discrete to continuous. In order to make the continuous functions orthonormal with reference to the distributed mass matrix, we use the Gram-Schmidt reorthonormalization technique, to yield what we call proper orthogonal modal functions (POMFs). We have used the linear natural modes (LNMs) of the cantilever for the curve fit. The motive is to identify the content of the natural modes in the dominant proper orthogonal modes.

In Figure 5 we present the orthonormalized POMFs obtained from the chaotic dynamics observed at the operating frequency of 41.1 Hz. In Table 1 we show the contribution of LNMs of the cantilever to these POMFs. In this case we need the first five LNMs to account for 99.99% energy in the first four POMFs and six LNMs to account for the same proportion of energy content in the fourth POMF. Thus we need six LNMs to describe the same dynamics as that of the four POMF model, showing that POMFs give a better reduction than LNMs.

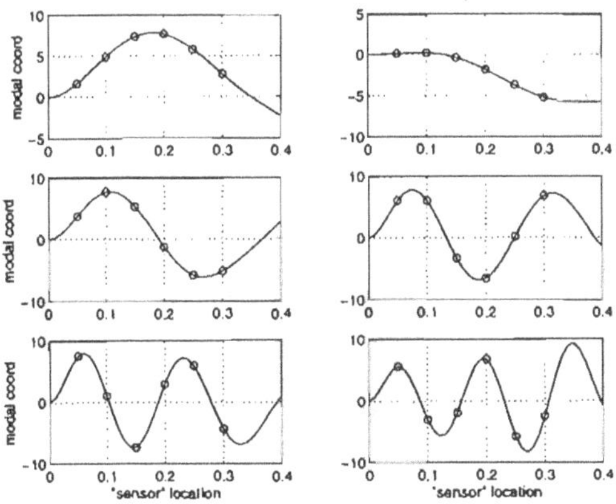

FIGURE 5. POMFs (obtained by orthonormalization of the curves fit on POMs) of chaotic dynamics at a driving frequency of 41.1 Hz.

TABLE 1. Contribution of LNMs to POMFs at 41.1 Hz.

POM #	LNM 1	LNM 2	LNM 3	LNM 4	LNM 5	LNM 6
1	0.2872	0.5748	0.1081	0.0216	0.0060	0.0023
2	0.7764	0.0443	0.1386	0.0326	0.0045	0.0038
3	0.1920	0.0937	0.5582	0.1386	0.0145	0.0030
4	0.1179	0.0819	0.0323	0.6254	0.1093	0.0331

This suggests that a model built from three POMFs would not be similar to the model built from three LNMs.

Ongoing studies, we have compared the reduced-order dynamics in a beam model projected onto dominant POMs and onto LNMs. The POMs tend to provide a good basis for modal projection of this nonlinear system.

4. References

[1] Berkooz, G., Holmes, P., and Lumley, J. L., 1993, The Proper Orthogonal Decomposition in the Analysis of Turbulent Flows, *Annual Review of Fluid Mechanics,* **25**, 539-575.

[2] Cusumano, J. P., Sharkady, M. T., and Kimble, B. W., 1993, Spatial Coherence Measurements of a Chaotic Flexible-Beam Impact Oscillator, *Aerospace Structures: Nonlinear Dynamics and System Response,* ASME

[3] Feeny, B., and Kappagantu, R., 1998, On the Physical Interpretation of Proper Orthogonal Modes in Vibrations, *Journal of Sound and Vibration* **211** (4) 607-616.

[4] Kappagantu, R., 1997, *An Optimal Modal Reduction of Systems with Friction Excitation,* PhD thesis, Michigan State University, East Lansing.

SESSION III–(A)

Analysis, Rigid Bodies, Multi-Body Systems, Mechanics, Contact Models

A Review of Some Existence Results for Quasistatic Contact Problems with Friction

L.-E. Andersson[1]

1 Elementary example of non-uniqueness and non-existence

When dealing with contact problems involving Coulomb friction, existence and uniqueness of solutions cannot be taken for granted. This has been known for more than a century through an example by Paul Painlevé for problems of rigid dynamics. However, it has seemingly remained unknown until 1980 that similar difficulties exist for static or quasistatic problems, some of which will be reviewed in this talk.

We will start by the following very simple example with two degrees of freedom, due to Klarbring, [5]. Consider a linearly elastic system consisting of a single particle which can move in a half-plane, *i.e.*, there is a rigid obstacle such for the displacement vector $u = (u_1, u_2)^T$ we have $u_1 \geq 0$. The external force $f(t)$ and the reaction force $p(t)$ are linearly coupled to u by the stiffness equation

$$\begin{bmatrix} k_{11} & k_{12} \\ k_{12} & k_{22} \end{bmatrix} \begin{bmatrix} u_1 \\ u_2 \end{bmatrix} = \begin{bmatrix} p_1 \\ p_2 \end{bmatrix} + \begin{bmatrix} f_1 \\ f_2 \end{bmatrix}$$

and the similar equation obtained by taking time-derivatives.

We now assume that at some time we have a reaction force p with $-p_1 = \mu p_2 > 0$, where μ is the coefficient of friction. Now the Coulomb friction model permits the following two possibilities (right hand derivatives).

- $-\dot{p}_1 < \mu\dot{p}_2$, in which case the node evolves into a *stick* case with $\dot{u}_1 = 0$ and $-p_1 < \mu p_2$.
- $-\dot{p}_1 = \mu\dot{p}_2$, in which case the node may evolve into a *slip* case with the rate vector in the opposite direction of p_1, *i.e.*, with $\dot{u}_1 \geq 0$.

Consequently we have the Linear Complementary conditions,

$$\dot{p}_1 + \mu\dot{p}_2 \geq 0, \quad \dot{u}_1 \geq 0 \quad \text{with at least one equality.}$$

[1]Department of Mathematics, Linköping University, S-581 83, Linköping, Sweden

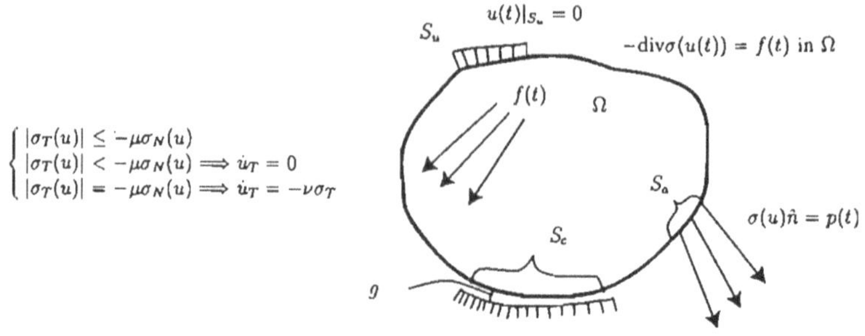

Figure 1: quasistatic frictional problem

Since $p_2 > 0$ we also have $\dot{u}_2 = 0$, which implies that

$$(k_{11}\dot{u}_1 = \dot{p}_1 + \dot{f}_1, \quad k_{12}\dot{u}_1 = \dot{p}_2 + \dot{f}_2.$$

Multiplying by μ and adding we get

$$(k_{11} + \mu k_{12})\dot{u}_1 = \dot{p}_1 + \mu\dot{p}_2 + \dot{f}_1 + \mu\dot{f}_2,$$

whence we conclude that there is

- exactly one solution if $k_{11} + \mu k_{12} > 0$,
- no solution if $k_{11} + \mu k_{12} < 0$ and $\dot{f}_1 + \mu\dot{f}_2 > 0$,
- no solution or infinitely many solutions if $k_{11} + \mu k_{12} = 0$, depending on the sign of $\dot{f}_1 + \mu\dot{f}_2 > 0$.

So if μ is too large, $\mu > k_{11}/(-k_{12})$, we are in trouble.

2 Classical formulation of the quasistatic frictional contact problem

We consider a linearly elastic elastic body occupying a bounded Lipschitz domain $\Omega \subset R^3$, subject to volume and traction forces $f(t)$ and $p(t)$ and obeying the equations of linearized elasticity, see the figure. $\sigma = \{\sigma_{ij}\}$ denotes the stress tensor and $\sigma_{ij} = E_{ijkl}\frac{\partial u_k}{\partial x_l}$, where $\{E_{ijkl}\}$ is the elasticity tensor. We also introduce the bilinear energy functional $a(u,v) = \int_\Omega E_{ijkl}\frac{\partial u_k}{\partial x_l}\frac{\partial v_i}{\partial x_j}$.

The classical Signorini-Fichera contact conditions are,

$$\sigma_N \leq 0, \quad u_N - g \leq 0, \quad \sigma_N(u_N - g) = 0.$$

These relations and those of the figure constitute our quasistatic frictional contact problem (no inertia terms). The non-penetration conditions of

Signorini-Fichera are sometimes replaced by the introduction of a so called normal compliance law $-\sigma_N = c_N(u_N - g)_+^{m_N}$, where c_N and m_N are positive parameters representing the physical characteristics of the interface. One can also generalize the friction law somewhat in the following way

$$|\sigma_T| \le c_T(u_N - g)_+^{m_T}, \quad \text{and}$$
$$|\sigma_T| < c_T(u_N - g)_+^{m_T} \Rightarrow \dot{u}_T = 0,$$
$$0 < |\sigma_T| = c_T(u_N - g)_+^{m_T} \Rightarrow \dot{u}_T = -\lambda\sigma_T, \quad \lambda \ge 0$$

where c_T and m_T are new physical parameters. One then retrieves Coulomb's friction law by taking $c_T = \mu c_N$ and $m_T = m_N$.

3 Variational formulations

We will now formulate a variational problem corresponding to the time dependent problem with the Signorini-Fichera conditions. Using Green's formula we are led to formulate the following time dependent variational problem.

VPQSF: Find a time dependent displacement field u such that $\sigma_N|_{S_c} \in L^2(S_c)$ and so that for all (or almost all) t we have

$$a(u, v - \dot{u}) - \int_{S_c} \mu\sigma_N(|v_T| - |\dot{u}_T|)\, dS \ge \int_\Omega f\cdot(v - \dot{u})\, dx + \int_{S_a} p\cdot(v - \dot{u})\, dS$$

for all $v \in K$. Further we require that $\sigma_N|_{S_c} \le 0$, and $\sigma_N(u_N - g) \ge 0$, almost everywhere on S_c. Here

$$V = \{u \in (H^1(\Omega))^3 : u|_{S_u} = \bar{u}\} \quad \text{and} \quad K = \{u \in V : u_N - g \le 0 \quad \text{on} \quad S_c\}.$$

In this generality no existence or uniqueness results are known, as far as we know. If, however, we regularize the friction term by introducing a bounded linear mapping $H^{-1/2}(\partial\Omega) \ni \sigma_N \mapsto \sigma_N^* \in L^2(\Omega)$ and by replacing σ_N by σ_N^* in the integral over S_c, then an existence result has been given by Cocu, Pratt and Raous, [3], for the case that the coefficient of friction is chosen sufficiently small. Their proof is based on an incremental formulation of the problem **VPQSF**.

For the case with a normal compliance model, we obtain by the aid of Green's formula the following time dependent quasistatic problem.

VPQNC: Find a time dependent displacement field $u \in K$ such that for all (or almost all) t we have

$$a(u, v - \dot{u}) + \varphi(u, v - \dot{u}) + j(u, v) - j(u, \dot{u}) \ge \langle F, v - \dot{u}\rangle$$

for all $v \in V$.

Here we have introduced the notation $\varphi(u,v) = \int_{S_c} (u_N - g)_+^{m_N} v_N \, dS$ and $j(u,v) = \int_{S_c} (u_N - g)_+^{m_T} |v_T| \, dS$, and $\langle F, u \rangle = \int_\Omega f \cdot v \, dx + \int_{S_0} p \cdot v \, dS$.

We will briefly review work by Klarbring, Mikelić and Shillor, [6, 7] and by Andersson, [1, 2] on the problem **VPQNC**.

The contact with friction was first considered in a static version, by Duvaut and Lions [4]. We will describe work on the static problem by Duvaut, Cocu and others, using various regularization of the friction term. The first result for the the static problem without any regularization of the friction term, was given by Nečas, Jarušek and Haslinger in [9], where the existence of a solution was proved, under the condition that the coefficient of friction was *small enough*.

4 Finite-dimensional problems

Martins *et al.* have studied the quasistatic (as well as the dynamic) Signorini problem for an elastic system with only two degrees of freedom in a series of articles, see for instance [8]. They consider the case with a large coefficient of friction, when, due to the examples of Klarbring, [5], one may expect discontinuities. Existence results are given for the quasistatic case, where the solution $u(t)$ may exhibit jump discontinuous as a function of time. They also show that these solutions may be obtained as limits of problems with inertia terms and viscosity terms, as the masses and the viscosity tend to zero.

References

[1] L.-E. Andersson, "A quasistatic frictional problem with normal compliance", *Nonlinear Analysis, Theory Methods & Applications* 16 (1991) 347-369.

[2] L.-E. Andersson, "A quasistatic frictional problem with a normal compliance penalization term", to appear in *Nonlinear Analysis, Theory Methods & Applications*.

[3] M. Cocu, E. Pratt, M. Raous, "Formulation and approximation of quasistatic frictional contact", *Int. J. Engng. Sci.* 34 (1996) 783-798.

[4] G. Duvaut, J.L. Lions, "Elasticité avec frottement", *J. Mécanique*, 10, (1971) 409-420.

[5] A. Klarbring, "Examples of non-uniqueness and non-existence of solutions to quasistatic contact problems with friction", *Ingenieur-Archiv* 60 (1990) 529-541.

[6] A. Klarbring, A. Mikelić, M. Shillor, "On friction problems with normal compliance", *Nonlinear Analysis, Theory, Methods & Applications* 13 (1989) 935-955.

[7] A. Klarbring, A. Mikelić, M. Shillor, "A global existence result for the quasistatic frictional contact problem with normal compliance", *International Series of Numerical Mathematics* 101 (1991) 85-111.

[8] J.A.C. Martins, F.M.F. Simões, F. Gastaldi, M.D.P.M. Marques, "Dissipative graph solutions for a 2 degree-of-freedom quasistatic frictional contact problem", *Int. J. Engng. Sci* 33 (1995) 1959-1986.

[9] J. Nečas, J. Jarušek, J. Haslinger, "On the solution of the variational inequality to the Signorini problem with small friction", *Bolletino U. M. I.* 5, (1980) 796-811.

Collision Between Rigid Bodies – Comparing Some Models

E. Cataldo[1]
R. Sampaio[2]

ABSTRACT There are in the literature some models that describe collision between rigid bodies. In modeling it is difficult to construct a model sufficiently simple that permits the simulation of a problem in real time and so sophisticated not to change the dynamics so that the preditions in time of observations are not unuseful. The aim of this paper is to describe some of the models more used and to compare them using numeric simulations and to show its limits of validity.

1 Introduction

To study the dynamics of rigid bodies with collision we must observe that for each collision the differential equation that describes the dynamics of the body must be reinitialized with appropriated new initial conditions. To determine this new initial conditions, obtained from the state of the body imediately before the collision, some models were developed. The classic models, developed in the literature, make symplifiying hypothesis that, in some cases, conduct us to facts that violate phisic laws. We use a problem to compare tehse different models and we visualize the different predictions. We discuss the Newton's law combined with Coulomb's law, Poisson's law combined with Coulomb's law and Smith's law. Each one of these laws make considerations about the coefficient of restitution and the coefficient of friction and in some cases different results can be obtained.

2 Basic Equations

The equations that describe the dynamics and the cinematics of collision are discussed in the following. We use A to indicate the situation imediately

[1]Department of Mechanical Engineering, PUC-Rio, Rio de Janeiro, Brazil
[2]Department of Mechanical Engineering, PUC-Rio, Rio de Janeiro, Brazil

before the collision, C to indicate the situation at the beginning of the compression phase and E to indicate the situation at the end of the expansion phase. Consider $\mathbf{q}$, the vector of generalized coordinates that shows the position of the rigid body. $\mathbf{D_N}$ is a vector in which the elements define the distance between the contact points of the bodies. $\mathbf{D_T}$ is a vector in which the elements define the tangential component of the distance between the contact points of the bodies. $\mathbf{I_N}$ is a vector in which the elements define the values of the normal impulses in the points of the bodies in contact. $\mathbf{I_T}$ is a vector in which the elements define the values of the tangential impulses in the points of the bodies in contact. W_N e W_T are matrices which elements are the coefficients of the vectors that define the normal impulses and the tangential impulses in the equations of the dynamics of the bodies. $\tilde{\mathbf{w}}_\mathbf{N}$ e $\tilde{\mathbf{w}}_\mathbf{T}$ are related to the rotation of the bodies.

In the compression phase we have

$$\begin{pmatrix} \dot{\mathbf{D}}_{NC} \\ \dot{\mathbf{D}}_{TC} \end{pmatrix} = \begin{pmatrix} W_N^T \\ W_T^T \end{pmatrix} \dot{\mathbf{q}}_C + \begin{pmatrix} \tilde{\mathbf{w}}_\mathbf{N} \\ \tilde{\mathbf{w}}_\mathbf{T} \end{pmatrix} \tag{1}$$

and

$$M(\dot{\mathbf{q}}_\mathbf{C} - \dot{\mathbf{q}}_\mathbf{A}) - \begin{pmatrix} W_N & W_T \end{pmatrix} \begin{pmatrix} \mathbf{I_{NC}} \\ \mathbf{I_{TC}} \end{pmatrix} = 0 \tag{2}$$

with the condition of complementarity

$$\mathbf{I_{NC}}\dot{\mathbf{D}}_\mathbf{NC} = 0 \ , \ \mathbf{I_{NC}} \geq 0 \ , \ \dot{\mathbf{D}}_\mathbf{NC} \geq 0.$$

In the expansion phase we have

$$\begin{pmatrix} \dot{\mathbf{D}}_\mathbf{NE} \\ \dot{\mathbf{D}}_\mathbf{TE} \end{pmatrix} = \begin{pmatrix} W_N^T \\ W_T^T \end{pmatrix} \dot{\mathbf{q}}_\mathbf{E} + \begin{pmatrix} \tilde{\mathbf{w}}_\mathbf{N} \\ \tilde{\mathbf{w}}_\mathbf{T} \end{pmatrix} \tag{3}$$

and

$$M(\dot{\mathbf{q}}_\mathbf{E} - \dot{\mathbf{q}}_\mathbf{C}) - \begin{pmatrix} W_N & W_T \end{pmatrix} \begin{pmatrix} \mathbf{I_{NE}} \\ \mathbf{I_{TE}} \end{pmatrix} = 0. \tag{4}$$

The condition of complementarity, for the cases that will be studied here, we have

$$\mathbf{I_{NE}}\dot{\mathbf{D}}_\mathbf{NE} = 0 \ , \ \mathbf{I_{NE}} \geq 0 \ , \ \dot{\mathbf{D}}_\mathbf{NE} \geq 0.$$

3 Comparing the models and simulations

We would like to compare Newton's law and Poisson's law. We are not going to describe these laws here but we will consider an example that can show when the usage of these laws can lead us to different results, that are, in fact, physically incorrect. Consider the collision of a rigid bar against a rigid barrier as it is showed in the figure 1. We consider the following

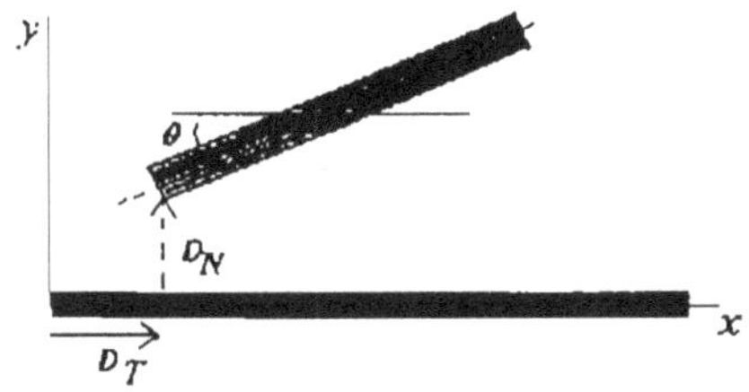

FIGURE 1. Collision of a bar against a rigid barrier.

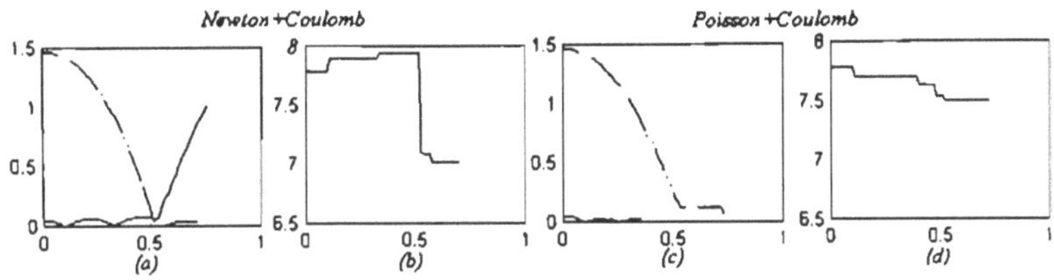

FIGURE 2. Case $\mu > \mu_d$ ($\mu = 1$). In the first collision: $\dot{D}_{NA} = -1$ and $\dot{D}_{TA} = 0.6$. (a) and (c) displacement of the left extremity (full line) and displacement of the right extremity (dasehed line) (b) e (d) Energy.

situation before the collision: $\dot{D}_{NA} = -1$ e $\dot{D}_{TA} = 0.6$, and $\mu > 0.6$. In this case, we have $\dot{D}_{NC} = 0$ e $\dot{D}_{TC} = 0$. Therefore, $I_{TC} = 0$ and $I_{NC} = 0.4$. Using Newton's law, we have $\dot{D}_{NE} = \epsilon$, $I_{TE} = 0.375\epsilon$ and $I_{NE} = 0.625\epsilon$ and using Poisson's law, we have $I_{TE} = 0.24\epsilon$, $I_{NE} = 0.4\epsilon$ and $\dot{D}_{NE} = 0.64\epsilon$.

When we use the Poisson's law, for collisions 2D, it is proved in [Pfeiffer and Glocker 1996] that $T_E - T_A \leq 0$ (T_E is the energy at the end of the expansion phase and T_A is the energy imediately before the collision). In the example that we have studied, we are concerned about the case in that the results given by Newton's law and Poisson's law are different. Using Newton's law we have $T_E - T_A = 0.3125\epsilon^2 - 0.2$. We observe that when $\epsilon > 0.8$, $T_E - T_A > 0$, therefore the law of conservation of energy is violated. Using Poisson's law, we have $T_E - T_A = 0.2\epsilon^2 - 0.2$. For $0 \leq \epsilon \leq 1 \Rightarrow T_E - T_A \geq 0$, and the law of the conservation of energy is not violated, as we could hope.

For the simulations we consider x, y and θ (the coordinates of the center of mass) the components of the vector of generalized coordinates, $2L$(length of the bar) $= 2$ and m (the mass of the bar) $= 1$. In order to compare with the data obtained previously we have simulated a situation so that imediately before the first collision the conditions discussed in the previous section were reached. We consider for the simulation, as initial conditions $x_0 = 0$, $\dot{x}_0 = 0.6$, $y_0 = 0.76$, $\dot{y}_0 = 0$, $\theta_0 = \frac{\pi}{4}$ and $\dot{\theta}_0 = 0$. In this case, the increase of energy is observed in the first collision when we use the Newton's law, but it doesn't occur when we use the Poisson's law. We

	Newton e Coulomb	Poisson e Coulomb	Smith
$\dot{D}_{TE}$	0	0	-0.4825
$\dot{D}_{NE}$	0.5	0.508	0.5
I_N	0.7125	0.6	0.5316
I_T	0.1875	0.12	-0.1141
$T_E - T_A$	-0.234	-0.234	-0.14

TABLE 1. Comparisson of the laws.

show the graphics of the simulations in the figure 2.

4 The Smith's Law

The Smith's law use the definition of the coefficient of restitution given by Newton and the normal and tangential impulses are related in the formula

$$I_T = -\mu I_N \frac{\| \dot{D}_{TA} \| \dot{D}_{TA} + \| \dot{D}_{TE} \| \dot{D}_{TE}}{\| \dot{D}_{TA} \|^2 + \| \dot{D}_{TE} \|^2}$$

with $I_T = I_{TC} + I_{TE}$ and $I_N = I_{NC} + I_{NE}$. We consider the case of the bar studied previously and we show the results in the table 1.

5 Conclusions

Some of the models considered, in the literature, to study the collision between the bodies can fail. They can violate physical laws. We simulate these models in order to show some of the characteristics of each one.

6 References

[1] F. Pfeiffer and C. Glocker (1996). Multibody Dynamics with Unilateral Contacts. Wiley series in nonlinear science.

[2] Anindya Chatterjee (1997). Rigid body collisions: some general considerations, new collision laws, and some experimental data. A PhD thesis presented to the faculty of the graduate school of Cornell Universitiy.

Realizability of Arbitrary Local Mass Matrices in Single-Point Rigid Body Collisions

Anindya Chatterjee[1]
Andy Ruina[2]

ABSTRACT The 3 × 3 local mass matrix fully characterizes a pair of colliding "rigid" bodies for many purposes. We prove here that arbitrary 3 × 3, symmetric positive definite matrices have physical realizations as local mass matrices for collisions of two bodies with finite dimensions and inertia. General collision models thus must be able to handle all such matrices.

1 Introduction

If two rigid bodies (labeled 1 and 2) collide at a point C, the impulse $\mathbf{P}$ exerted by body 1 on body 2 is related to the net change in $\mathbf{V}_c$, the velocity of the contact point on body 2 relative to the contact point on body 1, by

$$\mathbf{P} = \mathbf{M} \cdot \Delta \mathbf{V}_c \tag{0.1}$$

where $\mathbf{M}$ is a well known tensor, here called the "local mass tensor" [1, 6]. $\mathbf{M}$ can be interpreted as the anisotropic inertia experienced by a force pushing apart the contact points on the two bodies. If equal and opposite forces $\mathbf{F}$ and -$\mathbf{F}$ act on the two bodies at C, then the contact points' relative acceleration $\mathbf{a}_c$ satisfies (neglecting the effects of external forces and centripetal acceleration terms)

$$\mathbf{F} = \mathbf{M} \cdot \mathbf{a}_c \tag{0.2}$$

For two unconstrained rigid bodies $\mathbf{M}$ is a symmetric positive definite (SPD) second order tensor given by

$$\mathbf{M}^{-1} = \sum_{i=1}^{2} \left(\frac{1}{m_i}\mathbf{I} + \mathbf{S}^T(\mathbf{r}_{C/cm_i})\mathbf{J}_i^{-1}\mathbf{S}(\mathbf{r}_{C/cm_i}) \right) \tag{0.3}$$

[1]Engineering Science & Mechanics, Penn State University, University Park, PA 16802
[2]Theoretical & Applied Mechanics, Cornell University, Ithaca, NY 14853

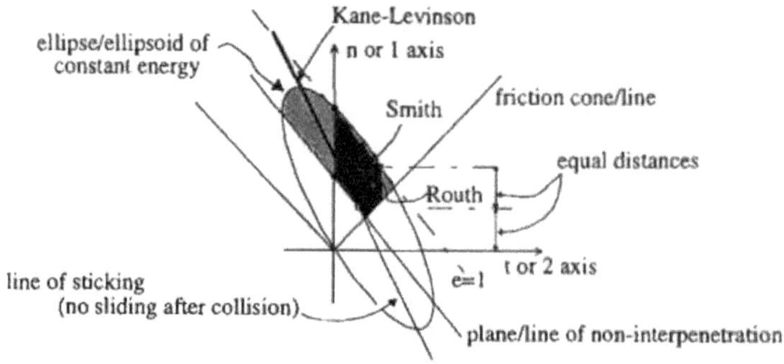

FIGURE 0.1. The region in impulse space accessible to generic "reasonable" collision. Also shown is the region accessible to some algebraic collision laws.

where $\mathbf{I}$ is the identity tensor, m_i is the mass of body i, $\mathbf{J}_i$ is the moment of inertia tensor of body i about its center of mass (c.m.), $\mathbf{r}_{C/cm_i}$ is the position vector of the contact point C with respect to the c.m. of body i, and the superscript T denotes transpose. $\mathbf{S}(\mathbf{r})$ is the cross product tensor of $\mathbf{r}$ satisfying $\mathbf{S}(\mathbf{r}) \cdot \mathbf{v} = \mathbf{r} \times \mathbf{v}$, for all $\mathbf{v}$. For calculations, we use the 3×3 matrix of components, $[\mathbf{M}]$, called the *local mass matrix.*

$\mathbf{M}$ fully characterizes the net dynamic interaction between the contacting rigid bodies. So, collision laws need only consider the family of all possible $\mathbf{M}$, and not the family of all pairs of possible contacting bodies or mechanisms. The set of all possible "reasonable" collisional impulses can be determined from $\mathbf{M}$ and the approach velocity of the about-to-contact points [1, 2]. These reasonable impulses are in a region of impulse space bounded by a non-interpenetration plane, a friction cone, and an energy-conserving ellipsoid, shown shaded light gray in the 2-D schematic Fig. 0.1. The coordinates are chosen with the 1 direction normal to the common contact-surface normal.

A geometric comparison is also presented in Fig. 0.1 between the range of possible predictions from Routh's law [4] (heavy dashed line), from Kane and Levinson's law [3] (heavy straight line), from Smith's law [6] (heavy curve) and a law proposed recently [2] (dark gray region). The geometric approach conveniently displays several features of the collision laws under study, such as the possibility of energy "creation" by Kane and Levinson's law, the impossibility here of a sliding collision as per Routh's law, etc.

Since all rigid body collision laws use $\mathbf{M}$ as part of their input (possibly implicitly) it is useful to know the set of all possible $\mathbf{M}$ (equivalently, all possible ellipsoids in impulse space, with some restrictions on their position [1, 2]) that need be considered as valid inputs for candidate collision laws.

As an analogy, consider the set of all possible rigid body moment of inertia tensors. These are SPD, but also have an additional restriction on

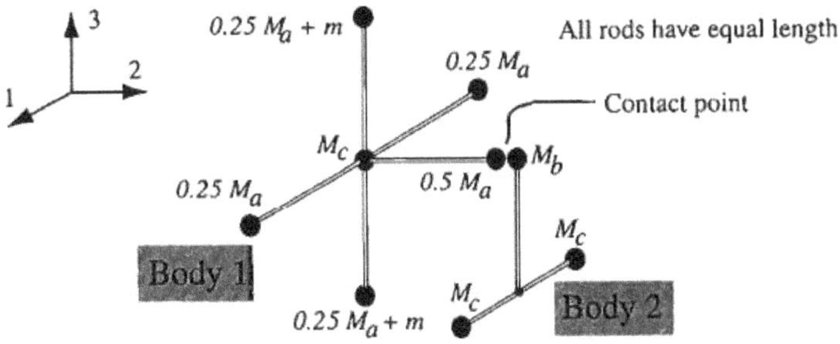

FIGURE 0.2. Physical realization of arbitrary mass matrices

the three eigenvalues: none of them is bigger than the sum of the other two. Are there any such restrictions on **M**?

2 Physical Realizability

Claim: Any 3×3, SPD matrix has a physical realization as [**M**] for two unconstrained rigid bodies of finite size, mass, and moments of inertia.

Proof: In general the contact tangent plane is independent of mass distribution and can have *any* orientation depending on the *local* shape of the colliding objects, near the contact point. Thus, we only need to demonstrate the realizability of arbitrary diagonal [**M**] with non-negative elements (eigenvalues) $(\lambda_1, \lambda_2, \lambda_3)$. Let $\lambda_1 \leq \lambda_3$.

Consider the 2-body system in Fig. 0.2. Body 1 has a mass distribution that is equivalent to six point masses as shown. Body 2 has a mass distribution equivalent to three point masses as shown. The rigid, massless rods are of equal length. We will show the existence of masses (M_a, M_b, M_c, m) that yield the eigenvalues $(\lambda_1, \lambda_2, \lambda_3)$.

Observe that as $M_c \to \infty$, we obtain a constrained system: body 1 has a ball and socket joint at point mass M_c (on body 1), while body 2 is hinged about the axis through the point masses M_c (on body 2). Now, by Eq. 0.2, **M** will have the required eigenvalues if

$$M_a = \lambda_1\,, \quad M_b = \lambda_2\,, \text{ and } M_a + 2m = \lambda_3\,, \text{ and } m = (\lambda_3 - \lambda_1)/2\,. \quad (0.4)$$

Thus with infinite masses, or with hinges, arbitrary **M** are obtainable.

For finite M_c, the mass matrix in the coordinate system shown in Fig. 0.2 (found using Eq. 0.3) is diagonal. Its eigenvalues (M_{11}, M_{22}, M_{33}) are functions of M_a, M_b, M_c and m (not reproduced here). We define $\epsilon := 1/M_c$, and find (using computer algebra) that for small ϵ,

$$M_{11} = M_a - \frac{11}{4} M_a^2\,\epsilon + \mathcal{O}(\epsilon^2)\,, \qquad (0.5)$$

$$M_{22} = M_b - M_b^2\,\epsilon + \mathcal{O}(\epsilon^2)\,, \tag{0.6}$$

$$M_{33} = M_a + 2m - \left(\frac{3}{4}M_a^2 + 4M_a m + 6m^2\right)\epsilon + \mathcal{O}(\epsilon^2)\,. \tag{0.7}$$

Setting $\epsilon = 0$ gives Eq. 0.4 as expected.

From Eqs. 0.5 through 0.7, the matrix of partial derivatives of (M_{11}, M_{22}, M_{33}) with respect to (M_a, M_b, m) evaluated at $\epsilon = 0$ is invertible. By the implicit function theorem (see e.g., [5]), if $0 < \lambda_1 \leq \lambda_3$ and $0 < \lambda_2$, then there is a finite M_{c_0} such that for $M_c > M_{c_0}$ there are nonnegative, finite $M_a(M_c), M_b(M_c), m(M_c)$ for which $[\mathbf{M}]$ has the specified eigenvalues $(\lambda_1, \lambda_2, \lambda_3)$. The demonstration presented here just shows one family of pairs of finite-inertia bodies associated with any possible positive definite $\mathbf{M}$. In practice, the minimum value of M_{c_0} may not be very large compared to the other masses.

3 Relevance to Collision Modeling

The physical realizability of arbitrary $[\mathbf{M}]$ shows the completeness of the geometric view of single-point rigid body collisions. That is, in drawing Fig. 0.1, we are assured in advance that a pair of bodies can be found to match *any* such figure. We can examine the nature of collision laws in relation to arbitrary impulse-space ellipsoids with the foreknowledge that all such ellipsoids correspond to some pair of realizable colliding rigid bodies.

4 References

[1] Chatterjee, A., *Rigid Body Collisions: Some General Considerations, New Collision Laws, and Some Experimental Data.* PhD thesis, Cornell University, 1997.

[2] Chatterjee, A. and Ruina, A., A new algebraic rigid body collision model with some useful properties. Accepted, subject to revisions, for publication in the *Journal of Applied Mechanics*, 1998.

[3] Kane, T. R. and Levinson, D. A., *Dynamics: Theory and Applications.* McGraw-Hill, New York, 1985.

[4] Routh, E. J., *Dynamics of a System of Rigid Bodies.* Macmillan and Co., London, sixth edition, 1897.

[5] Rudin, W., *Principles of Mathematical Analysis.* McGraw-Hill, New York, third edition, 1976.

[6] Smith, C. E., Predicting rebounds using rigid-body dynamics. *ASME Journal of Applied Mechanics*, 58:754–758, 1991.

Control of Joint Forces of Planar Mechanism with Clearances at Joints for Reducing Vibration and Impact at Joints

B. Feng[1]
N. Morita[1]
T. Torii[1]

ABSTRACT This paper presents a new dynamic design method of reducing vibration and impact by optimizing the magnitude and direction of the joint forces of planar linkage. It is shown that the change of the joint force of the optimized linkage becomes smooth, and the relative vibrations between pairing elements reduce.

1 Introduction

A link mechanism at high operating speed produces large inertia forces, which make the magnitude and directions of joint forces acting on the pairs change sharply. Because a small joint clearance exists between the pairing elements, the sharp changes of joint force will cause the relative vibration or impact between the elements. Therefore, it is important for a mechanism with clearances at joints to control the joint forces for reducing the relative vibration or the impact. The problem to control the joint forces is to control the inertia forces of moving links. This paper presents a new optimum design method to control the joint forces by redistributing masses of moving links, without adding counter-weight. The dynamic optimum designs to control the joint forces for three pairs of four-bar mechanism with clearances are carried out. To form the shape of links using the optimized parameters, the small element method is proposed and used. The numerical example is given and it is shown that the method presented is efficient to control the joint forces and to reduce the relative vibrations of elements.

---

[1] Department of Mechanical Engineering, Shizuoka University, Johoku 3-5-1, Hamamatsu, 432-8561, Japan

2. Optimum Design of Joint Force of Linkage with Clearances

Consider a planar four-bar linkage with clearances r_2, r_3 r_4 at the pairs *B*, *C*, *D*, as shown in Fig. 1. The model of continuous contact is used in the kinetic analysis of the linkage, in which it is assumed that the pin is always in contact with the socket in each pair [1]. It is known that this model is applicable to great part of region of the joint force, excepting the joint force, which is equal to the gravitational force. The clearances are assumed as the massless links. The joint forces F_{12}, F_{23} and F_{43} and the direction α_2 at the pair *B* are taken as the objective functions shown in Eq. (1)~(3).

$$\min f_1(\mathbf{x},\varphi_1) = W_1\sum_{i=1}^{n}\left[\alpha_{2i}(\mathbf{x},\varphi_1)-\alpha_{2i}(\varphi_1)\right]^2 + W_2\sum_{i=1}^{n}\left[F_{12i}(\mathbf{x},\varphi_1)-F_{12-id}\right]^2 \quad (1)$$

$$\min f_2(\mathbf{x},\varphi_1) = \sum_{i=1}^{n}\left[F_{23i}(\mathbf{x},\varphi_1)-F_{23-id}\right]^2 \quad (2)$$

$$\min f_3(\mathbf{x},\varphi_1) = \sum_{i=1}^{n}\left[F_{43i}(\mathbf{x},\varphi_1)-F_{43-id}\right]^2 \quad (3)$$

Where, $\alpha_{2i}(\varphi_1)$ and $F_{ij\text{-}id}$ are the ideal values. They are given by analyzing the changes of joint forces and directions of basic linkage. W_1 and W_2 are the weighting coefficients. The mass, the center of mass and the mass moment of inertia of coupler and lever are taken as the design variables.

$$\begin{aligned}\mathbf{x} &= \begin{bmatrix}x_1 & x_2 & x_3 & x_4 & x_5 & x_6 & x_7 & x_8\end{bmatrix}^T \\ &= \begin{bmatrix}m_2 & u_2 & v_2 & I_2 & m_3 & u_3 & v_3 & I_3\end{bmatrix}^T\end{aligned} \quad (4)$$

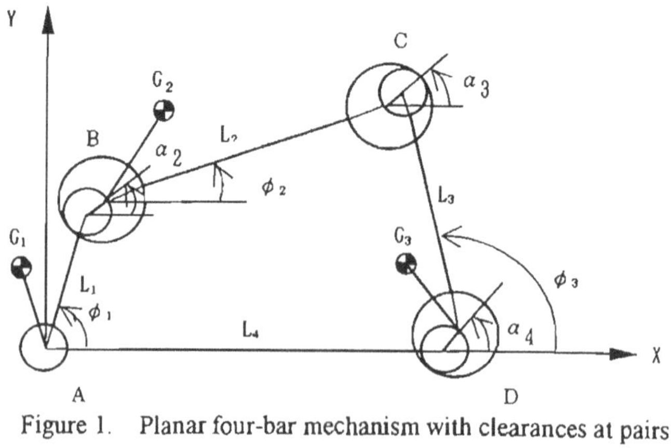

Figure 1. Planar four-bar mechanism with clearances at pairs

The maximum dynamic stresses $\sigma_{i\max}$ of links and the minimum and maximum values of design variables are taken as the constraints.

$$\sigma_{i\max} < [\sigma]_i \qquad i = 1,2,3 \tag{5}$$

$$x_{j\min} < x_j < x_{j\max} \qquad j = 1,2,\cdots,8 \tag{6}$$

The presented problem is an optimum problem of multi-objective functions. This paper uses the method of multi-times optimization to control three joint forces and the direction α_2. At first, the optimal process of objective function (1) is carried out. Then, optimized joint force is taken as a constraint, and the optimal process of objective function (2) is carried out. Lastly, two optimized forces are taken as the constraints, and the optimal process of objective function (3) is carried out.

3 Numerical Examples

The dynamic optimum design is carried out for the linkage, which lengths and thickness are shown in Table 1, and the crank speed are $600\,rpm$. The weighting coefficients are taken as W_1=1 and W_2=0.5. The optimized parameters of linkage are shown in Table 2. It is seen that the optimized center of mass of the coupler is close to the crank and the center of mass of lever is close to the frame. Then, the magnitude and the direction of joint forces are calculated, and they are shown in Fig. 2 and Fig.3. From Fig. 2, it is clear that

Table 1 Parameters of original linkage

	L_i (m)	h_i (m)	t_i (m)	u_i (m)	v_i (m)	m_i (kg)	I_i (kgm^2)	Clearance (m)
Crank	0.030	0.030	0.005	0.015	0.0	0.0122	1.82×10^{-6}	0.0001
Coupler	0.180	0.030	0.005	0.090	0.0	0.0737	2.02×10^{-4}	0.0001
Lever	0.120	0.030	0.005	0.060	0.0	0.0490	6.19×10^{-5}	0.0001
Frame	0.200	–	–	–	–	–	–	–

Table 2 Parameters of optimized linkage

	L_i (m)	t_i (m)	u_i (m)	v_i (m)	m_i (kg)	I_i (kgm^2)	Clearance (m)
Crank	0.030	0.005	0.015	0.000	0.0122	1.82×10^{-6}	0.0001
Coupler	0.180	0.005	0.025	0.0023	0.0911	2.71×10^{-4}	0.0001
Lever	0.120	0.005	0.015	-0.005	0.0602	6.02×10^{-5}	0.0001
Frame	0.200	–	–	–	–	–	–

the magnitude of optimized joint forces is approximately uniform, and the relative vibrations are vanished. From Fig. 3, the optimized direction of joint force F_{12} is approximately the ideal straight line. To form the shape of the link, the small element method is used. It is shown in Fig. 4.

4 Conclusion

The joint forces of planar linkage with clearances are optimized with respect to the magnitude and direction. The results are summarized as follows:

(1) The optimized joint forces obtained by redistributing masses of links are approximately the ideal values, and the relative vibrations between pairing elements reduce.

(2) The shapes of the optimized links are designed by using the small element method.

5 Reference

[1] T. Furuhashi, N. Morita , and M. Matsuura, Research on Dynamics of Four-Bar Linkage with Clearances at Turning Pairs, *Bull. of JSME* 21, 518-523 (1978).

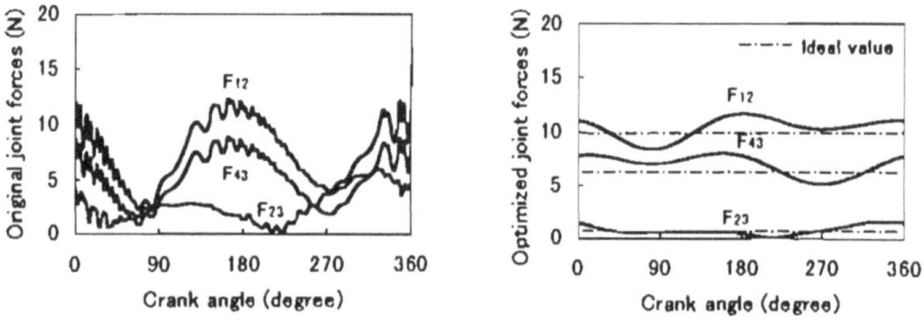

Figure 2 Joint forces of planar mechanism with clearances at joints

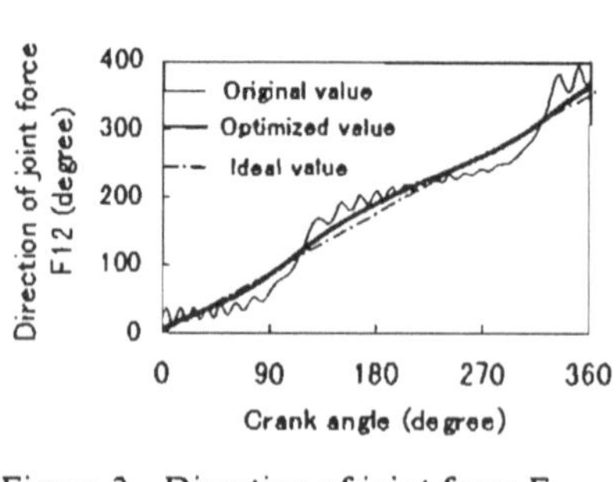

Figure 3 Direction of joint force F_{12}

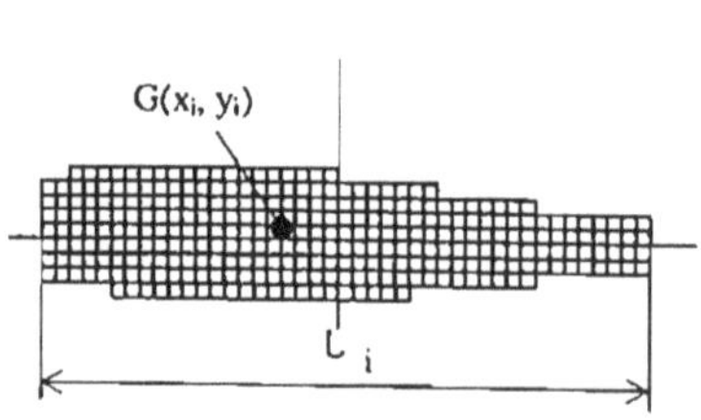

Figure 4 Small element method

Optimization of Shaking Force and Shaking Moment of Planar Linkages for Vibration Reduction Using Genetic Algorithms

G. Guo[1]
N. Morita[1]
T. Torii[1]

ABSTRACT This paper proposes a new mixed mass redistribution method, uses genetic algorithms to optimize shaking force/moment of linkages for reducing vibration , and give a design example to verify the effectiveness and efficiency of the new method.

1 Introduction

The shaking force/moment produced at high operating speed of link mechanism will cause the vibration and noise of a mechanism. So it is important to reduce the shaking force/moment as more as possible in order to decrease the vibration.

This paper proposes a new mixed mass redistribution method , investigates the optimum balance of shaking force /moment using genetic algorithms(GA). The results calculated by mass redistribution method and counter-weight method and new method are compared using GA. To verify the efficiency of GA , the complex method of conventional optimization is also used for the same design model. Lastly, the small element method[1] is used to form the shape of links according to the optimized design parameters.

2 Dynamic Characteristics and Optimum Design Model of Linkages

A 4-bar linkage shown in Figure 1 is taken as an example. F_{SX} and F_{SY} are the X and Y component of shaking force and can be obtained by D'Alembert's principle [2]. M_S is the shaking moment which acts on the fixed frame and is given in [2].

In establishing the optimum design model, we propose a new mixed mass redistribution method, that is, the method mixed both the method of mass redistribution

[1]Department of Mechanical Engineering, Shizuoka University, Johoku 3-5-1,Hamamastu 432, Japan

and that of adding counter-weight . The design variables are shown as follows.

$$\overline{X} = \left[m_2\ S_2\ \theta_2\ I_2\ m_3\ S_3\ \theta_3\ I_3\ m_1^*\ S_1^*\ \theta_1^*\ m_3^*\ S_3^*\ \theta_3^*\right]^T \tag{1}$$

In order to reduce the vibration of linkages, we hope to reduce the shaking force / moment as more as possible, therefore, we constitute the objective function shown below.

$$\min f(\overline{X}) = \min\{W_{b1} \times B_{\max}(F_{SX}) + W_{r1} \times |F_{SX\max}| + W_{b2} \times B_{\max}(F_{SY}) + W_{r2} \times |F_{SY\max}| + W_{b3} \times B_{\max}(M_S) + W_{r3} \times |M_{S\max}|\} \tag{2}$$

Where, W_{bi} and W_{ri} are the weighted coefficients for the components of objective function, respectively. $B_{max}(G\{t\})$ expresses the difference between the maximum and the minimum of a function G{t}.

Considering the material strength, shape and manufacture easiness of the links, we establish the constraint condition equations given in equation (3).

$$g_{ui}(\overline{X}) = a_i - x_i\ ,\ g_{li}(\overline{X}) = x_i - b_i\ ,\ (i = 1,2,\cdots 14) \tag{3}$$

3 Optimum Design of Linkages Based on GA

The fitness for this link mechanism is defined as the reciprocal of objective function value multiplied by a constant.

The GA used in this paper is based on a simple GA[3]. A series of treatments, such as , selection, reproduction, crossover and mutation is repeated to get converged solutions.

4 Design Example

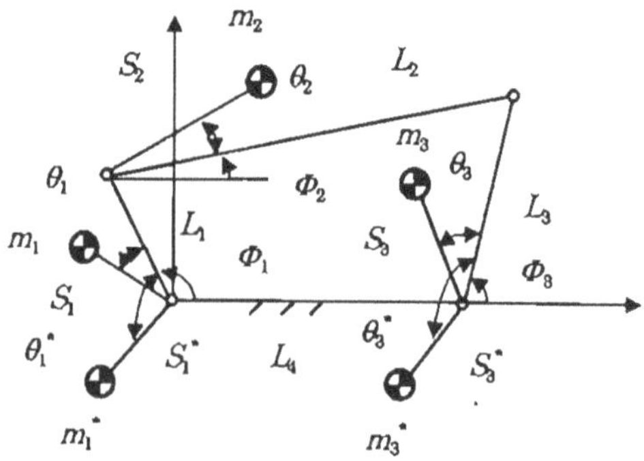

Figure 1 Dimensions and mass parameters of 4-bar linkage

Table 1 Parameters of original 4-bar linkage

	L_i(m)	m_i (kg)	S_i(m)	θ_i (deg)	I_i(kgm^2)
Link1	0.06	0.024	0.03	0.0	3.68e-6
Link2	0.36	0.294	0.18	0.0	2.32e-3
Link3	0.24	0.196	0.12	0.0	8.58e-4
Link4	0.40	–	–	–	–

Table 2 Parameters of optimized 4-bar linkage by GA

	L_i(m)	m_i (kg)	S_i(m)	θ_i (deg)	I_i(kgm^2)
Link1	0.06	(0.142)	(0.116)	(199.1)	–
Link2	0.36	0.388	0.155	327.5	7.0e-3
Link3	0.24	0.183 (0.224)	0.089 (0.106)	297.9 (145.4)	3.0e-3
Link4	0.40	–	–	–	–

Table 3 Parameters of optimized 4-bar linkage by complex method

	L_i (m)	m_i (kg)	S_i(m)	θ_i (deg)	I_i(kgm^2)
Link1	0.06	(0.123)	(0.203)	(165.4)	–
Link2	0.36	0.263	0.223	194.7	3.76e-3
Link3	0.24	0.587 (0.387)	0.166 (0.154)	144.6 (220.8)	1.39e-3
Link4	0.40	–	–	–	–

Taking the 4-bar linkage shown in Figure 1 as a design example, which original parameters are given in Table 1, we performed the optimization of the shaking force/moment using GA. Table 2 shows the parameters of the linkages that have the best fitness, and the values written into parentheses in Table 2 gives the results corresponding to the counter-weight case.

Then, the linkages parameters optimized by GA, shown in Table 2, are used to calculate the shaking force/moment of the mechanism, and the results are illustrated in Figure 2.

For the linkages shown in Table 1, the model set by mixed mass redistribution method is also optimized by the complex method (CP, in short) to investigate the effectiveness and efficiency of GA. The result is given in Table 3, and its dynamic characteristics is shown in Figure 3.

Lastly, to design the feasible shape of links according to the optimized design parameters, the small element method developed in this research division[1] is applied. Figure of the linkages after shape design is omitted because of space limit.

5 Conclusion

In summary, we can attain following conclusions.

(1) To optimize the shaking force/moment and decrease the vibration of the linkages, a new mixed mass redistribution has been proposed, the effectiveness of the new method

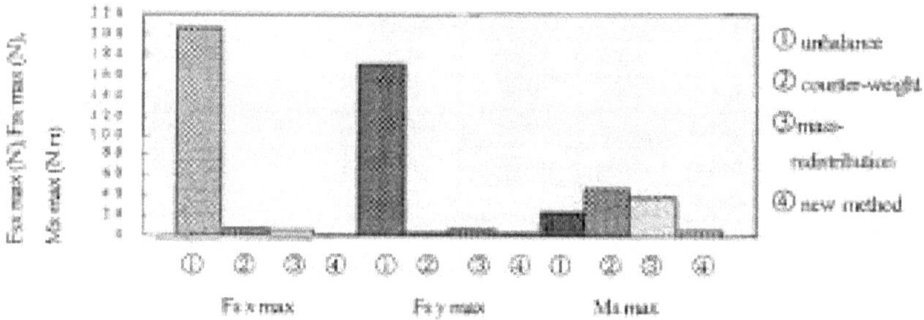

Figure 2 Dynamic characteristics of 4-bar linkage by different balance methods

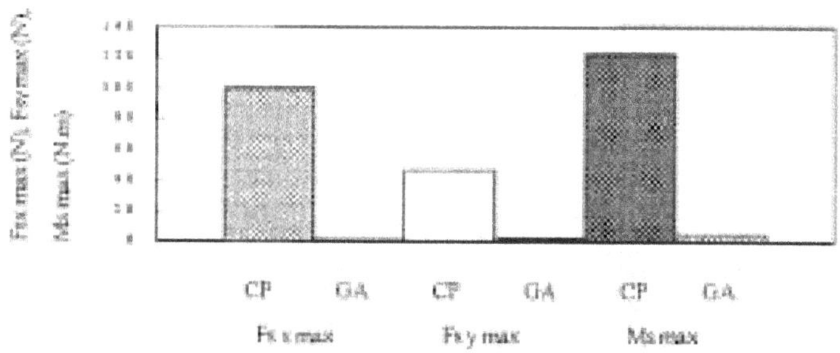

Figure 3 Dynamic characteristics of 4-bar linkage by GA and complex method

is confirmed.

(2) Optimum dynamic design of planar linkages using genetic algorithms has been investigated and the design system has been set up.

(3) For the same design model, it is affirmed that the genetic algorithms can obtain the optimum balance results more efficiently than the complex method of traditional optimization approaches under the same conditions.

(4) By computing the dynamic characteristics of the linkages after optimum design ,the shaking force/moment has been greatly decreased ,so the vibration of the linkages can also be reduced.

6 References

[1] Bin Feng, Nobuyoshi Morita, Hidehisa Sakai and Takao Torii, Optimum Design of Linkages, *Trans. JSME(in Japanese). Series C*, Vol. 61, No. 585, 1995, 2014-2019.

[2] Beliang Guan, Takeshi Furuhashi and Nobuyoshi Morita, Control of Shaking Force and Shaking Moment in Link Mechanism, *Bulletin of JSME*, Vol. 28, No. 243, 1985, 2078-2083.

[3] Hiroshi Tachiya and Yoshiya Tsujimura, Synthesis of Planar Multilink Mechanisms Based on Genetic Algorithms, *Jpn. Soc. Mech. Eng. Intl. J. Series C*, Vol. 39,No. 4, 1996, 828-836.

Modeling of Multiple Impacts

A.P. Ivanov[1]

ABSTRACT Phenomenon of multiple collision and methods of its analysis are discussed. The principal conclusion is: multiple impacts can not be considered as just a set of simple "pair" impacts, since they possess some special features. Thus, special techniques are to be developed. Some of them are presented below.

1 Introduction

In mechanics, an impact is considered as a short-term interaction of bodies which leads to essential changes in their velocities. Impacts are accompanied with small deformations causing very big contact forces. In practice, we usually meet impact as a result of collision of two or more interconnected bodies, for instance, in vehicle dynamics, robotic dynamics, offshore structures, and so forth. The simplest case is pair impact (where two free bodies collide), then triple impact, and so on. We will call pair impacts *simple*, while all the more complicated cases will be called *multiple* impacts.

The theory of mechanical collision involves several branches [4]. The most popular are classical stereomechanics where deformations of the colliding solids are not accounted and the wave impact theory where the history of these deformations is investigated. In addition, various intermediate approaches exist which are based on certain models of rigid bodies with deformable elements [4,10,12,13,7]. To solve a particular problem one should choose such approach that unites reliability of results with numerical efficiency. In particular, the wave theory is used for the analysis of colliding pivots and plates while the classical theory is convenient for the description of low-speed collisions of spherical particles.

It should be pointed out that all the aforementioned methods were initially used for analysis of simple collisions where they were shown to be efficient. However, it is not much evidence on they validity in the case of multiple impact. One of sparse examples of reliable solution of the problem is known Boussinesq solution of triple collision problem (of a rigid body,

[1]Department of Mathematical Modeling, Moscow State Academy of Device and Informatics, Moscow, Russia

a pivot and a massive wall) by means of the wave theory. The classical theory of impact was used in [14,11,5,6], for the description of multiple collision in a wide class of multibody systems. However, the authors have not cosidered such important topics as correctness and reliability of their models. Another approach was used by the author [8,9], who computed the impulsive reaction with the help of certain strain-stress relations.

We consider several typical examples of multiple collision and discuss various methods of solution of impact problem.

2 Triple Collision

2.1 Collinear Impact

Examine the system of three balls with masses $m_1, m2$, and m_3, moving along a line without rotation. The balls m_2 and m_3 are originally at rest and touch one another, while the ball m_1 strikes against the ball m_2 with non-zero approach velocity V_1. The problem is to determine the velocities V_1', V_2', V_3' after collision.

The law of conservation of total impulse provides unique equation for the three unknowns. In the case of simple collision such uncertainty is usally removed with the help of the coefficient of restitution $e \in [0,1]$. Similarly, we can try to set the restitution coefficient for each of two contact pairs. As a result we conclude that $V_2' = 3DV'3$ (as $V_2 = 3DV_3$) which is realistic for plastic collisions only (i.e. $e = 3D0$).

An attempt to reduce the triple impact to a sequence of simple collisions is fruitful provided the balls are identical, i.e. three billiard balls or coins. However, if they are made of different materials, such approach becomes inapplicable. For instance, if we take two iron balls (or coins) and one rubber ball (or an eraser), then the system behavior differs in an essential way from theoretical predictions. Namely, if the rubber ball is situated opposite to the striking one, then after collision it rebounds from the central ball while other two balls collapse. Such result is similar to collision of three coins and is in good agreement with the idea of successive simple impact. On the contrary, in the case where the rubber ball is the striker, the result is quite different: the iron balls do not separate at all (or rather slightly), while the rubber ball rebounds from the central one (Fig.1,b). Hence, no impact between the harder balls takes place.

This simplest example shows inapplicability of classical methods to the problem of multiple impact: an account of deformations is needed to obtain realistic solution. The simplest model was proposed as early as in 18-th century by d'Alembert [1], who suggested to place imaginary strings at contact points between the balls. Such approach an be extended by using various strain-stress functions. The triple collision is described by the following system of ordinary differential equations:

$$\begin{aligned} \ddot{\delta}_1 &= 3Dm_2^{-1}R_{23}(\delta_2, \dot{\delta}_2) - (m_1^{-1} + m_2^{-1})R_{12}(\delta_1, \dot{\delta}_1), \\ \ddot{\delta}_2 &= 3Dm_2^{-1}R_{12}(\delta_1, \dot{\delta}_1) - (m_2^{-1} + m_3^{-1})R_{23}(\delta_2, \dot{\delta}_2) \end{aligned} \quad (1)$$

where $\delta_{1,2}$ are normal contact deformations. The system (1) is to be computed under initial conditions $\delta_1^0 = 3D\delta_2^0 = 3D0$, $\dot{\delta}_1^0 = 3DV_1, \dot{\delta}_2^0 = 3D0$ and under final conditions $\delta_{1,2}^k = 3D0, \dot{\delta}_{1,2}^k \le 0$. Numerical results, obtained from this model, agree with experimental data (see [8],[3]).

2.2 *Bernoulli Problem*

Along with the above considered case of triple collision, there exist more complicated cases, including non-collinear collisions. Clearly, the classical theory is inapplicable for their analysis. It is possible to describe impact by means of a system similar to (2). However, as was shown in [8], such system is ill-posed in general.

The reason is impossibility to determine the initial conditions for integration with the accuracy of contact deformations: we measure velocities of balls with error of millimeters per second, while these deformations are of order of micrometers and impact duration is of order of milliseconds or even microseconds. Due to such indetermination there exist several scenarios of multiple impact, which lead to quite different outcomes. Therefore, it is natural to consider multiple impact as multivalued random function. This situation is similar in essence to a thrown coin, which behaviors in two ways. An exceptional regular case exist when the solution is single-valued: where so called orthogonality conditions are satisfied [8].

A simple demonstration of such multi-valuedness is a problem of Iohann Bernoulli, who studied [2] collision of a moving ball upon two symmetrically posed identical balls. Bernoulli predicted that the symmetry preserves after impact. In fact, such outcome is rather impropable, as it is impossible to ensure the simultaneity of the simple impacts. In experiments, we observe two equiprobable outcomes where one of the formerly resting balls rebounds much faster than another one.

2.3 *Pendulum Hitting a Wall*

Another example of triple impact is a spherical pendulum, hitting a wall. Here the pendulum is suspended to an immovable base, which represents the third body. There exist just two cases where the classical impact theory may be used: (i) the impulsive reaction of the suspension vanishes (so called *centre de percussion*), or (ii) the pendulum is much harder than the wall. Otherwise one should use some more complicated model. For instance, the author considered a model o type (1) where $\delta_{1,2}$ are vectors [9]. This approach can be extended to the case of impact with friction. For this purpose an elsto-plastic [13] or visco-elastic [7] model of friction might be used.

3 Acknowledgments

This work is supported in part by the Russian Foundation for Basic Researches (96-01-01440) and by INTAS (96-2138).

4 References

[1] D'Alembert, J. *Traité de Dynamique*, David, Paris, 1743.

[2] Bernoulli, I. *Opera omnia*, Lausannae, 1742.

[3] Brogliato, B. *Nonsmooth Impact Mechanics*, Springer-Verlag, Berlin, Heidelberg, New York, 1996.

[4] Goldsmith, W., *Impact. The theory and physical behavior of colliding solids*, Arnold, London, 1960.

[5] Hurmuzlu Y. and Marghitu D., Rigid Body Collisions of Planar Kinematic Chains With Multiple Contact Points, *International J. of Robotics Research*, 13 (1), 82-92, 1994.

[6] Marghitu D. and Hurmuzlu Y., Three-Dimensional Rigid-Body Collisions With Multiple Contact Points, *ASME J. of Applied Mechanics*, 62 , 725-732, 1995.

[7] Ivanov, A., Visco-Elastic Approach to Impact With Friction, *Impact, Waves and Fracture* (ed.Batra, Mal, MacSithigh) ASME AMD-205, 115-127, 1995.

[8] Ivanov, A., On the Problem of Multiple Collision. *Prikladnaya Matematika i Mekhanika*, 59 (6), 930–946, 1995.

[9] Ivanov, A., The Problem of Constrained Impact. *Prikladnaya Matematika i Mekhanika*, 61 (3), 341–353, 1997.

[10] Johnson, K., *Contact Mechanics*, Cambridge, University Press, 1985.

[11] Pfeiffer, F. and Glocker Ch., *Multibody dynamics with unilateral contacts*, New York, Wiley, 1996.

[12] Bodies, *International J. of Impact Engineering*, 15 (4), 435-450, 1994.

[13] Stronge, W., Theoretical Coefficient of Restitution for Planar Impact of Rough Elasto-Plastic Bodies, *Impact, Waves and Fracture* (ed.Batra, Mal, MacSithigh) ASME AMD-205, 351-362, 1995.

[14] Wittenburg, J., *Dynamics of Systems of Rigid bodies*, Teubner, Stuttgart, 1977.

Analysis of Frictional Impact in Constrained Multibody Mechanical Systems

H. Lankarani[1]
S. Ahmed[1]
M. Pereira[2]

ABSTRACT This paper presents a formulation for the frictional impact analysis in constrained multibody mechanical systems. It is shown here that by considering the correct mode of impact, and by using Poisson's definition of coefficient of restitution the energy gains observed by the conventional analysis methods can be avoided.

1 Introduction

The inclusion of friction in the rigid body model of impact for a mechanical system adds to the complexity of an already complex problem. Previous studies shows that the conventional method may predicts energy gains when friction is taken into account (Reference [1]). Further investigation revealed that the use of Newton's hypothesis, for the definition of coefficient of restitution as well as the phenomenon of reverse sliding are responsible for these energy gains. Hence a correct model of frictional impact must take into account this reverse sliding, and must be based on the Poisson's instead of Newton's hypothesis. Such a model exists for the impact between two free and unconstrained bodies. This paper presents the extension of the methodology for an impact in constrained or jointed mechanical system.

The formulation presented here is developed using a canonical form of the systems' equation of motion in terms of joint coordinates and joint momenta. The joint coordinates reduces the equations of motion to a minimal set, and eliminate the complications arisen from the kinematic constraint equations. The canonical form of equations are solved for the change in joint momenta using Routh's graphical method and considering the correct mode of impact. The velocity jumps are then calculated balancing the accumulated momenta of the system during the impact process.

2 Velocity Jumps Due to Impact in Multibody Systems

The relative velocity vector $\mathbf{v}_r$ between the contact points P_i and P_j (Figure 1) can be written in terms of the vector of joint velocities θ as

$$\mathbf{v}_r = \mathbf{R}\dot{\theta} \tag{1}$$

[1]Department of Mechanical Engineering Wichita State University, Wichita KS 67208
[2]DIDMEC/IST, Technical University of Lisbon, 1096 Lisboa, Portugal

where $\mathbf{R}=\mathbf{D}_{ij}\mathbf{B}$. Here $\mathbf{B}$ is the velocity transformation matrix for the relation between joint velocities $\dot{\theta}$ into absolute velocities $\mathbf{v}$, and $\mathbf{D}_{ij}$ is an incidence matrix and it is a function of the coordinates of bodies 'i' and 'j' only. Using Equation 1 the velocity jumps in the normal ($\mathbf{n}$) and tangential ($\mathbf{t}$) directions (shown in Figure 1) as a result of impact can be written as

$$v_n^+ = v_n^- + \mathbf{c}_n^T \Delta\dot{\theta} \quad 2$$

$$v_t^+ = v_t^- + \mathbf{c}_t^T \Delta\dot{\theta} \quad 3$$

where $\mathbf{c}_n=\mathbf{R}^T\mathbf{n}$ and $\mathbf{c}_t=\mathbf{R}^T\mathbf{t}$. Here '-' and '+' signs denote the state immediately before and after the impact respectively. The canonical form of the equations of motion using joint coordinates is derived in Reference [2] as

$$\mathbf{p} = \mathbf{M}\dot{\theta} \quad 4$$

$$\dot{\mathbf{p}} = \mathbf{f} + \dot{\mathbf{M}}\dot{\theta} \quad 5$$

where $\mathbf{M}$ is the system generalized mass matrix, and $\mathbf{p}$ is a vector of system generalized momenta. Equation 4 can be used in Equation 2 and 3 to replace $\Delta\dot{\theta}$ with $\Delta\mathbf{p}$. Also if π_n and π_t are components of impulse associated with the change in momenta $\Delta\mathbf{p}$ in the normal and tangential directions respectively, then Equation 2 and 3 can be written as

$$v_n^+ = v_n^- + m_{nn}\pi_n + m_{nt}\pi_t \quad 6$$

$$v_t^+ = v_t^- + m_{nt}\pi_n + m_{tt}\pi_t \quad 7$$

where $m_{nn} = \mathbf{c}_n^T\mathbf{M}^{-1}\mathbf{c}_n$, $m_{nt} = \mathbf{c}_n^T\mathbf{M}^{-1}\mathbf{c}_t$, and $m_{tt} = \mathbf{c}_t^T\mathbf{M}^{-1}\mathbf{c}_t$. These three coefficients m_{nn}, m_{nt}, and m_{tt} can be evaluated knowing the inertia properties and initial configuration of the impacting bodies. It is clear from Equation 6 and 7 that once we know π_n and π_t the velocity jumps due to impact can be calculated.

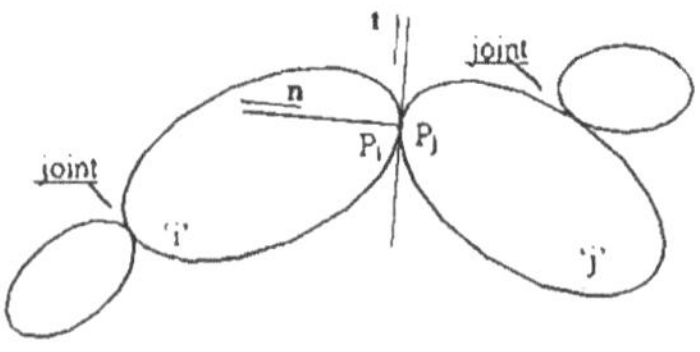

Figure 1. Impact of Multibody Systems

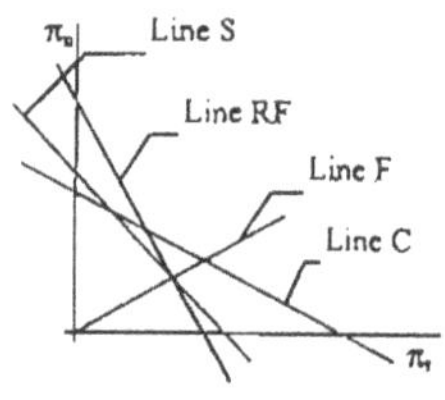

Figure 2. Impact Process diagram

3 Impulse Calculations using Routh's Method

The expressions for π_n and π_t are obtained by following the impact path on the impact process diagram (Figure 2). The technique is developed by Routh in 1891. The impact paths depends on the modes of impact, i.e., sliding, sticking, or reverse sliding.

3.1 Sliding, Sticking, and Reverse Sliding

If a sliding contact occur during the impact then the impact will follow Line F shown in Figure 2. The Line F is obtained from Coulomb's law as

$$\pi_t = i\mu_k \pi_n \qquad 8$$

where $i = v_t^- / |v_t^-|$. For the sticking contact, i.e., $v_t^+=0$, the impact path can be obtained by using Equation 7

$$v_t^- + m_{nt}\pi_n + m_{tt}\pi_t = 0 \qquad 9$$

Equation 9 is plotted as Line of sticking or Line S in Figure 2. In reverse sliding the tangential velocity changes its direction and so does the tangential impulse. The equation for reverse sliding path (Line RF in Figure 2) is obtained by reversing the slope of Line F and making it passing through the intersection of Line F and Line S.

$$\pi_t - \pi_{ts} = -i\mu_k(\pi_n - \pi_{ns}) \qquad 10$$

where π_{ts} and π_{ns} correspond to the intersection of Lines S and Line F.

3.2 Poisson's Hypothesis

With Poisson's definition of coefficient of restitution the total normal impulse at the end of the impact π_n can be expressed as

$$\pi_n = (1+e)\pi_n(t_c) \qquad 11$$

where e is the coefficient of restitution and $\pi_n(t_c)$ is the normal impulse at the end of the compression phase. The end of the compression phase give us another relation between π_n and π_t which is obtained by putting $v_n^+=0$ in Equation 6. The relationship is plotted as Line C in Figure 2.

$$v_n^- + m_{nn}\pi_n(t_c) + m_{nt}\pi_t(t_c) = 0 \qquad 12$$

Thus, once the correct mode of impact is determined, i.e., sliding, sticking, or reverse sliding, one can solve either of Equation 8, 9, or 10 with Equation 12 to obtain the values of $\pi_n(t_c)$ and $\pi_t(t_c)$ at the end of compression phase. After that Equation 11 can be used to calculate the total normal impulse π_n generated due to impact. The corresponding total tangential impulse π_t is obtained by substituting π_n in the equation of the line on which the impact ends. Once π_n and π_t are known velocity jumps can be calculated using Equation 6 and 7.

4 Impact Mode Prediction

The sliding and sticking modes depend on the satisfaction of Coulomb's law at the time of impact. Since differential impulse is a force, Coulomb's law can be expressed in terms of $d\pi_t$ and $d\pi_n$. By differentiating Equation 9, the ratio of $d\pi_t$

to $d\pi_n$ can be expressed in terms of the generalized coefficients m_{nt} and m_{tt}, and thus Coulomb's law become

$$|d\pi_t| / d\pi_n = |m_{nt}| / m_{tt} \leq \mu_s \qquad 13$$

Since m_{nt} and m_{tt} depends only on the pre-impact condition and the inertia properties, Coulomb's law and hence sliding or sticking can be checked before solving for π_n and π_t. Reverse sliding occurs when the impacting bodies initially have a non-zero relative tangential velocity, and when the differential tangential impulse can overcome the stiction force ($d\pi_t > \mu_s d\pi_n$) once this relative tangential velocity becomes zero. Again using Equation 13 reverse sliding can be predicted at the onset of impact.

5 Results & Conclusions

The developed formulation is used to solve two classical problems. The first problem, which is the impact of a falling rod with ground, is used to verify the developed formulation by comparing the results with the previous solutions. The solution of the second problem, which is the impact of a double pendulum (Figure 3) with ground, showed no energy gain for any value of e and μ (Figure 4). The impact of a closed loop multibody system is also analyzed without observing any energy gains. Thus it is concluded that by using Poisson's hypothesis, and by considering the correct mode of impact the energy gains from frictional impact analysis can be avoided.

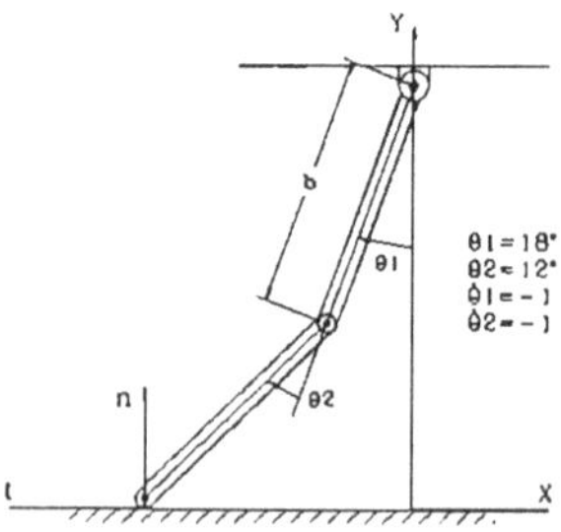

Figure 3. Double Pendulum

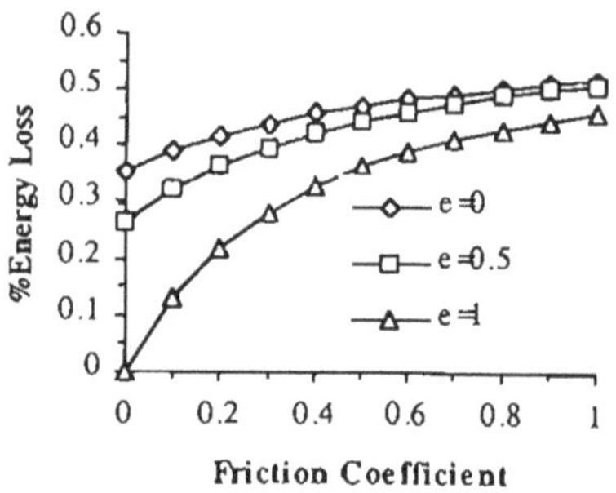

Figure 4. % Energy Loss

6 References

[1] Kane, T.R., 1984, "A Dynamics Puzzle," Stanford Mechanics Alumni Club Newsletter, pg.

[2] Nikravesh P.E., and Gim G., 1989, "Systematic Construction of the Equations of Motion for Multibody System Containing Closed Kinematic Loops," Paper No. 89-DAC-58, Montreal, Canada, Sept. 17-20

An Equation of Volterra in the Impact of Two Rigid Bodies

R. Souchet[1]

1 - INTRODUCTION

Impact with friction is a very old problem in Rigid Body Dynamics, Phillips (1849), Darboux (1880). Recently several authors have revisited the basic equations of this theory. The main hypotheses are (1) the validity of motion laws and friction laws throughout a fictitious duration (τ^-, τ^+) of the impact process, (2) the introduction of a coefficient of restitution e $(0 \le e \le 1)$ in order to model the internal (normal) dissipation during collision.

Due to assumptions (1), the unknown variables, viz impulses and relative velocities, are obtained from initial conditions and geometrical data ; that is the more difficult problem. Assumptions (2) provides only the final time of impact τ^+, once the first problem is solved. Here we will consider the first problem with a non-null initial slip velocity, recalling that a zero initial slip velocity leads to a « linear solution » of the problem, Darboux (1880).

Some notations are introduced concerning the collision of a rigid body (B) against a fixed plane : friction coefficient f, inverse mass matrix a_{ij} at the contact point, slip velocity (u, v), tangential impulse (X, Y), normal impulse Z, time-like independant variable $\{s = f Z\}$.

2 - STATEMENT OF THE PROBLEM

The problem is an initial value problem for the following system of differential equations $(\dot{X} = dX / ds)$

$$u = u_0 + a_{11} X + a_{12} Y + a_{13}^* s \quad , \quad a_{13}^* = a_{13} / f$$
$$v = v_0 + a_{12} X + a_{22} Y + a_{23}^* s \quad , \quad a_{23}^* = a_{23} / f \qquad (1)$$

$$\dot{X} = -u (u^2 + v^2)^{-1/2} \quad , \dot{Y} = -v \; (u^2 + v^2)^{-1/2} \qquad (2)$$

$$u(0) = u_0 > 0 \quad , \quad v(0) = v_0 \quad , \quad X(0) = 0 \quad , \quad Y(0) = 0$$

(1) Laboratoire de Modélisation Mécanique et de Mathématiques Appliquées, BP 179 - 86960 FUTUROSCOPE - France

The slip direction $\{x = v/u = \dot{Y}/\dot{X}\}$ plays a central role. From (2) and $\{u_0 > 0\}$, it results

$$\dot{X} = -(1+x^2)^{-1/2} \quad , \quad \dot{Y} = -x(1+x^2)^{-1/2} \tag{3}$$

The variation of x(s) follows from

$$\frac{dx}{ds} = (u^2+v^2)^{-1/2}\, g(x)$$
$$g(x) = a_{12}(x^2-1) + (a_{11}-a_{22})x + (\dot{a}_{23} - \dot{a}_{13}x)(1+x^2)^{1/2} \tag{4}$$

If the initial value x_0 is a root of the quartic equation $\{g(x)=0\}$, then it is seen that a motion with $\{x = x_0\}$ is a possible event. It is known that $\{g(x)=0\}$ has two or four real solutions, Darboux (1880). In the sequel we suppose that x_0 is not a root of g(x), by example that $\{g(x_0)>0\}$. So $\{x = v/u\}$ is an increasing function from x_0 until some value x_1 (x_1 is a root of g(x) eventually). It must be noted that, if body B is a sphere, $\{a_{11} = a_{22}\ ;\ a_{12} = 0\ ;\ a_{13} = a_{23} = 0\}$, then $\{g(x)=0\}$; so we have $\{x = x_0\}$ and solution is not relevant of the actual procedure.

3 - VOLTERRA INTEGRAL EQUATION

Since x is an increasing function for $\{s \geq 0\}$, it is convenient to choose x as the independant variable. From $\{x = \dot{Y}/\dot{X}\}$, $\{X(0)=0\ ,\ Y(0)=0\}$, it results

$$X(x) = \int_{x_0}^{x} X'(t)\,dt \quad , \quad Y(x) = \int_{x_0}^{x} t\,X'(t)\,dt \tag{5-1}$$

(notation : $X' = dX/dx$)

Then from (2)

$$\dot{X}^2 + \dot{Y}^2 = 1 \quad , \qquad \text{i.e.} \quad ds^2 = (1+x^2)\,dX^2 \quad ;$$

so we have

$$\dot{X} = \frac{dX}{ds} = -(1+x^2)^{-1/2}$$

since the sign of X is the sign of (-u), i.e. $(-u_0)$.

So
$$s(x) = -\int_{x_0}^{x} (1+t^2)^{1/2}\, X'(t)\, dt \tag{5-2}$$

Formulae (1) give u and v, as functions of x, by means of linear functionals of X (cf 5-1, 5-2) :

$$u(x) = u_0 + \int_{x_0}^{x} \left\{ a_{11} + a_{12}\, t - a_{13}^{*} (1+t^2)^{1/2} \right\} X'(t)\, dt$$

$$v(x) = v_0 + \int_{x_0}^{x} \left\{ a_{12} + a_{22}\, t - a_{23}^{*} (1+t^2)^{1/2} \right\} X'(t)\, dt \tag{6}$$

Finally the constraint $\{v - x\, u = 0\}$ may be written

$$\int_{x_0}^{x} G(x,t)\; X'(t)\, dt = v_0 - u_0\, x \tag{7}$$

$$G(x,t) = a_{12}\,(x\,t - 1) + (a_{11}\, x - a_{22}\, t) + (a_{23}^{*} - a_{13}^{*}\, x)\,(1+t^2)^{1/2}$$

(7) is a linear integral equation of Volterra of first kind. Performing an integration by parts, (7) is equivalent to the following linear integral equation of Volterra of second kind, Delves (1985),

$$g(x)\, X(x) = \int_{x_0}^{x} K(x,t)\; X(t)\, dt + (v_0 - u_0\, x)\;,\quad g(x) = G(x,x)\;,$$

$$K(x,t) = (a_{12}\, x - a_{22}) + (a_{23}^{*} - a_{13}^{*}\, x)\, t\, (1+t^2)^{-1/2} \tag{8}$$

Such a formulation of impact problems was never proposed in the past. The linearity of (8) is due to the choice of the independant variable $x = v/u$ in place of $s = f\, z$. In return, a Volterra equation seems natural concerning an initial value problem for differential equations.

4 - APPLICATIONS

A numerical solution procedure is generally used in order to solve (8). We refered to Delves (1985) that proposes various quadrature methods : the integral term in (8) is approximated by means of a quadrature rule of weights w_{ij}, the fixed steplength being h :

$$\int_{x_0}^{x_i} K(x_i,t)\; X(t)\, dt = h \sum_{j=0}^{i} w_{ij}\; K(x_i, x_j)\, X(x_j)\;,\quad x_i = x_0 + i\, h$$

Here we consider a solid of revolution,

$$a_{12} = 0 \quad , \quad a_{13} = 0 \quad , \quad a_{23} \neq 0 \, ,$$

and we choose $\{u_0 > 0 \, , \; v_0 > 0\}$. Then

$$g(x) = (a_{11} - a_{22})x + \overset{*}{a}_{23} (1 + x^2)^{1/2}$$

$$K(x,t) = -a_{22} + \overset{*}{a}_{23} \, t (1 + t^2)^{-1/2}$$

Since kernel K (x, t) does not depend on x, solution X (x) is given by

$$\mathrm{Log}\left|u_0 + a_{11} \, X(x)\right| = h_0 + \frac{p}{p^2 - q^2} \mathrm{Log}\left(x + \sqrt{1 + x^2}\right)$$

$$- \frac{q}{p^2 - q^2} \mathrm{Log}\left(qx + p\sqrt{1 + x^2}\right)$$

$$p = -\overset{*}{a}_{23} / a_{11} \; , \; q = (a_{22} / a_{11}) - 1 \quad , \quad h_0 \; : \; \text{initial constant}$$

So an explicit solution is available, Souchet (1996). It must be noted that this last example concerns the impact of a spherical pendulum against a plane, Stronge (1994).

5 - REFERENCES

BATTLE J.A., 1996, The sliding velocity flow of rough collisions in multibody systems, J. Appl. Mech., Vol.63, 804-809.

DARBOUX G., 1880, Etude géométrique sur les percussions et le choc des corps, Bulletin des Sciences Math., 126-160.

DELVES L.M. and MOHAMED J.L., 1985, Computational methods for integral equations, Cambridge University Press.

PHILLIPS E., 1849, Sur les changements instantanés de vitesse qui ont lieu dans un système de points matériels, J. de Math. Pures et Appliquées, tome XIV, 300-336.

SOUCHET R., 1996, An illustrative impulse pattern in rigid body collisions, Euromech 351, Systems with Coulom friction, Vadstena, Sweden, August 5-7.

STRONGE W.J., 1994, Swerve during three-dimensional impact of rough rigid bodies, J. Appl. Mech., vol.61, 605-611.

A Variational Principle for Equation of Hertz's Theory of Impact

T. M. Atanackovic[1]
A. Guran[2]
D. T. Spasic[1]

ABSTRACT The differential equation of classical Hertz's impact is written in the form suitable for variational formulation. By using special type of Ritz procedure and approximate solution is obtained from which duration of impact is estimated. Also a generalized problem is studied where the global deformability of the sphere is taken into account according to the method proposed by Villaggio.

1 The classical impact model

The differential equation describing classical Hertz's impact of two bodies of revolution is [1]

$$\frac{d}{dt}\left(m_1 \frac{m_2}{m_1+m_2}\frac{d\alpha}{dt}\right) = -P \tag{1}$$

where m_1 and m_2 are masses of the impacting bodies, α is the relative displacement of the centers of mass of the two bodies and P is the force between the bodies. By using the assumption the static solution is applicable, the force P may be written as

$$P = k_2\alpha^{3/2} \tag{2}$$

The initial conditions corresponding to the system (1), (2) are

$$\alpha(t=0) = 0; \qquad \frac{d\alpha(t=0)}{dt} = v_0. \tag{3}$$

In (2) k_2 is a constant. By solving (2) for α we obtain

[1]University of Novi Sad, Yugoslavia
[2]Institute for Structronics, Canada

$$\alpha = kP^{2/3}, \tag{4}$$

where $k = (1/k_2)^{2/3}$. Let $M = m_1 m_2/(m_1 + m_2)$. By introducing a dimensionless time τ and dimensionless displacement u by the relations (see [2] for a motivation)

$$t = \tau \frac{M^{2/5} k^{3/5}}{(v_0)^{1/5}}; \qquad u = \frac{\alpha}{k^{3/5}(M v_0^2)^{2/5}}, \tag{5}$$

the system (1), (2) becomes

$$\frac{d^2u}{d\tau^2} + u^{3/2} = 0; \qquad u(\tau = 0) = 0; \quad \frac{du(\tau = 0)}{d\tau} = 1. \tag{6}$$

The initial value problem (6) we solved numerically and determined the point $\tau = \tau^*$ such that $u(\tau^*) = 0$. This corresponds to (dimensionless) duration of the impact. The results of our calculations show that $\tau^* = 3.21805$. We note that the energy integral of (6) reads

$$E = \frac{1}{2}\left(\frac{du}{d\tau}\right)^2 + \frac{2}{5}u^{5/2} = \frac{1}{2} \tag{7}$$

From (7) the maximal dimensionless approach of the spheres is obtained as

$$u_{\max} = \left(\frac{5}{4}\right)^{2/5} = 1.093362074. \tag{8}$$

To obtain an analytical approximation of the duration of impact we use the least squares type of variational principle for (6). Namely, consider the functional

$$I = \int_0^{\tau^*} \left[\frac{d^2u}{d\tau^2} + u^{3/2}\right]^2 d\tau \tag{9}$$

It is easy to see that the generalized variation of (9), taking into account the fact that τ^* is unknown, reads

$$\Delta I = \int_0^{\tau^*} \left[\frac{d^2u}{d\tau^2} + u^{3/2}\right] \delta\left\{\left[\frac{d^2u}{d\tau^2} + u^{3/2}\right]\right\} d\tau + \left[\frac{d^2u}{d\tau^2} + u^{3/2}\right]_{\tau=\tau^*} \Delta\tau, \tag{10}$$

where $\delta(\cdot)$ denotes the classical Lagrange variation operator. Thus $\Delta I = 0$ on the solution of (6). We assume an approximate solution of (6) in the form

$$u = \frac{(1 + a)}{b} \sin b\tau - \frac{a}{3b} \sin 3b\tau, \tag{11}$$

where a and b are constants to be determined. The function u given by (11) satisfies $(6)_{2,3}$. Note this function has a maximum at $\tau = \tau^*/2 = \pi/(2b)$. We choose constant a in such a way that (8) holds. Thus,

$$a = \frac{3}{4}\left[\left(\frac{5}{4}\right)^{2/5} b - 1\right]. \tag{12}$$

By substituting (11), (12) into (9) and by following the classical Ritz procedure, we obtain $b = 0.9753306692$ so that $\tau^*/2 = 1.610526949$. The value τ^* obtained from the Ritz solution agrees well (the error is less than 2×10^{-3}) with the value obtained by numerical solution. The solution (11) thus becomes

$$u = 1.076 \sin 0.975\tau - 0.017 \sin 2.926\tau. \tag{13}$$

We show the solution (13) in the next Figure.

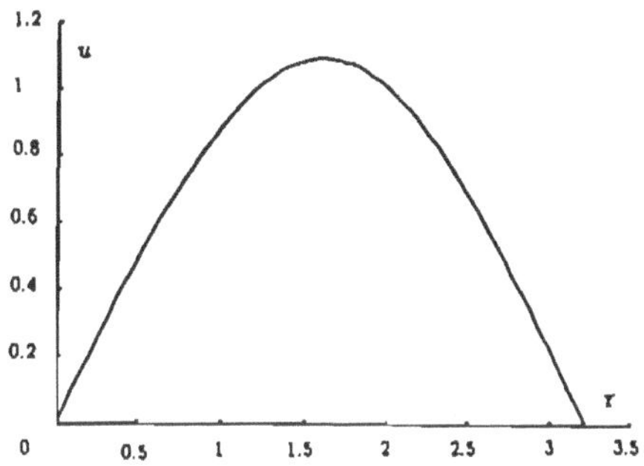

The difference between (13) and the numerical solution of (6) is less then 5×10^{-3}. Also the coefficient of restitution (should be equal to 1) in the case of solution (13) is 0.9974.

2 The generalized model

Recently Villaggio in [3] proposed a generalization of the classical Hertz's impact model by using the solution of Bondareva [4] for a contact problem of a heavy sphere and a rigid plane. Thus, instead of (6), for the case of the impact between an elastic sphere and rigid plane Villaggio considers (in our notation) the problem

$$\frac{d^2u}{d\tau^2} + \psi(u)u^{3/2} = 0; \qquad u(\tau = 0) = 0; \qquad \frac{du(\tau = 0)}{d\tau} = 1, \tag{14}$$

where

$$\psi(u) = \left[1 - \tfrac{2}{3}\tfrac{(1-2\sigma)^2}{(1-\sigma)}\tfrac{1}{\pi}(\beta u)^{1/2}\left\{\ln 2 - \tfrac{2}{3} + \ln(\beta u)^{1/2}\right\} + \tfrac{1}{2}\tfrac{(1-2\sigma)}{(1-\sigma^2)}\tfrac{1}{\pi}(\beta u)^{1/2}\right]^{-1}. \tag{15}$$

In (15) we use σ to denote the Poisson's ratio and

$$k = \left(\frac{3}{4E^* R_0^{1/2}}\right)^{2/3}; \qquad E^* = \frac{E}{1-\sigma^2}; \qquad \beta = \frac{\alpha}{R_0} = \frac{k^{3/5}\left(Mv_0^2\right)^{2/5}}{R_0}. \tag{16}$$

In (16) we used E to denote modulus of elasticity and R_0 the radius of the impacting sphere. Note that β is the ratio of the relative displacement of the centers of mass of the sphere with respect to plane ant its radius. Note that for incompressible materials (i.e., $\sigma = 1/2$) the problem (14) reduces to (6). We solved (14) for $\sigma = 1/3$ and for few values of β. The results are shown in the next Table

β	0	0.00025	0.1	0.2
$u_{\max}$	1.0934	1.0950	1.1129	1.1189

As could be seen from the table by taking into account deformation of the sphere outside the contact region by the method proposed in [3] the maximal deformation increases with increasing β.

3 References

[1] A. E. H. Love, *A Treatise on the Mathematical Theory of Elasticity*, Dover, New York, 1944.

[2] N. A. Klichevski, *Dynamical contact of solid bodies. Impact.* Kiev, Naukova Dumka, 1976.

[3] P. Villaggio, The Rebound of an Elastic Sphere Against a Rigid Wall. *Journal of Appl. Mechanics (transactions of ASME)*, 259, **63**, 1996.

[4] V. F. Bondareva, Contact problem for an elastic sphere. *P.M.M.* 37, **35**, 1970.

Analytical Mechanics with Multivalued Constraints and Elastic Impacts – The Convex Case

P.D. Panagiotopoulos[1]

1 Introduction

Multivalued constraints (e.g. inequality constraints, friction constraints etc.) have not been studied until now in Analytical Mechanics with the same intensity and rigour as they have been studied in Mechanics of Deformable Bodies and in Continuum Physics. Concerning this last subject we refer to the book by Duvaut and Lions [1] and to the books by Panagiotopoulos [2],[3] where large classes of continua subjected to this type of constraints have been studied, and to the references given there. It is today well-known that multivalued constraints lead in the case of convexity (resp. of nonconvexity) of the energy to variational inequalities (resp. to hemivariational inequalities) which are variational expressions for multivalued differential equations. This area has been called by the author ([1] p.374) Nonsmooth Mechanics (see also [4],[5]). Note that the case of monotone (resp. nonmonotone) constraints is a special case of monotone (resp. nonmonotone) superpotentials related to the notion of subdifferential (resp. generalized gradient) of Nonsmooth Analysis.

Shocks, or more generally impact effects, can be taken into consideration in the equations of Mechanics by two methods:
i) The method proposed by Moreau [4] and extensively developed by M. Marques (cf. [6] and the references given there) which is based on the formulation of the differential inclusions, or variational inequalities, governing the behaviour of the mechanical system under consideration. Then the search for the solution takes place in appropriate measure spaces permitting

[1]Department of Civil Engineering, Aristotle University, GB-54006 Thessaloniki, Greece; Faculty of Mathematics and Physics, RWTH, D-52056 Aachen, Germany

the velocity discontinuity.

ii) In [7][8] P. D. Panagiotopoulos has introduced in continuum problems a new class of variational expressions including the impact effect. These effects introduce two terms in the variational expressions, the first including the variation of the impact time and the second the variation of the unit tangential vector on the obstacle at the point of impact. It should be noted that the variation of time has already been considered by Kozlov and Treshchëv, for a first time, together with the variation of the unit tangential vector, in the theory of Billiards [9], but only partially. The variational principle, must hold in inequality form, (i.e. $\delta I \geq 0$), and not in equality form, due to the inequality constraints [10].

2 Arbitrary Variations of the Displacements

We consider a discrete system of n material particles m_i with $3n$ degrees of freedom in $\mathbb{R}^3$ if the system is unconstrained. The system is moving in a convex closed subset $C \subset \mathbb{R}^{3n}$ and is subjected to the convex superpotential Φ. Let us assume that

$$C = \{x | x \in \mathbb{R}^{3n} \ f_j(x) \leq 0 \ j = 1, ..., r\}, \tag{1}$$

where f_j are continuously differentiable functions on $\mathbb{R}^{3n}$ and $r < 3n$. Thus the system has at least $f = 3n - r$ degrees of freedom. We assume, moreover, that the contact of each material particle with the boundary of C is without frictional forces. The kinetic energy of the system is (no summation with respect to i in $m_i \dot{x}_i^2$ and in all analogous cases)

$$T = \frac{1}{2} \sum_{i=1}^{3n} m_i \dot{x}_i^2 \tag{2}$$

where $m_i = m_{i+1} = m_{i+2}$ denotes the mass corresponding to the i-th particle and x_i its velocity in the i-th direction. On the system act the forces F_i, $i = 1, 2, ..., 3n$ and the reactions R_i and R_i'. We assume that F_i are continuous functions of t. Thus we may write the relation

$$R_i' + R_i + F_i = m_i \ddot{x}_i \quad i = 1, ..., 3n \ , \tag{3}$$

where

$$-\{R_i\} = \sum_{j=1}^{r} \lambda_j \operatorname{grad} f_j, \ \ \lambda_j \geq 0, \ f_j \leq 0, \ \lambda_j f_j = 0, \tag{4}$$

$$\text{and} - \{R_i'\} \in \partial\Phi(x).$$

Here $\{R_i\}$ denotes the vector with components R_i and Φ is a convex, l.s.c., proper functional. Henceforth the λ_j denote the various Lagrange multipliers appearing in this paper. Then fullfil some inequalities due to the inequality conditions in (4). Note that R_i is a classical reaction force, if no impact occurs, otherwise R_i is an impulsive force.

Equivalently ([2] p.43) relation (4) is written as

$$-\{R_i\} \in \partial I_C(x) = N_C(x), \quad -\{R'_i\} \in \partial\Phi(x). \tag{5}$$

Here $N_C(x)$ denotes the normal cone (in the sense of convex analysis) to C at x and $\partial I_C(x)$ denotes the subdifferential of the indicator I_C of C. We recall that $I_C(x) = \{0 \text{ if } x \in C,\ \infty \text{ if } x \notin C\}$. Thus the equation of motion becomes

$$\{F_i - m_i\ddot{x}_i\} \in \partial I_C(x) + \partial\Phi(x) \tag{6}$$

which by means of the definition of the subdifferential is equivalent to the variational inequality (summation convention)

$$x \in \mathbb{R}^{3n}, \quad I_C(x^*) - I_C(x) + \Phi(x^*) - \Phi(x) \geq \sum_{i=1}^{3n}(F_i - m_i\ddot{x}_i)(x_i^* - x_i) \tag{7}$$

$$\text{for all } x^* \in \mathbb{R}^{3n}\ ,$$

or to the inequality: Find $x \in C$ such as to satisfy the relation

$$\sum_{i=1}^{3n}(m_i\ddot{x}_i(x_i^* - x_i) + \Phi(x^*) - \Phi(x) - F_i(x_i^* - x_i)) \geq 0 \tag{8}$$

$$\text{for all } x^* = \{x_i^*\} \in C.$$

This relation holds for every time t (with the exception of the impact time $\tilde{t}$) and we assume that x and x^* are functions of t. This inequality expresses the d'Alembert "principle" in inequality form. The inequality arises from the fact that certain variations $x_i^* - x_i$ are irreversible or unilateral, i.e. if $x_i^* - x_i$ is admissible $-x_i^* + x_i$ will not be admissible, because $x_i^*, x_i \in C$. Note in (8) that $\ddot{x}_i$ is not defined at the time $\tilde{t}$ of impact. Only $\dot{x}_i^+$ and $\dot{x}_i^-$ are defined at this time and $\ddot{x}_i$ is a measure. Thus, in order to consider the impact at $t = \tilde{t}$, one should replace in (8) $m_i\ddot{x}_i$ by $d(m_i\dot{x}_i)$ (cf. in [4] the Chapter 1 by J. J. Moreau). We assume further that, for $t \neq \tilde{t}$, $\dot{x}$ and $\ddot{x}$ are continuous functions and for simplicity we assume that only for $t = \tilde{t}$ impact occurs. The following theory holds also if impact occurs at $t = \tilde{t}$, $t = \tilde{t}_1$, $t = \tilde{t}_2$ etc. Then obvious modifications are needed in the following expressions.

Thus we will look for solutions of (8) such that $x \in C^0(0T, \mathbb{R}^{3n})$, $x(t) \in C$, $\dot{x} \in BV(0T, \mathbb{R}^{3n})$ and $\ddot{x} \in M(0T, \mathbb{R}^{3n})$, where C^0, BV, M denote respectively the spaces of continuous functions, of the bounded variation functions and of the Radon-measures. Here $(0T)$ denotes the time interval. Now we have ([10]) that $(x^\star \in C^0(0T, \mathbf{R}^{3n}))$

$$d[\dot{x}_i(x_i^\star - x_i)] = (x_i^\star - x_i)d\dot{x}_i + \frac{1}{2}(\dot{x}_i^+ + \dot{x}_i^-)d(x_i^\star - x_i) \tag{9}$$

because $t \to \dot{x}(t)$ is a BV-function.

Combining (8) with (9) and integrating from t_1 to t_2, (Lebesgue-Stieltjes integral), where $t_1, t_2 \in (0,T)$ and $(0,T)$ includes the impact time $\bar{t}$, we obtain (assuming that $\left|\int_{t_1}^{t_2} F_i(x_i^\star - x_i)dt\right| < \infty$) that (we omit $\sum_{i=1}^{3n}$ and we consider further the summation convention with respect to a repeated index (exception the products by the mass i.e. $m_i\ddot{x}_i$, $m_i\dot{x}_i$ etc; further we put $\sum_{i=1}^{3n}$ only in order to avoid a misinterpretation of the formulas)

$$\int_{t_1}^{t_2} d[m_i\dot{x}_i(x_i^\star - x_i)] - \int_{t_1}^{t_2} \frac{1}{2}m_i(\dot{x}_i^+ + \dot{x}_i^-)d(x_i^\star - x_i)dt - \\ - \int_{t_1}^{t_2} \{F_i(x_i^\star - x_i) + \Phi(x^\star) - \Phi(x)\}dt \geq 0 \qquad \forall\, x^\star \in C. \tag{10}$$

Since $d[m_i\dot{x}_i(x_i^\star - x_i)]$ is a measure at $t = \bar{t}$ we obtain

$$[m_i\dot{x}_i(x_i^\star - x_i)]\Big|_{t_1}^{\bar{t}^-} + [m_i\dot{x}_i(x_i^\star - x_i)]\Big|_{\bar{t}^+}^{t_2} + \{m_i\dot{x}_i^+(x_i^{+\star} - x_i^+) - m_i\dot{x}_i^-(x_i^{-\star} - x_i^-)\} - \\ - \int_{t_1}^{t_2} \frac{1}{2}m_i(\dot{x}_i^+ + \dot{x}_i^-)d(x_i^\star - x_i) - \int_{t_1}^{t_2} \{F_i(x_i^\star - x_i) + \Phi(x^\star) - \Phi(x)\}dt \geq 0 \\ \text{for all } x^\star \in C. \tag{11}$$

This inequality leads to the following two equivalent formulations of the problem, where T denotes the kinetic energy (cf [10]).

Problem 1. Find $x \in C^0(t_1t_2, \mathbf{R}^{3n})$, $t \to x(t) \in C$, $\dot{x} \in BV(t_1t_2, \mathbf{R}^{3n})$, $\ddot{x} \in M(t_1t_2, \mathbf{R}^{3n})$ such that

$$m_i\dot{x}_i(x_i^\star - x_i)(t_2) - m_i\dot{x}_i(x_i^\star - x_i)(t_1) - \\ - \int_{t_1}^{\bar{t}^-} \mathrm{grad}T(\dot{x})_i(\dot{x}_i^\star - \dot{x}_i)dt - \int_{\bar{t}^+}^{t_2} \mathrm{grad}T(\dot{x})_i(\dot{x}_i^\star - \dot{x}_i)dt - \\ - \int_{t_1}^{t_2} \{F_i(x_i^\star - x_i) + \Phi(x^\star) - \Phi(x)\}dt \geq 0 \text{ for all } x^\star(t) \in C \tag{12}$$

and

Problem 2. Find $x \in C^0(t_1 t_2, \mathbf{R}^{3n}), t \to x(t) \in C$, $\dot{x} \in BV(t_1 t_2, \mathbf{R}^{3n})$, $\ddot{x} \in M(t_1 t_2, \mathbf{R}^{3n})$ such that

$$\int_{[t_1,\bar{t}-)\cup(\bar{t}+,t_2]} m_i \ddot{x}_i (x_i^\star - x_i)dt + \{m_i \dot{x}_i^+ (x_i^{+\star} - x_i^+) - m_i \dot{x}_i^- (x_i^{-\star} - x_i^-)\}_{t=\bar{t}} -$$

$$- \int_{t_1}^{t_2} \{F_i(x_i^\star - x_i) + \Phi(x^\star) - \Phi(x)\}dt \geq 0 \qquad \text{for all } x^\star(t) \in C. \tag{13}$$

Until now we have not introduced any specific impact law in our problem. We have merely derived two variational expressions holding if the velocity is a BV-function, with a discontinuity at $t = \bar{t}$. We may state the following theorem.

Theorem 1. If the velocity $\dot{x}$ of a material system, such that $x \in C$, is a BV-function, then on an interval (t_1, t_2), including the time $\bar{t}$ of the discontinuity of the velocity, relation (12) holds. Similarly relation (13) holds.

Note that the procedure of the proof of the above theorem can be inversed and that the following theorem holds:

Theorem 2. Relation (12) describes completely the dynamics of the inequality constrained system, for $t \in (t_1, t_2) \subset [0, T]$. The same holds for relation (13).

Proof. (Analogous to the proof in [10])
Further let us assume that the variations $x^\star(t) - x(t) = \delta x$, where δx is small enough in the sense of the classical calculus of variations. Moreover let

$$\delta x = 0 \ \text{ at } \ t = t_1, \ t = t_2. \tag{14}$$

We use further the notation

$$p_i = m_i \dot{x}_i$$

and we assume that

$$-F_i = \frac{\partial V}{\partial x_i}, \quad \text{and let } \tilde{\delta}\Phi = \Phi(x + \delta x) - \Phi(x)$$

i.e. that the applied forces are derived by a potential $x \to V(x)$. Thus we may write instead of (12) the relation: Find $x \in C^0(t_1 t_2, \mathbf{R}^{3n})$, $x(t) \in C$, $\dot{x} \in BV(t_1 t_2, \mathbf{R}^{3n})$ such that

$$\int_{[t_1,\bar{t}-)\cup(\bar{t}+,t_2]} -\delta T(\dot{x})dt + \int_{t_1}^{t_2} \{\delta V(x) - \tilde{\delta}\Phi\}dt \geq 0 \tag{15}$$

$$\text{for all } \delta x \text{ such that } x + \delta x \in C$$

and instead of (13) the relation: Find $x \in C^0(t_1 t_2, \mathbb{R}^{3n}), t \to x(t) \in C$, $\dot{x} \in BV(t_1 t_2, \mathbb{R}^{3n})$, $\ddot{x} \in M(t_1 t_2, \mathbb{R}^{3n})$ such that

$$\int_{[t_1, \bar{t}-) \cup (\bar{t}+, t_2]} m_i \ddot{x}_i \delta x_i dt + \{p_i^+ \delta x_i^+ - p_i^- \delta x_i^-\}_{t=\bar{t}} + \int_{t_1}^{t_2} (\delta V - \bar{\delta}\Phi) dt \geq 0$$

$$\text{for all } \delta x \text{ such that } x + \delta x \in C. \tag{16}$$

Note that the above expressions are independent of the impact law. We have assumed only for their derivation that $t \to \dot{x}(t)$ is a BV-function. Further we shall introduce the impact laws. Note that if the impact law implies that

$$\int_{\bar{t}-}^{\bar{t}+} \delta T dt = 0 \,, \tag{17}$$

then instead of (28) we can write the inequality

$$-\int_{t_1}^{t_2} \delta(T - V) + \bar{\delta}\Phi dt = -\int_{t_1}^{t_2} \{\delta L(x, \dot{x}) + \bar{\delta}\Phi\} dt \geq 0$$

$$\text{for all } \delta x \text{ such that } x + \delta x \in C. \tag{18}$$

Here $L = T - V$ is the expression of the action. Accordingly we have proved the following results.

Theorem 3. Suppose that a material particle system satisfies the equations of motion (3) and is constrained to move, subjected to a convex superpotential, in a convex closed set C having frictionless boundary.

i) Then the Hamilton-type variational inequalities (12) (with the two first terms equal to zero), or (13), are satisfied for arbitrary variations of any perturbed motion satisfying the conditions (14) at $t = t_1$ and $t = t_2$.

ii) Analogously for appropriately small variations $x_i^* - x_i = \delta x$ the Hamilton type variational(δ)-inequality (15), or (16), is satisfied for any perturbed motion satisfying the conditions (14) at $t = t_1$ and $t = t_2$. If, moreover, (17) holds, then the variational (δ)-inequality (18) is valid.

3 The Impact Law and Hamilton's "Principle" for Impacts

Up to now the impact law was not considered in its complete form in our system. Now we will introduce the impact equations for non-dissipative collisions in the sense of Newton. An extension for the case of dissipative collisions is possible, with minor modifications of the following procedure. The equations of motion (3) must be replaced at the time $\tilde{t}$ of impact by the equation of momentum (cf. [10] eq. (32)). We follow the same procedure as in [10] and we prove the following result.

Theorem 4. Let us consider a time interval (t_1, t_2) including one elastic impact on a regular point (i.e. not a corner point) at the boundary of C (eq. (1)). A trajectory $t \to x(t)$ which is assumed generally to be a continuous function $C^0(0T; \mathbb{R}^{3n})$ is a motion of the system, if and only if, it is a solution of the variational (δ) - inequality: Find $x \in C$ such that

$$-\delta \int_{t_1}^{t_2} L(x, \dot{x}) dt + \int_{t_1}^{t_2} [\Phi(x + \delta x) - \Phi(x)] dt \geq 0 \tag{19}$$

for all δx such that $x + \delta x \in C$.

In the above inequality we assume that the variation δx satisfies (14).

Proof. The proof is analogous to the proof of Theorem 4 in [10].

4 References

[1] Duvaut, G., Lions, J.L.: Un Probléme d' élasticité avec Frottement, *J. de Mécanique* **10** (1971) 409-420.

[2] Panagiotopoulos, P.D.: *Inequality Problems in Mechanics and Applications. Convex and Nonconvex Energy Functions*. Birkhäuser Verlag, Basel, Boston 1985. Russian Translation MIR Publ. Moscow 1989.

[3] Panagiotopoulos, P.D.: *Hemivariational Inequalities*. Springer Verlag Berlin, Heidelberg, 1993.

[4] Moreau, J.J., Panagiotopoulos, P.D., Strang, G.(eds): *Topics in Nonsmooth Mechanics*. Birkhäuser Verlag, Basel, Boston 1988.

[5] Moreau, J.J., Panagiotopoulos, P.D.(eds): Nonsmooth Mechanics and Applications, *CISM* Vol. **302**, Springer Verlag, Wien 1988

[6] Monteiro Marques, M.D.P.: *Differential Inclusions in Nonsmooth Mechanical Problems. Shocks and Dry Friction*. Birkhäuser Verlag, Basel, 1993.

[7] Panagiotopoulos, P.D.: Variational Principles for Contact Problems Including Impact Phenomena. In: Contact Mechanics (ed. by M. Raous et al.), *Plenum Press*, New York 1995.

[8] Panagiotopoulos, P.D.: Modelling of Nonsmooth Energy Problems. Dynamic Hemivariational Inequalities with Impact Effects. *J. Comp. Appl. Math.* **63** (123-138), 1995.

[9] Kozlov, V.V., Treshchëv, D.V.: A Genetic Introduction to the Dynamics of Systems with Impacts, *Tranl. of Math. Monographs* **89**, A.M.S., Providence, 1991.

[10] Panagiotopoulos, P.D., Glocker Ch.: Analytical Mechanics Addendum I: Inequality Constraints with Elastic Impacts. The Convex Case, *ZAMM* **78** (1998), 219-229.

SESSION III–(B)

Applications, Rotating Systems

Neck Dynamics and Injury Mechanisms Following Head Impact

B. S. Myers[1]
R. W. Nightingale[1]

ABSTRACT The clinical presentation of cervical fractures following head impact is often complex, particularly when multiple non-contiguous fractures are present. Based on the results of 22 human cadaver head-neck impact experiments, a biomechanical framework is developed in which these injuries may be better understood. This includes the effects of head rebound, head and neck decoupling, cervical spine buckling, cervical injury mechanisms, and the effects of padding. On the basis of these observations, we hypothesize that impact injury may be modeled as the dynamic response of two large masses coupled by a segmented curved beam-column comprised of seven small masses with interposed nonlinear flexibility elements.

Introduction

Catastrophic cervical spinal injury has remained among the most difficult and socially significant impact injury problems in structural biomechanics. While these injuries occur through a variety of mechanisms, head contact resulting in compression-bending neck loading remains among the most common mechanisms of injury. The volume of literature devoted to cervical spinal impact injury is large, and has been recently been reviewed in detail [Myers and Winkelstein, 1995]. Despite this collection of writing, considerable confusion remains as to the basic mechanisms which result in catastrophic cervical spinal injuries including: the effects of end condition; the relationships between head motion and injury mechanisms; the effects of head, neck, and torso inertia; the role of buckling in injury, and the effects of the initial orientations of the head, neck, and torso relative to the

* Department of Biomedical Engineering, Division of Orthopaedic Surgery, Duke University, Durham, NC, 27705

impact surface. Although previous studies have examined these variables and provided invaluable insights into the dynamic behavior of the cervical spine, few have had the statistical power to obtain significance.

Of equal importance is the prevention of injury. It is widely assumed that energy absorbing devices which reduce the severity and incidence of head injury also reduce the risk for neck injury. Despite this belief, reduction of neck injury through the use of padding has never been demonstrated in any experimental or epidemiological study. On the contrary, a number of studies have shown that padded surfaces have no effect on neck injury risk [Alem et al., 1984; Nusholtz et al., 1983], and others have presented evidence suggesting that they may increase the risk of neck injury [Hodgson and Thomas, 1980; Myers et al., 1991; Yoganandan et al., 1986]. Despite these arguments, there have been no studies which have examined the effects of padding statistically. Clearly, if we are to become more effective in preventing these injuries, a cogent understanding of how they occur must first be achieved. It is therefore, the purpose of this paper to examine catastrophic head impact neck injury using a human cadaver impact model to provide insight into cervical dynamics, and neck injury prevention.

Methods

An experimental apparatus was designed to model cervical spine injury resulting from impacts near the head vertex with the torso following. An effective torso mass of 16 kg was coupled to the first thoracic vertebra of 22 unembalmed human head cervical spine preparations. The specimen preparations were mounted to a vertical drop tower in an inverted position with the C7-T1 intervertebral disc oriented at 25° to horizontal to preserve the lordosis (curvature) of the cervical spine. Multiaxis transduction fully quantify the forces, moments, and accelerations of the head and neck during the impact event. Impact velocity was recorded using an optical sensor. The transducer data were sampled at 62.5 kHz and digitally filtered using an SAE J211b 1000 Hz filter. Photographic targets inserted on the vertebrae were imaged at 1000 Hz (Figure 1).

The impact surface angle was varied about the y-axis between -15° (posterior head impact) and +30° (anterior head impact). The anvil was covered with 3 mm of lubricated Teflon sheet to simulate impacts onto a rigid, frictionless surface (n=10). Impacts onto a padded surface were simulated by attaching foams to the anvil with duct tape using either an expanded polystyrene foam or a less stiff, open cell polyurethane foam (n=12).

Break-away sutures maintained the lordosis of the cervical spine and applied a preload equal to the weight of the head and neck. Each specimen was mechanically stabilized and dropped from 0.53 m. Conventional radiographs and dissection were used to document injuries.

The impulse of the compressive force at T1 was calculated for all the impacts. Differences in axial impulse between the padded and rigid tests were

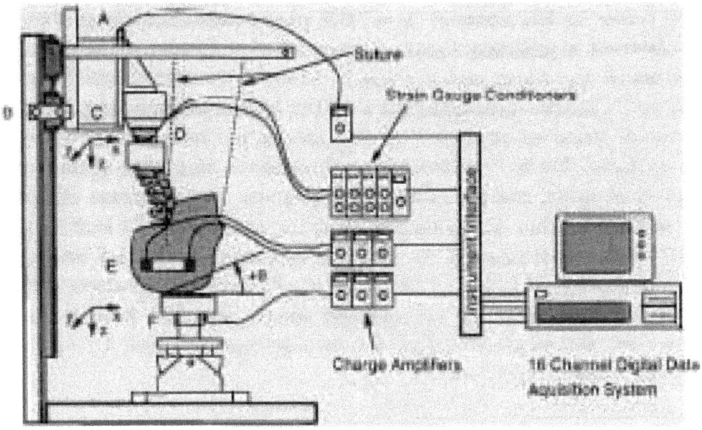

FIGURE 1. A diagram of the test apparatus showing the accelerometer on the torso mass (A), the optical velocity sensor (B), the carriage and torso mass (C), the six-axis load cell at T1 (D), the head accelerometers (E), and the anvil and three-axis load cell (F).

evaluated using two-way ANOVA. The effect of padding on the frequency of injury was examined using a χ^2 comparison of proportions for two independent samples with a continuity correction. The effects of padding on the peak head and neck forces were examined using t-tests.

Results

Using this test system, basicranial and cervical spine injuries were produced in 16 of 22 tests. Injuries included anterior disc tears, anterior longitudinal ligament ruptures, Jefferson fractures, Hangman's fractures, odontoid fractures, burst fractures, facet dislocations, posterior element fractures, and basilar skull fractures.

The impact dynamics of the head and neck are bimodal and reflect the vibrations of a two mass system with an interposed poorly coupled viscoelastic spring striking an impact surface (Figure 2). For the rigid impacts, Mode 1 is attributed almost entirely to stopping the head and had a duration of 4.3±1.6 milliseconds. During the first half of this head inertial loading mode, the head impact force reached a maximum with no concomitant neck force. For the padded impacts, the head contact times during Mode 1 were significantly increased. Therefore, Mode 1 contained loading by the torso in addition to the force required to stop the head. The inertia and compliance of the head mass caused rebound loading. Head rebound forces on the order of 10 to 35% of the peak head force were commonly observed, indicating that in impact, the neck must manage both the momentum of the torso and the head. Mode 1 durations for the padded impacts could not be calculated because the increased coupling of the head and cervical spine resulted in less separation between modes. For both impact surfaces, Mode 2 represents loading of the impactor surface by the head, cervical spine, and the effective torso mass. The duration of this neck impact surface loading mode for the rigid impact was 27.3±14.3 ms. In all the tests there was a delay in the onset of measured neck load with respect to the head load. This lag in response at T1 provides evidence that the head and cervical spine are not coupled during the first half of the head impact mode, and was 1.6±0.3 ms for the rigid impacts and 4.7±1.3 ms for the padded impacts ($p<0.001$) .

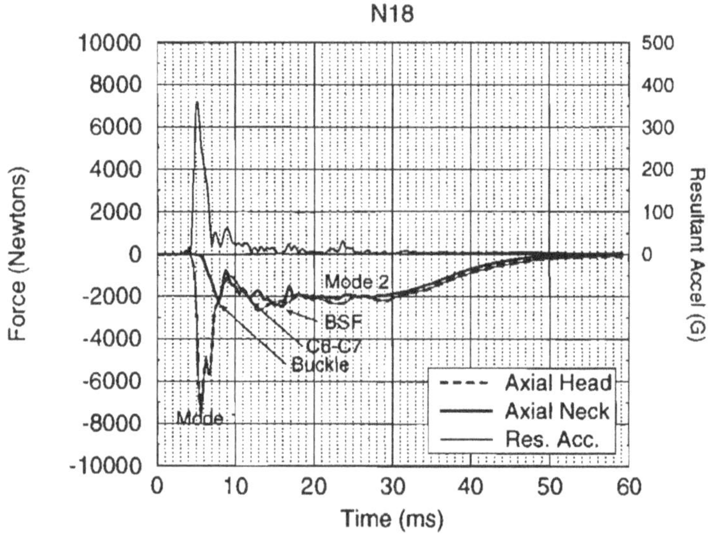

FIGURE 2. Magnitude of the head and neck axial forces, and the head center of gravity acceleration for an impact into a rigid surface oriented at +15°. A bimodal response consisting of a head inertial deceleration mode (Mode 1), and a neck impact surface loading mode (Mode 2) are seen. Kinematic analysis revealed a stable buckle followed by a cervical and a basicranial injury.

Head motion was not related to injury mechanism. Cervical spine injuries were produced within 9 ms following head contact with a rigid surface and in less than 20 ms in all but one of the padded surface impacts. In contrast, significant head motions, as characterized by rotations greater than 20°, did not occur until between 20 ms and 100 ms following head impact (larger times in the padded impacts and shorter times in the rigid impacts). The final head motions, the kind which would be observable on a 60 Hz recording, were directly related to the point of impact. Injuries produced prior to the head flexion motion included eight extension injuries, one compression injury, and one flexion injury. Injuries produced prior to the head extension motion included, nine extension injuries, four compression injuries and two flexion injuries. Of those impacts without significant head motions, compression, extension and flexion injuries were produced. Further, head motion was unable to explain the four cases of multiple noncontiguous injuries by with different injury mechanisms.

Buckling occurred 3 to 8 ms following head impact (Figure 2). The buckle was a rapid transition from a compression mode of deformation to a bending mode of deformation (Figure 3). Because the buckle was observed in uninjured specimens, and because the injured spines demonstrated post-bucking stability prior to injury, we concluded that the buckle did not result in material failure. Deforming as a fixed-pinned precurved member, the spine underwent a characteristic postbuckled configuration with midcervical extension and lowermost cervical flexion, while the upper cervical spine flexed, extended, or compressed. Unlike head motion, the post-buckled deformation and resultant force location were able to explain the mechanism of injury.

Padding was found to significantly increase the risk for injury and to increase the severity of injury. All nine of the padded impacts and four of the nine rigid impacts produced cervical spine injury. Analysis of variance showed that the frequency of injury in the padded impacts was significantly greater than the frequency of injury in the rigid impacts ($p=0.0375$). In the same context, only two of the rigid impacts were found to be clinically unstable, while seven of the nine padded impacts were unstable. For the rigid impacts, the axial impulses were 41.0 ± 14.2 N-s. For the padded impacts, the axial impulses were 60.2 ± 21.9 N-s. The impulses in the padded impacts were significantly larger than the impulses in the rigid impacts ($p=0.00023$) and also varied as a function of impact angle ($p<0.0001$) with the posterior impacts producing the smallest impulses and the anterior and vertex impacts producing larger impulses. Padding significantly lowered the risk for head injury. The average HIC for the padded impacts, 136 ± 32, was significantly lower than the average HIC for the rigid surface, 1010 ± 534. Average axial neck force to produce the first neck injury was -1948 ± 666 N and did not vary with impact surface.

Discussion

Catastrophic head impact neck injury remains a significant societal problem. These data illustrate the ease with which the neck may be injured, with only 40% of the 50th percentile torso mass following the neck at only

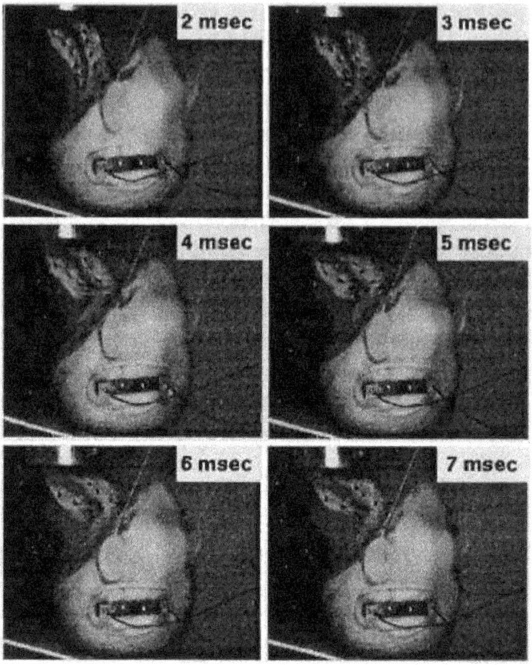

FIGURE 3. High speed digital images of a rigid impact at -15° illustrating buckling kinematics. The images show the normal lordotic curve (2 ms), followed by a transient higher order mode (3 ms) which decays quickly (4 ms) to a first order buckle (5 ms).

3.1 m/s. The lag between the neck load and the head load indicates that the torso and neck are poorly coupled despite the presence of a compressive preload. The presence of significant forces owing to head rebound from the impact surface demonstrates the importance of head inertia in the dynamics. The sensitivity of injury risk to the impact surface properties and the production of injury in the complete absence of constraints imposed by the impact surface, show that in addition to managing the momentum of the torso, the neck must also accelerate the head out of the path of the torso.

As in prior studies, head motion was unrelated to injury mechanism [Nusholtz et al., 1983]. This occurred because head motion lags the applied forces which produce injury. These experiments show that the spine acts as a slender, curved beam column with a fixed base and pinned apex. Once a critical load is reached, this structure snap-through buckles with a rapid transition from a predominantly compression mode of deformation to a predominantly bending mode of deformation. As in prior static studies [Myers et al., 1991], the cervical spine shows post-buckling stability, as it can be buckled without injury. Thus, the buckle is a structural failure and not a material failure (a neck injury). While buckling is not injury, the post-buckled mode shape does explain the mechanism of these injuries. The buckled deformation exhibits regions of extension in the middle cervical spine and flexion in the lower cervical spine. As a result, the neck load vector passes behind the middle cervical spine and causes compression-extension, and in front of the lower cervical spine and causes lower cervical compression-flexion.

While able to explain the injuries seen in this study, the first order buckle does not explain the production of midcervical (C3 to C6) flexion injuries. Yet clinical studies show that this is a common injury. Previous research has demonstrated that preflexion may be one mechanism for producing midcervical compression-flexion injuries [McElhaney et al., 1983]. Another, mechanism of midcervical flexion injuries may be the development of higher order buckling modes. In these experiments, the higher order mode shape, which produces midcervical flexion decayed prior to injury. Mechanistically, the development of a higher order mode is expressed because of the inertial forces developed in each of the small vertebral masses during the snap-through motion. It is possible that at higher impact velocities, in which the torso loads the neck more rapidly, injury may occur during this higher order mode, and result in midcervical compression-flexion injuries.

The effects of padding were examined using the extremes of boundary conditions. The pads used were more compliant than those paddings and liners that are currently in use by the automotive and other safety industries. Analysis of the injury data shows that there was a significantly greater frequency of injury in the padded impacts than in the rigid impacts. This shows that pads do not protect the neck and supports the hypothesis that padded surfaces may increase the risk of cervical spine injury. In the experiments where a padded surface was used, the deformed pad opposed head motion, increased the contact time of the head with the impact surface and thereby delayed head escape. As a result, the neck was subjected to significantly more of the torso momentum. Thus, pads force the neck to stop the moving torso, and sustain more injuries which were more severe. Importantly, the padding was able to significantly reduce the risk for head injury. Therefore, these experiments suggest that highly deformable padded contact surfaces should be employed carefully in environments where there is the risk for cervical spine injury. In addition, the results suggest that the orientation of the head, neck, and torso relative to the impact surface is of equal if not greater importance in neck injury risk.

Conclusions

1. Neck injuries are commonly produced at velocities of 3.1 m/s with an effective dynamic mass of only 40% of the total torso mass.
2. Head and neck dynamics include a bimodal response of a head impact mode, and a neck impact surface loading mode. Characteristic features

of this response include the poor coupling of the head and neck owing to an initial low neck stiffness, and head rebound.
3. Head motion is unrelated to the neck injury mechanism as neck injuries occur prior to significant head motions.
4. Cervical spine buckling is not an injury, however, the post-buckled mode shape is able to explain the distribution of neck injuries, and the relationship between head forces and neck injuries.
5. Constraint of head motion due to pocketing in a highly deformable padded surface may increase the risk of cervical spine injury; however, the orientations of the head, neck, and torso relative to the impact surface are of equal if not greater importance. These pads significantly reduce peak head forces and therefore reduce head injury risk.

References

[1] Alem, N. M., Nusholtz, G. S., Melvin, J. W., Head and Neck Response to Axial Impacts, *Proceedings of the 28th Stapp Car Crash Conference*, 1984, 927-940.

[2] Hodgson, V. R., Thomas, L. M., Mechanisms of Cervical Spine Injury During Impact to the Protected Head, *Proceedings of the 24th Stapp Car Crash Conference*, 1980, 17-42.

[3] McElhaney, J. H., et al., Biomechanical Analysis of Swimming Pool Neck Injuries, *Proceedings of the 23rd Stapp Car Crash Conference*, 1979, 494-500.

[4] Myers, B. S., et al., The Role of Head Constraint in the Development of Lower Cervical Compression Flexion Injury, *Proceedings of the 35th Stapp Car Crash Conference*, 1991, 391-399.

[5] Myers, B. S., Winkelstein, B. A., Epidemiology, Classification, Mechanism and Tolerance of Human Cervical Spine Injuries, *Critical Reviews in Biomedical Engineering*, 23, 1995, 307-409.

[6] Nusholtz, G. S., et al., Cervical Spine Injury Mechanisms, *Proceedings of the 27th Stapp Car Crash Conference*, 1983, 179-197.

[7] Yoganandan, N., et al., Experimental Spinal Injuries with Vertical Impact, *Spine*, 11, 1986, 855-860.

Acknowledgments

Research has been provided by the DOT NHTSA Grant, DTNH22-94-Y-07133, and the DHHS CDC Grant, R49/CCR402396-10. The views and opinions expressed in this paper are those of the authors and not necessarily those of the granting institutions.

Development of an Intelligent Vibro-Impact Machine for Ground Moling Based on Vibro-impact Dynamics

R. D. Neilson[1]
A. A. Rodger[1]

ABSTRACT This paper provides an overview of the research undertaken to date on the development of a vibro-impact mole concentrating on machine development and computer modelling. In particular the model of a cam driven mole is put forward and some results from simulation of the model are presented, showing the possibility of predicting penetration rates. The conclusion of the work is that further work is required to produce a model in which the soil parameters change depending on the motion of the mole itself.

1. Introduction

Over the past decade there has been a vast increase in the range and flexibility of techniques available for the trenchless installation of cables, pipes and ducts [1]. Although becoming more widely used conventional impact moles can: induce deleterious ground movement, fail to penetrate in certain ground conditions and may damage underground services and foundations unexpectedly encountered in their path [2]. A research programme conducted at the University of Aberdeen since 1989 has led to the invention of a new, potentially more efficient, and intelligent form of ground moling machine based on vibro-impact dynamics. The dual mode of excitation associated with the vibro-impact driver enables it to relate the form of the driving energy to the type of soil encountered by the driven element. This results in efficient penetration employing predominant vibration with cohesionless soils and predominant impact with cohesive soils. Due to the general heterogeneous nature of most soil profiles, such an ability to adapt the driving mode to the soil conditions encountered, can increase penetration depths and rates, when compared with pure impact or vibration driving systems. As well as a potential increase in penetration, the self-adjusting capabilities of a vibro-impact device also reduce the disturbance to the surrounding soil, and, allow remote identification of the material being encountered by the mole using neural network techniques.

[1] Department of Engineering University of Aberdeen, Scotland, United Kingdom

2. Background

The initial research work was undertaken on a pneumatically driven 100mm diameter device, in which the source of vibration was a pneumatic oscillator driven by a high pressure supply switched by solenoid valves [3]. Unfortunately, this configuration was found to be restricted to very low excitation frequencies (of the order of 5Hz) by the flow rate through the valves and therefore did not provide enough vibration component to reduce the soil resistance appreciably. The machine tended to run in impact mode only and was therefore not viable. Further research suggested that a cam driven device powered through a flexible shaft might allow higher frequencies and higher associated body accelerations to be generated. A design of a 50mm diameter model was undertaken, and the device was shown to have a speed of up to 10mm/s in sand. This is appreciably better than existing pure impact devices in this type of soil. Unfortunately poor penetration was found in cohesive (clay like). soils. As a consequence a dynamic model has been derived which should allow optimisation of the design to be achieved to allow penetration in cohesive soils. This model is based around the techniques described in [4]. This is reported in the following sections.

3. Equations of Motion

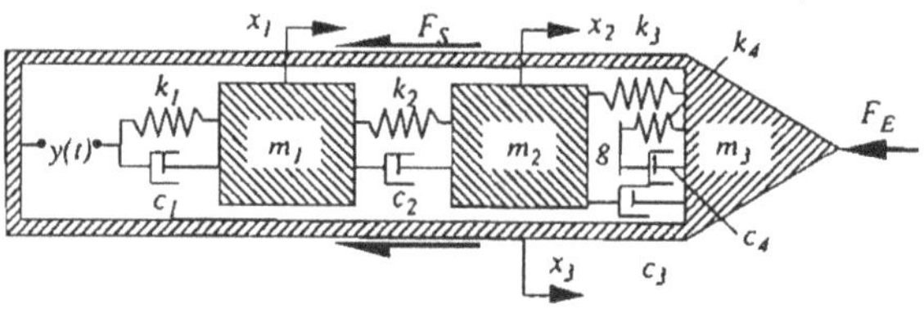

Figure 1 Dynamic model of the cam driven mole

Figure 1 shows the dynamic model of the mole including end and side resistance forces F_E and F_S. The equations of motion for this system may be readily obtained and are as follows:

$$m_1\ddot{x}_1 + c_2(\dot{x}_1 - \dot{x}_2) + k_2(x_1 - x_2) + k_1(x_1 - x_3 - y) = 0 \tag{1}$$

$$\begin{aligned} m_2\ddot{x}_2 + c_2(\dot{x}_2 - \dot{x}_1) + c_3(\dot{x}_2 - \dot{x}_3) + c_4 < \dot{x}_2 - \dot{x}_3 > + k_2(x_2 - x_1) \\ + k_3(x_2 - x_3) + k_4 < x_2 - x_3 + g >= 0 \end{aligned} \tag{2}$$

$$\begin{aligned} m_3\ddot{x}_3 + c_3(\dot{x}_3 - \dot{x}_2) + c_3\dot{x}_3 + c_4 < \dot{x}_3 - \dot{x}_2 > + k_3(x_3 - x_2) \\ + k_1(x_1 - x_3 - y) + k_4 < x_3 - x_2 - g >= -F_E - F_S \end{aligned} \tag{3}$$

where the soil end resistance is of the form shown in Figure 2 and given by equations (4) and (5)

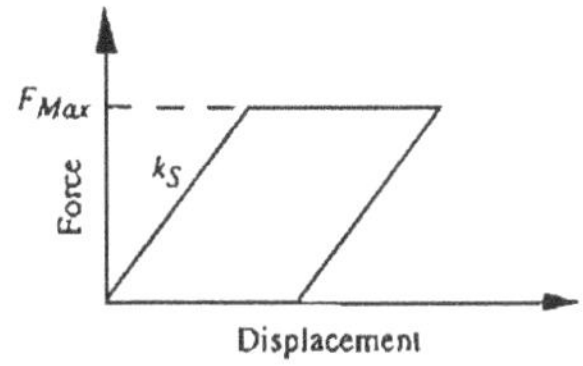

Figure 2 Force displacement characteristic

$$F_E = k_s(x_3 - x_4) \quad \text{if} \quad F_E < F_{max} \tag{4}$$

and

$$F_E = F_{Max} \quad \text{if} \quad F_E > F_{max} \tag{5}$$

where x_4 is the soil displacement, and

$$F_S = F_S sign(\dot{x}_3) \tag{6}$$

This soil model is a simplified version of that proposed by Spektor [5].

These equations are non-smooth and therefore have to be solved carefully by numerical techniques. In this case, the equations were written in Matlab code and solved using the ode15s stiff solver.

4. Results

The mole has been run for a number of configurations. Figures 3 and 4 show some of the responses elicited. As can be seen the penetration of the mole into the soil can be predicted if the soil parameters are chosen correctly. Figure 3 shows the motion of the mole and the associated impact force as would be measured behind the nose cone for driving the mole into a cohesionless soil (sand). In contrast, Figure 4 shows the reduced penetration rate and slightly higher impact force as the mole attempts to penetrate a "soft" clay which has a higher side resistance gripping the body of the mole than the sand. Further simulation with a "stiff" clay showed responses in which the mole failed to penetrate the soil and the impact force waveform had a maximum value around-600N. This reduction in force at the nose cone has also been found in experiments in which the side resistance is high compared to the end resistance. The soil parameters for these cases were chosen to be representative of the soils under the action of vibration. In reality the soil parameters will change depending on the motion of the mole. This needs further research.

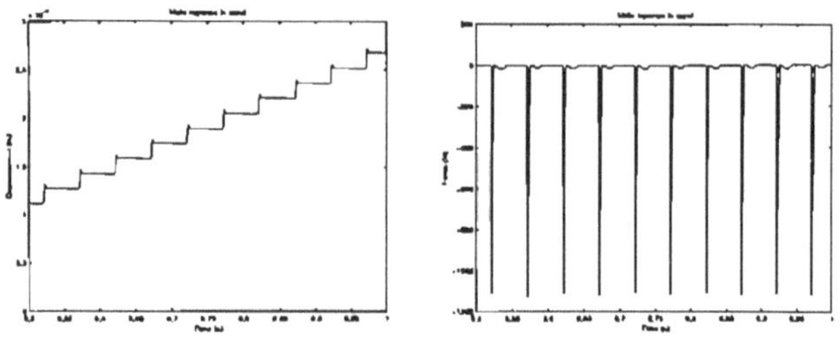

Figure 3 Response of the mole at 20Hz in sand

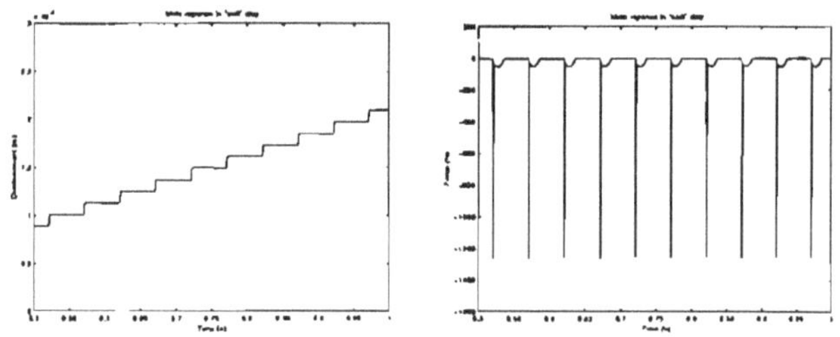

Figure 4 Response of the mole at 20Hz in "soft" clay

5. Conclusions

The work present shows that dynamic models of vibro-impact moling devices can be used to predict penetration rates if the soil parameters are chosen correctly. Further development of the model is required to close the feedback mechanism between soil resistance and the actual mole motion.

6. References

1. Binnie & Partners (1985), Trenchless Construction for New Pipelines - A Review of Current Methods and Development, Report No. ER 167E.
2. Etherton, P T (1985) Development of Impact Moling Techniques in the UK: Design, Use and Cost Considerations, Paper 3.1, No-Dig 85, London.
3. Stevenson, R.G., An investigation into the use of vibro-impact techniques in the design of a ground moling system., PhD Thesis, University of Aberdeen, 1993.
4. Neilson,R.D., Rodger,A.A .&Stevenson, R.G. Development of a Computer Based Model of Vibro-Impact Driving, Machine Vibration.
5. Spektor, M., Minimization of the Energy Consumption of Soil Deformation, Journal of Terramechanics, 1980, 17, 2, 63-77.

Towards a Model for Self-Excited Solitary Waves in a Spinning Flexible Disk

D. Boulahbal[1]

ABSTRACT Recent experiments reported by the author showed that a flexible disk, spinning against a stationary base-plate on an air-film, exhibits both harmonic as well as solitary self-excited traveling waves. The harmonic waves are very well understood and linear models exist that properly describe and predict them. The solitary waves are the manifestation of non-linear mechanisms linked to the dynamics of both the flexible disk and the air-film. In this paper, the non-linearities associated with large transverse deflections of the disk are investigated. The coupling between the in-plane centrifugally induced stresses in the disk and the disk's transverse displacement is formulated and yields coupled nonlinear equations that are solved numerically. The dependence of the disk's in plane stresses upon its transverse displacement provides the necessary ingredient for the establishment of solitary waves.

Introduction

Spinning disks are widely used in applications such as computer floppy disks, large circular saws in the timber industry, and automotive disk brakes. In a series of experiments aimed at understanding the mechanisms responsible for unstable operation of a spinning disk, it was discovered [1] that, depending on the spin rate of the disk, the latter supports harmonic as well as solitary waves. Photos of these waves are given in Figure 1.

FIGURE 1. Harmonic and Solitary Waves in a Spinning Disk.

Most published studies on rotating disks reported the existence of harmonic waves whose behavior is very well understood. Maher and Adams [2] inves-

[1]Department of Mechanical Engineering, University of Waterloo, Waterloo, Ontario, Canada, N2L 3G1

tigated the equilibrium position of a disk spinning next to a baseplate and the effects of displacement-dependent membrane stresses. In the present work, we extend on the results of [1, 2] and calculate the distribution of in-plane stresses in the disk. This calculation reveals a key ingredient towards understanding the mechanism responsible for triggering the solitons.

The Linear Model

Let us consider the flexible disk shown in Figure 2, spinning at a constant rate Ω on a thin air-film next to a rigid back-up plate.

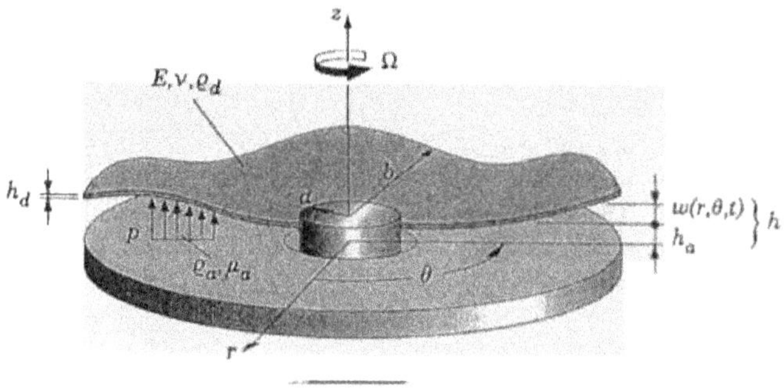

FIGURE 2. Disk Spinning on an Air-Film next to a Back-Up Plate

With respect to stationary coordinates, small transverse deflections of the disk are governed by the partial differential equation

$$\varrho_d h_d \frac{D^2 w}{Dt^2} + D_b \nabla^4 w - \frac{h_d}{r}\frac{\partial}{\partial r}\left(\sigma_r r \frac{\partial w}{\partial r}\right) - \frac{h_d}{r^2}\sigma_\theta \frac{\partial^2 w}{\partial \theta^2} = p - \varrho_d h_d g \quad (1)$$

where $D/Dt = \partial/\partial t + \Omega \partial/\partial \theta$ is the substantial derivative and ∇ is the two-dimensional differential operator. The disk has Young's modulus E, Poisson's ratio ν, mass density ϱ_d, uniform thickness h_d, and bending rigidity $D_b = Eh_d^3/12(1-\nu^2)$. The last two terms on the left side of equation 1 account for the stiffening introduced by the centrifugally induced membrane stresses σ_θ and σ_r which, for a full disk ($a \ll b$), are given by

$$\sigma_r = \frac{3+\nu}{8}\varrho_d \Omega^2 \left(b^2 - r^2\right) \quad , \quad \sigma_\theta = \frac{3+\nu}{8}\varrho_d \Omega^2 \left(b^2 - \frac{1+3\nu}{3+\nu} r^2\right) \quad (2)$$

The air-flow in the air-film is governed by a *Reynold's* lubrication equation with an added term due to centrifugal loading.

$$\frac{\partial}{\partial r}\left(r h^3 \frac{\partial p}{\partial r}\right) + \frac{1}{r}\frac{\partial}{\partial \theta}\left(h^3 \frac{\partial p}{\partial \theta}\right) = 12\mu_a r \left(\frac{Dh}{Dt} - \frac{\Omega}{2}\frac{\partial h}{\partial \theta}\right) + \frac{3}{10}\varrho_a \Omega^2 \frac{\partial}{\partial r}\left(r^2 h^3\right) \quad (3)$$

where $h = h_a + w$ is the total thickness of the air-film, and μ_a and ϱ_a are respectively the kinematic viscosity and air density inside the air-film, taken to be incompressible. Perhaps the strongest assumption in the linear model is that the stress-field (2) is calculated for a *flat* disk. In the equilibrium configuration, the disk is pulled towards the stationary plate as shown in Figure 3 and thus, it is *not* flat.

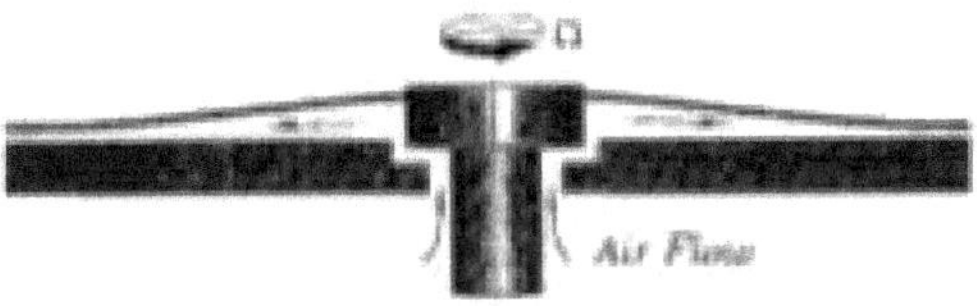

FIGURE 3. Steady-State Deflection of a Spinning Disk

To circumvent this limitation we now examine a model that captures the coupling between the disk's in-plane stresses and transverse displacement.

The Nonlinear vonKarman Model

If we denote by u the in-plane radial displacement of the disk and ignore the circumferential deformations v, the radial and circumferential strains, assuming moderately large transverse deflections w, are given by

$$\epsilon_r = \frac{\partial u}{\partial r} + \frac{1}{2}\left(\frac{\partial w}{\partial r}\right)^2 \qquad \textit{and} \qquad \epsilon_\varphi = \frac{u}{r} + \frac{1}{2}\left(\frac{1}{r}\frac{\partial w}{\partial \theta}\right)^2 \tag{4}$$

and yield the disk's differential equilibrium equation

$$\frac{\partial^2 u}{\partial r^2} + \frac{1}{r}\frac{\partial u}{\partial r} - \frac{1}{r^2}u + \left(\frac{\partial^2 w}{\partial r^2} + \frac{1-\nu}{2r}\frac{\partial w}{\partial r}\right)\frac{\partial w}{\partial r} + (1-\nu^2)\frac{\varrho_d \Omega^2}{E} r = 0 \tag{5}$$

The first three terms in (5) reflect the linear model. The fourth term captures the effect of transverse displacement and the last term accounts for the centrifugal loading. In the equilibrium case, (1) and (3) reduce to

$$D_b\left(\frac{d^2}{dr^2} + \frac{1}{r}\frac{d}{dr}\right)^2 w - \frac{h_d}{r}\frac{d}{dr}\left(\sigma_r r \frac{dw}{dr}\right) = p - (\varrho_d h_d)\, g \tag{6}$$

$$\frac{d}{dr}\left(r h^3 \frac{dp}{dr}\right) = \frac{3}{10}\varrho_a \Omega^2 \frac{d}{dr}\left(r^2 h^3\right) \tag{7}$$

The radial tensile stress in (6) is calculated from

$$\sigma_r = \frac{E}{1-\nu^2}\left\{\frac{du}{dr} + \frac{1}{2}\left(\frac{dw}{dr}\right)^2 + \nu\frac{u}{r}\right\} \tag{8}$$

The equilibrium configuration $w(r)$ of the disk and the pressure $p(r)$ in the air-film, along with the radial stress $\sigma_r(r)$ and displacement $u(r)$, are obtained by solving the coupled set of nonlinear equations 5–8, with the corresponding boundary conditions.

Numerical Results and Discussion

Equations 5–8 are solved iteratively using the finite difference technique. The initial guesses for $p(r)$, $u(r)$, and $\sigma_r(r)$ are those of the *flat* disk. The displacement $w(r)$ is chosen so as to resemble the shape of Figure 3. Once the iterative scheme has converged, the circumferential component of the disk's in-plane stress is calculated according to

$$\sigma_\theta = \frac{E}{1-\nu^2}\left\{\nu\left[\frac{du}{dr}+\frac{1}{2}\left(\frac{dw}{dr}\right)^2\right]+\frac{u}{r}\right\} \tag{9}$$

Figure 4 shows the distribution of $\sigma_\theta(r)$, normalized by $\varrho_d\Omega^2b^2$, for the cases of the linear model 2, and the nonlinear model for low and high spin rates and the same conditions that yielded the waves of Figure 1.

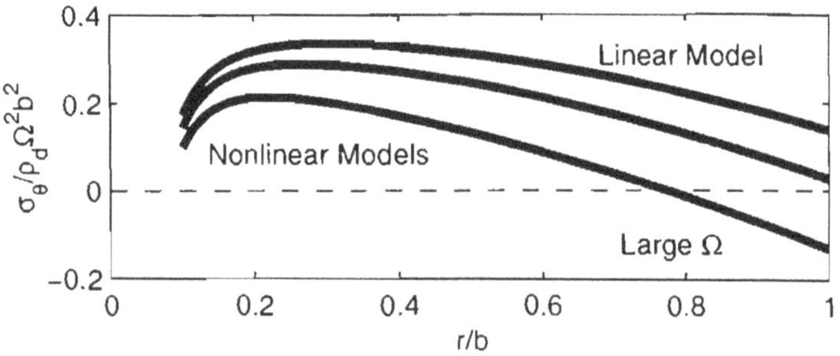

FIGURE 4. Circumferential Stress Distribution in Spinning Disk

In the linear model, the circumferential stress σ_θ is always positive and the disk is always stretched. There is a decrease in the stresses predicted by the nonlinear model, for low rotation rates. For sufficiently large Ω, the disk is pulled closer to the baseplate and the stresses become negative near the disk's outer edge. The disk gets thus compressed in the circumferential direction and since it cannot sustain such compressive loads, it locally "buckles" and gives rise to the observed solitary humps.

References

[1] Boulahbal, D. and Crandall, S.H., Self-Excited Harmonic and Solitary Waves in a Spinning Disk (With Video Presentation). *Proceedings of ASME DETC'97*, Sacramento, CA, September 1997.

[2] Maher, J.F. and Adams, G.G., Effect of displacement-dependent membrane stresses on the axisymmetric configuration of a spinning flexible disk. *STLE Tribology and Mechanics of Magnetic Storage Systems*, SP-29:123–129, 1990.

Applications of Friction to Vibration Control of Civil and Aerospace Structures

G. Ahmadi[1]

ABSTRACT Passive frictional vibration control of structures and their applications to civil and aerospace structures is reviewed. Attention is given to the base isolation methodology and passive frictional control techniques for earthquake resisting design. Several base isolation devices for seismic protection of buildings and vibration control of payload and space structures is described. The active control methodologies for vibration control is also discussed. It is concluded that the acceleration transmitted to compact stiff structures during an earthquake can be effectively reduced by using properly designed base isolation systems. One the other hand, the active control methodologies may be used for protection of tall buildings against earthquake. In addition, these latter techniques may become useful for seismic rehabilitation of existing structures.

1 Introduction

The conventional technique for aseismic design of structures is to strengthen the structural members in order to protect them against strong earthquakes. However, this strengthening strategy inevitably leads to higher masses and hence higher seismic forces. A structure designed in this way may survive a strong earthquake, while there could result intolerable damages to its members, as well as, to its sensitive internal equipment. Furthermore, the economical consideration limits the construction of a completely safe structure within the bounds of traditional design methodology. In the past two decades, significant progress has made in developing an alternative and attractive design strategy. The new approach is to use passive and active control mechanisms to control the vibration of structure during the earthquake strong motion. Thereby, the structure will become lighter and, eventually, can be constructed at lower cost. Similarly, considerable attention has been given to use of passive and active vibration control techniques

[1]Department of Mechanical and Aeronautical Engineering, Clarkson University, Potsdam, NY 13699-5725

(including smart materials) in aerospace structures.

In this paper, the recent development on vibration control of civil engineering and aerospace structures is reviewed. Particular emphasis is placed on the recent advances in frictional base isolation technology. Recent studies on passive dampers for dissipating vibration energy of structure and the new development in active and hybrid vibration control methodologies are also described.

2. Base Isolation

The passive base isolation techniques attempts to decouple the structure from the base excitations (earthquake, spacecraft vibration and etc.). The approach relays on minimizing the excitation energy that enters the structure, by partly reflecting the energy and partly dissipating it at the point of attachment (foundation level). As a result, the base isolation methodology appears to have considerable potential in preventing earthquake damages to structures and their internal equipment, as well as protecting sensitive equipment during space shuttle lift-off. Reviews of recent developments on base isolation and active control against earthquakes were provided by Kelly [1], Skinner et al. [2]. Housner et al. [3] and Ahmadi [4]. Recent studies on vibration control of aerospace structures were reported by Lee-Glauser et al. [5,6].

2.1 Frictional Base Isolation System

The Resilient-Friction Base Isolation system (R-FBI) for buildings was developed by Mostaghel and co-workers [7,8]. This isolator consists of several layers of Teflon coated friction plates with a central core of rubber. The rubber provides the restoring force for the system and hence control the relative displacement while energy is dissipated by the friction forces. An extensive study of the responses of a five-story building isolated by the R-FBI system was provided by Mostaghel and Khodaverdian [8].

Another frictional system which is used for base isolation of nuclear power plants in regions of high seismicity was developed undeer the auspices of Electricite de France (EDF) [9]. The Kroeberg nuclear power plant in South Africa relies on the EDF base isolation system for protection against earthquake. An EDF base isolator unit consists of a laminated (steel-reinforced) neoprene pad topped by a lead-bronze plate which is in frictional contact with a steel plate anchored to the base raft of the structure. Whenever there is no sliding in the friction plate, the EDF system behaves as a LRB and the flexibility of the neoprene pad provides isolation for the structure. The presence of the friction plate serves as an additional safety feature for the system. Whenever the ground acceleration becomes very large, sliding occurs which dissipates energy and limits the acceleration transmitted to the superstructure.

A base isolation concept which combines the desirable features of the R-FBI and the EDF systems was proposed by Su et al. [10]. This system which is referred to as the Sliding Resilient-Friction (SR-F) base isolation system is a R-FBI unit with an additional upper friction plate. Whenever there is no sliding in the upper friction plate, the SR-F base isolator behaves as a R-FBI unit. For high ground accelerations, sliding in the upper friction plate occurs which provides an additional mechanism for energy dissipation and increases the overall effectiveness of the isolation system.

2.2 Comparative Study

Su et al. [11] carried out several comparative studies of different base isolation devices for rigid and shear beam structures. Fan et al. [12,13] considered the performances of various base isolation systems for a multi-story building under a horizontal sinusoidal, as well earthquake ground motions. In these works, performances of different base isolation systems subject to various earthquake excitations were studied. The peak absolute acceleration of each floor, the peak base displacement, and the peak structural deflection for various base isolation systems were evaluated and the results were compared with each other and with those of the fixed-base structure. The acceleration responses both in time and frequency domains were examined.

2.3 Random Excitations

In [14-20] a number of studies on performance of various base isolation systems to random models of earthquake excitations. Stochastic earthquake response of secondary systems in base isolated structures were analyzed in [20]. In these studies, the equivalent linearization method were used for analyzing responses of structures with nonlinear frictional base isolation systems. It was shown that this method leads to reasonable peak response statistics for the base-isolated structures. More advance procedure for random response analysis of nonlinear systems were described in the literature. These include the Gaussian and non-Gaussian cumulant-neglect method developed by Noori and Davoodi [21], and Wiener-hermite functional expansion technique proposed by Jahedi and Ahmadi [22] and Orabi and Ahmadi [23,24]. These techniques, however, have not been used for response analysis of base isolated structures as yet.

3. Active Control

Active vibration control systems have been successfully utilized for aircraft, spacecraft, mechanical devices and various structures [25-30] in the last two decades. Extensive reviews on active control methodology were provided by Soong [31,32] and Meirovitch [33]. An extensive review of literature on active vibration control and future research need was provided by Housner et al. [3]. Recently, Lee-Glauser et al. [5] studied the effectiveness of an active vibration absorber (AVA) in conjunction with the use of a model independent active control strategy. Currently, significant progress is being made in active control of structures. In particular, the active control appear to have considerable potential for vibration control of tall buildings and for protection of existing structures.

4. Hybrid Control

There has been considerable recent interest in developing hybrid (combined passive and active) control systems for vibration control and structural protection against earthquake. The hybrid system which combines the advantages of passive and active systems could, in principal, be highly effective. Houser et al. [3] described the recent efforts in this direction. Tadjbakhsh and Rofooei [34] and Luco et al [35] performed computer simulation of the performance of hybrid systems. Active control of frictional isolation system was studied in [36]. Application of hybrid vibration control of aerospace structures were reported by Lee-Glauser et al. [6]. These studies clearly showed the feasibility of using a combination of passive and active control systems for optimum performance.

5. Concluding Remarks

Recent advances in passive, active and hybrid vibration control of civil and aerospace structures is described. Various frictional base isolation systems and active and hybrid control systems are also discussed. It is concluded that the compact and relatively stiff structures during an earthquake can be effectively protected by using a properly designed base isolation system. The active control methodologies is most suitable for protection of tall buildings against earthquake. The hybrid combination of passive and active control strategies may be designed for optimal performance.

6. References

[1] Kelly, J.M., "Aseismic base isolation: Review and bibliography", *Solid Dyn. Earthquake Engng.*, Vol. 5, pp. 202-216, 1986.

[2] Skinner, R.I., Robinson, W.H. and McVerry, G.H., *An Introduction To Seismic Isolation*, John Wiley & Sons, New York, 1993.

[3] Housner, G.W., et al., "Structural control: Past, present and future", *J. Engng. Mechanics*, Vol. 123, pp. 897-971, 1997.

[4] Ahmadi, G., "Overview of base isolation, passive and active vibration control strategies for aseismic design of structures", *Scientia Iranica*, Vol. 2, pp. 99,11-6, 1995.

[5] Lee-Glauser, G., Ahmadi, G. and Horta, L., "Passive/active vibration absorber for multistory buildings", *J. Struct. Engng.*, Vol. 123, pp. 499-504, 1997.

[6] Lee-Glauser, G., Ahmadi, G. and Layton, J.B., "Satellite active and passive vibration control during lift-off", *J. Spacecraft Rockets*, Vol. 33, pp. 428-432.

[7] Mostaghel, N. and Tanbakuchi, J., "Response of sliding structures to earthquake support motion", *Earthquake Engng. Struct. Dyn.*, Vol. 11, pp. 729-748, 1983.

[8] Mostaghel, N. and Khodaverdian, M., "Dynamics of resilient-friction base isolator (R-FBI)", *Earthquake Engng. Struct. Dyn.*, Vol. 15, pp. 379-390, 1987.

[9] Gueraud, R., Noel-Leroux, J.-P., Livolant, M. and Michalopoulos, A.P., "Seismic isolation using sliding-elastomer bearing pads", *Nuclear Engng. Design*, Vol. 84, pp. 363-377, 1985.

[10] Su, L., Ahmadi, G. and Tadjbakhsh, I.G., "Performance of a sliding resilient-friction base isolation system for a shear beam structure", *ASCE J. Struct. Engng.*, Vol. 117, pp. 165-181.

[11] Su, L., Ahmadi, G. and Tadjbakhsh, I.G., "A comparative study of performances of various base isolation systems, Parts I & II", *Earthquake Engng. Struct. Dyn.*, Vol. 18, pp. 11-12, 1989 and Vol. 19, pp. 21-33, 1990.

[12] Fan, F.G., Ahmadi, G. and Tadjbakhsh, I.G., "Multi-story base-isolated buildings under a harmonic ground motion, Parts I & II", *Nuclear Engng. and Design*, Vol. 123, pp. 1-6, 1990, and Vol. 123, pp. 17-26, 1990.

[13] Fan, F.G., Ahmadi, G., Mostaghel, N. and Tadjbakhsh, I.G., "Performance analysis of aseismic base isolation systems for a multi-story building", *Solid Dynamics and Earthquake Engineering*, Vol. 10, pp. 152-171, 1991.

[14] Su, L. and Ahmadi, G., "Response of frictional base isolation systems to horizontal-vertical random earthquake excitations", *Probabilistic Engineering Mechanics*, Vol. 3, pp. 12-21, 1988.

[15] Su, L., Ahmadi, G. and Tadjbakhsh, I.G., "A probabilistic comparative study of various base isolation systems", *Mechanics Struct. Machines*, Vol. 18, pp. 107-133, 1990.

[16] Su, L., Ahmadi, G. and Tadjbakhsh, I.G., "Responses of base-isolated shear beam structures to random excitations", *J. Probabilistic Engng. Mech.*, Vol. 5, pp. 35-46, 1990.

[17] Fan. F.G. and Ahmadi, G., "Random response analysis of a frictional base isolation system", *ASCE J. Engng. Mech.*, Vol. 116, pp. 1881-1901, 1990.

[18] Chen, Y. and Ahmadi, G., "Wind effects on base-isolated structures", *ASCE J. Engng. Mech.*, Vol. 118, pp. 1708-1727, 1992.

[19] Ahmadi, G. and Su, L., "Probabilistic responses of base-isolated structures to El Centro 1940 and Mexico City 1985 earthquakes", *Engineering Struct.*, Vol. 14, pp. 217-230, 1992.

[20] Chen, Y. and Ahmadi, G., "Stochastic earthquake response of secondary systems in base-isolated structures", *Earthquake Engng. Struct. Dyn.*, Vol. 21, pp. 1039-1057, 1992.
[21] Noori, M. and Davoodi, H., "An Ito-based general approximation method for random vibration of hysteretic systems, Part I: Gaussian analysis", *J. of Sound and Vibrations*, Vol. 127, No. 2, December 1988.
[22] Jahedi, A. and Ahmadi, G., "Application of Wiener-Hermite expansion to nonstationary random vibration of a duffing oscillator", *J. Appl. Mech. Trans. ASME*, Vol. 50, pp. 436-442, 1983.
[23] Orabi, I.I. and Ahmadi, G., "A functional series expansion method for response analysis of nonlinear systems subjected to random excitations", *Int. J. Nonlinear Mech.*, Vol. 22, pp. 451-465, 1987.
[24] Orabi, I.I. and Ahmadi, G., "Nonstationary response analysis of a duffing and duffing oscillator by the Wiener-Hermite expansion method", *J. Appl. Mech. Trans. ASME*, Vol. 54, pp. 434-440, 1987.
[25] Meirovitch, L., Van Landingham, H.F., and Oz, H., "Control of spinning flexible spacecraft by modal synthesis", *Acta Astronautica*, 4, pp. 985-1010, 1977.
[26] Meirovitch, L. and Oz, L., "Modal space control for distributed gyroscopic systems", *J. Guidance Control*, 3, pp. 140-150, 1980.
[27] Meirovitch, L., Baruh, H. and Oz, L., "A comparison of control techniques for large flexible systems", *J. Guidance Control*, 6, pp. 302-310, 1983.
[28] Lightner, E.L., "Structural dynamics and control of large space structures", *NASA Conference Publ. 2187*, Langley Research Center, Hampton, VA, 1981.
[29] Johnson, T.L., "Progress in modeling and control of flexible spacecraft", *J. Franklin Institute*, 315, pp. 495-520, 1983.
[30] Plump, J.M., Hubbard, J.E. and Bailey, T., "Nonlinear control of a distributed system: Simulation and experimental results", *J. Dynamic Systems Measurement Control Trans. ASME*, 109, pp. 133-139, 1987.
[31] Soong, T.T., "State-of-the-art review: Active structural control in civil engineering", *Engineering Structures*, Vol. 10, pp. 78-84, 1988.
[32] Soong, T.T., *Active Structural Control: Theory and Practice*, Longman, London and Wiley, New York, 1990.
[33] Meirovitch, L., *Dynamics and Control of Structures*, Wiley-Interscience, New York, 1990.
[34] Tadjbakhsh, I.G. and Rofooei, F., "Optimal hybrid control of structures under earthquake excitation", *Earthquake Engng. Struct. Dyn.*, Vol. 21, pp. 233-252, 1992.
[35] Luco, J.E., Wong, H.L. and Mita, A., "Active control of seismic response of structures by combined use of base isolation and absorbing boundaries", *Earthquake Engng. Struct. Dyn.*, Vol. 21, pp. 525-541, 1992.
[36] Feng, M.Q., "Application of hybrid sliding isolation system to buildings", *ASCE J. Engng. Mechanics*, Vol. 119, pp. 2090-2108, 1993.

The Effect of Friction on the Dynamic Behavior of Spindle Motor in Data Storage Devices

J. H. Wang[1]
C. H. Chang[1]

ABSTRACT. The dynamic behavior of a spindle motor with porous bearings in the radial direction and a frictional plate in the axial direction was investigated. Due to the gyroscopic effect, the friction forces in the axial and radial directions have strong effect on the dynamic behavior of the spindle. From the results, some design criteria were proposed.

1 Introduction

Brushless small dc motors are widely used in computer data storage deices, such as hard disk drives, CD-ROM or DVD drives, et. al.. The spindles of the motors are generally supported with either ball bearing, hydrodynamic bearing or sintered porous plain bearing [1,2]. Due to the low cost and easy installation, sintered porous plain bearings are widely used in hard disk and CD-ROM driving motors. Generally, a fully lubricated fluid film can not be sustained in porous braring, and partial metal contact between the spindle and the bearing is expected. A typical motor used in CD-ROM is shown in Fig.1. The spindle is supported by two porous bearings in the radial direction and by a frictional plate as a trust bearing in the axial direction. As discussed above, one can find that the spindle may be subjected to friction forces both in the axial and radial directions simultaneously. The dynamic behavior of a rotor with rubbing has been discussed by many papers, for instance, [3, 4]. However, most of the reports disscussed the rotor with rubbing only in the radial direction, and the rotor was generally horizontal. The spindle used in the hard disk or CD-ROM drives is vertical. Due to the gyroscopic effect, the frictional force in the axial direction has a strong effect on the dynamic

[1] Department of power Mechanical Engineering, National Tsing Hua University, Hsinchu, Taiwan

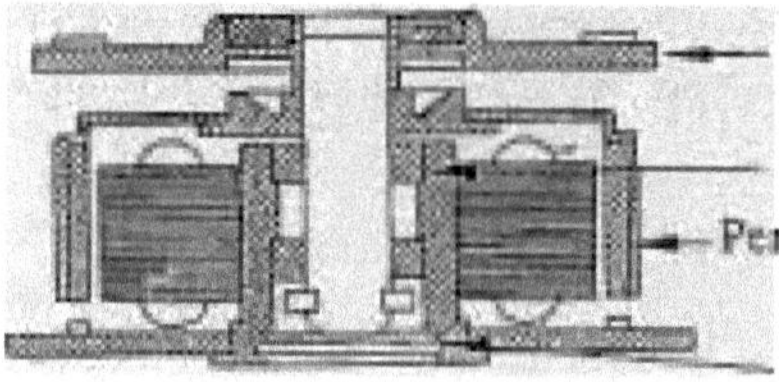

Figure 1 Schematic of the spindle motor

behavior of the spindle.

In this work, the dynamic behavior of a vertical spindle motor with sintered porous bearings in the radial direction and a frictional plate in the axial direction was investigated.

2 Theoretical Formulation

The schematic of the spindle motor investigated in this work is shown in Fig. 1. The motor is designed to have a maximum speed of 8000 rpm. Because the interest range of frequency is below 150Hz, the rotating part (spindle) is considered as rigid. In other words, a model of rigid spindle is considered in this work.

The fundamental equations of motion of a rigid body are well known as:

$$\sum F = ma \tag{1}$$

$$\sum M = \dot{H}_o \tag{2}$$

where H_o represents the angular momentum of the spindle with respect to the mass center. In order to solve the equations (1) and (2), one shall know all the external forces and moments acting on the rigid body. The external forces and moments acting on the rigid body include: the unbalance force due to mass eccentricity, the elastic and damping forces of the radial bearings, the normal contact force and friction force at the radial bearings, the normal contact and friction forces at the thrust bearing, and the gravitational force. Note that all the above mentioned forces also produce moments with respect to the mass center.

Due to space reason, the detailed derivation of equations (1), (2)can not be given here. Due to the gyroscopic effect and the frictional forces at the bearings, equations (1), (2) are highly nonlinear. Especially, the direction of the friction force can be known only when the velocity of the rigid body is known. Therefore, a special iteration procedure was developed in this work to cooperate with the Runge-Kutta method to solve the equations.

3 Results of Numerical Simulation

The system parameters of the simulated spindle are summarized here: polar moment of inertia=0.0156Kgcm^2; transverse moment of inertia=0.011 Kgcm^2; rotating speed=6000rpm; spindle diameter=3mm; bearing clearance=0.003mm; radial bearing stiffness=10^4N/m and damping=10^2N·s/m; axial preload=2N; mass unbalance=0.25g.cm.

As mentioned, the main object of this work is to investigate the effect of the friction forces on the behaviors of the motor. Therefore, the friction coefficients of the thrust bearing (μ_1) and the radial bearings (μ_2) are the main variables in the following simulations.

Example 1

In order to study the effect of the friction force of the thrust bearing , only the μ_1 was considered in this example. Figs.2 and 3 show the vibration of the mass center due to different values of μ_1. The results show that the friction coefficient is not the smaller the better. This is due to the fact that the friction force at the thrust bearing can help the gyroscopic effect to rise the inclined spindle to a vertical position. As best as we know, in the past no report has been published to discuss the design problem of the frictional plate. Most of the people believe that the friction coefficient of the plate is the smaller the better.

Example 2

In this example, both the $\mu_1$and μ_2 were considered. The results of Figs. 4 and 5 show that the friction coefficient of the radial bearing should be as small as possible. The radial rubbing force can not only enhance the wear, but also reduce the avaiable damping of the system because the direction of the friction force is opposite to that of the viscous damping force.

4 Conclusions

The dynamic behavior of a spindle motor with porous bearings in the radial direction and a frictional plate in the axial direction was investigated. From the results, some conclusions can be drawn:

(1)The friction coefficient of the frictional plate is not necessary the smaller the better. A reasonable range of the friction coefficient is about 0.1 to 0.2.

(2)The friction coefficient of the porous bearings should be as small as possible. Special attention should be paid to the roughness of the contact surface of the bearing.

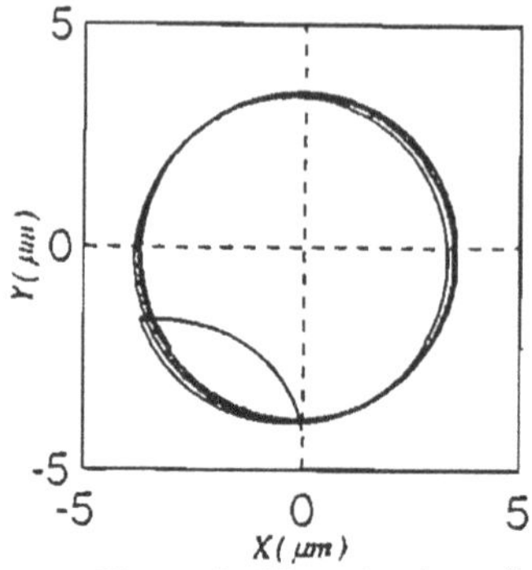

Figure 2 . The vibration of mass center $\mu_1 = 0.01$

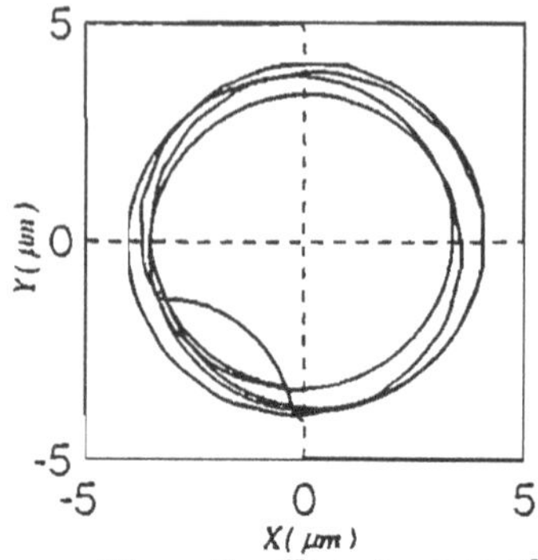

Figure 3. The vibration of mass center $\mu_1 = 0.3$

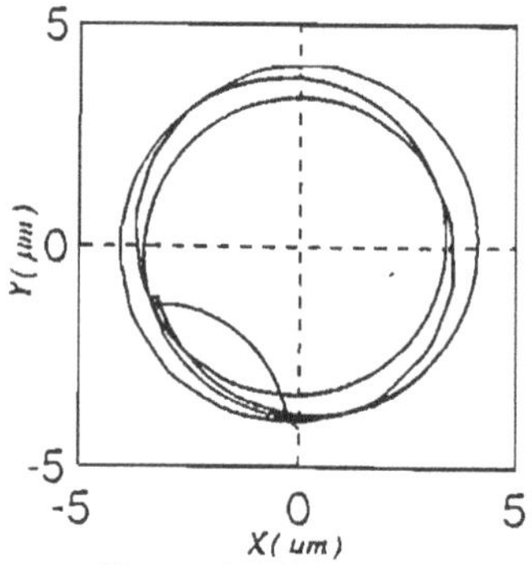

Figure 4. The vibration ot mass center, $\mu_1 = 0.3, \mu_2 = 0.01$

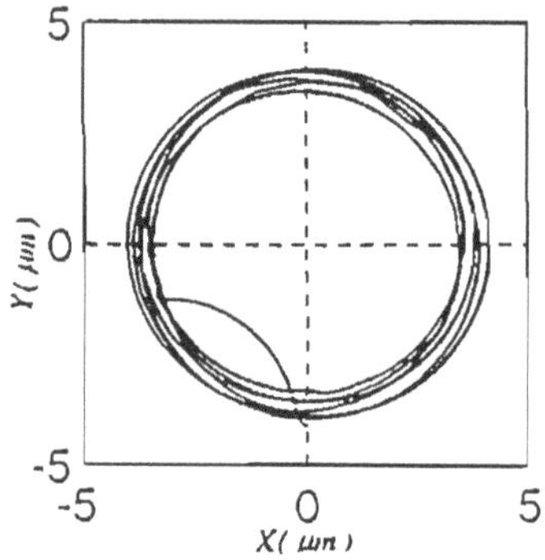

Figure 5. The vibration of mass center. $\mu_1 = 0.3, \mu_2 = 0.3$

5 References

[1]Nagarathnam, L., Leuthod, H. and Jennings D.," Mechanical Vibration Caused by Ball Bearings in Spindle Motors", Proceedings os the 24th Annual IMCSD Symposium, San Jose, CA, June 6-8, 251-255, 1995.

[2]Leuthold, H. and Murthy S., "Impact Testing of Spindle Motor with Hydrodynamic Bearings", Proceedings of the 25th Annual IMCSD Symposium, 93-101, 1996.

[3]Ehrich, F. F., "The Dynamic Stability of Rotor/Stator Radial Rubs in Rotating Machinery", Trans. of ASME, J. of Engineering for Industry, Vol. 91, No. 4, 1025-1028, 1969.

[4]Choy, F. K. And Padovan, J., "Non-linear Transient Analysis of Rotor-Casing Rub Events", J. of Sound and Vibration, Vol.113, 529-545, 1987.

Intelligent Control of Man-machine System for Stability of Industrial Vehicle

K. Tsujioka[1]
A. Nagamatsu[1]

ABSTRACT A new approach on intelligent control of man-machine system is proposed for its safety, in which the control system judges driver's operation autonomic and adaptively using the neural network.

1. Introduction

A new approach on intelligent control of man-machine system is proposed in this report, in which the control system judges driver's operation autonomic and adaptively using the neural network. The forklift vehicle is taken as a typical example of man-machine system. This control system devotes itself to promote comfortable riding by suppressing shock and vibration, obeying driver's operation docilely, while the control system judges that safety is maintained. But, when it happens that the vehicle may fall down, the control system devotes itself to prevent it irrespective of driver's operation.

The proposed control system is composed of the four elements as the human driver, the controller for preventing fall down and for vibration control, the system neural network for estimating state variables at one sampling time later and the distributor for judging driver's operation. The validity and the usefulness of the proposed approach are demonstrated by experiment.

2. Vehicle Model

Figure 1 shows a simplified model of the actual forklift vehicle, in which body mass M_b is 3.08t, mast mass M_m is 1.88t, mast height is 3.20m and body length is 2.13m. The mathematical model of this forklift is as follows.

$$M_m \ddot{x}_m + M_b \ddot{x}_b = u \tag{1}$$

$$M_m \ddot{z}_m + M_b \ddot{z}_b = -(M_m + M_b)G + F_{1z} + F_{2z} \tag{2}$$

$$I_m \ddot{\theta}_m = -F_{12} \sin\theta_g l_{x,12,m} + F_{12} \cos\theta_g l_{z,12,m} - F_{11x} l_{z,11,m} + F_{11z} l_{z,11,m} \tag{3}$$

$$\begin{aligned} I_b \ddot{\theta}_b = {} & F_{1z} l_{x,1,b} + F_{2z} l_{x,2,b} - F_{11z} l_{x,11,b} + F_{11x} l_{z,11,b} - u l_{z,1,b} \\ & - F_{10} \sin\theta_g l_{x,10,b} + F_{10} \cos\theta_g l_{z,10,b} \end{aligned} \tag{4}$$

---

[1] Department of Mechanical Engineering Tokyo Institute of Technology, Meguro-ku, Tokyo 152, Japan

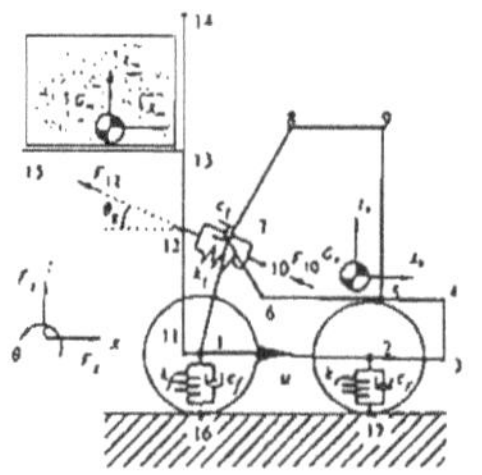

Fig.1 Vehicle Model

where u is input force (including control force), G is gravity, F is inside force, I is moment of inertia, θ is angle, l is distance, and lower suffixes explain mast center m, body center b, number of nodal point in Fig. 1 and coordinate axis (x y z). Giving the initial and the constraining conditions, Eqs. (1)~(4) are solved numerically. When the body inclines to the front direction and the rear wheel separates from the earth surface, F_{2z} is given to be zero.

3. Control System

Figure 2 shows the outline of the block diagram of the newly proposed control system which is composed of four kinds of elements, namely the human driver, the controller, the system neural network and the distributor. The largest strong point of this control system is that it contains the human driver, in order to observe and to check his operation every time from the viewpoint of safety.

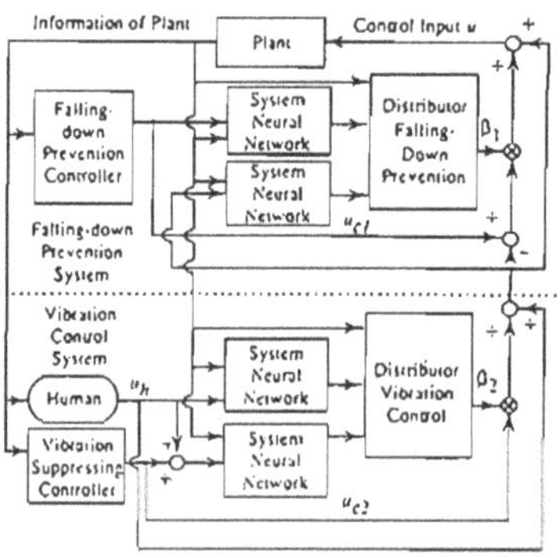

Fig.2 Block Diagram of Control System

3.1. Controller

Two controllers are used. One is for preventing fall down, and the other is for suppressing vibration. Both of them are designed using the linear optimum control theory. It is assumed in design that the two points, 10 and 12 in Fig. 1 are connected rigidly, so the body and the mast become the one body whose rotating angle is θ_s and the gravity coordinate is $(x_s \quad z_s)$. Thus, the state variable vector of this rigid vehicle is

$$q_s = \begin{bmatrix} \dot{\theta}_s & \theta_s & \dot{x}_s & x_s & \dot{z}_s & z_s \end{bmatrix}^T \quad (5)$$

The cost functions of these two controllers are written as follows.

$$J = \int_0^T \left\{ q_s^T Q q_s + u^T R u \right\} dt \quad (6)$$

where u is the control input force in Fig. 1. Q and R are the weighting functions. Q of the former controller is determined so that it becomes overwhelmingly sensitive to θ_s, because fall down can be prevented by suppressing the rotating angle of the vehicle. Q of the later controller is determined so that it becomes overwhelmingly sensitive to $\dot{\theta}_s$, because vibration control can be done mainly by suppressing rotating angular velocity. R of the former controller is given to be ten times smaller than R of the later one. This is because the full power is used to prevent fall down, but vibration control must been performed while moving the vehicle as the driver operates.

3.2. System neural network

Four system neural networks of the same structure shown in Fig. 3 are used in the control system of Fig. 2. These all have acquired the same dynamics of the vehicle plant. The inputs are all state variables except the displacement of x direction and the input force, and the outputs are the differences of the angle, the angular velocity, the displacement and the velocity of x direction and the displacement of z direction of the body.

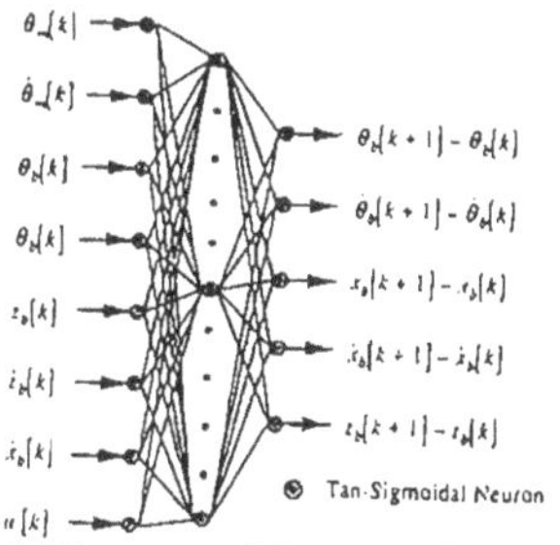

Fig. 3 Diagram of System Neural Network

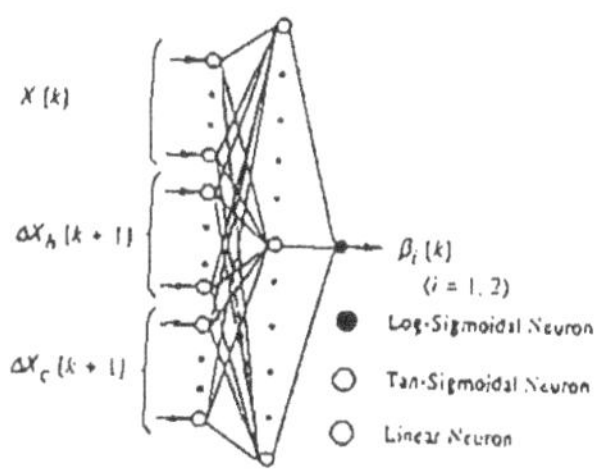

Fig.4 Distributor

3.3. Distributor

Two distributors of the same neural network structure shown in Fig. 4 are used in the control system of Fig. 2. One is for vibration control and the other is for prevention of fall down. Inputs to the distributor are the state value $X[k]$ at the sampling time k, the estimated difference of state value at the one sampling time later $\Delta X_h[k+1]$ by only driver's operation and the one $\Delta X_c[k+1]$ by only the controller's output Output from the distributor is the distributing ratio β_i, where i=1 for prevention of fall down and i=2 for vibration control. β_i takes a value between 0 and 1.

The final value of the control force u to the plant becomes as follows.

$$\left.\begin{aligned} u &= u_{hc2} + \beta_1\left(u_{c1} - u_{hc2}\right) \\ u_{hc2} &= u_h + \beta_2 u_{c2} \end{aligned}\right\} \tag{7}$$

4. Experimental Result

Figures 5 and 6 are the experimental results. The shows that the forklift falls down when the control system is cut out, but does not fall down when it operates.

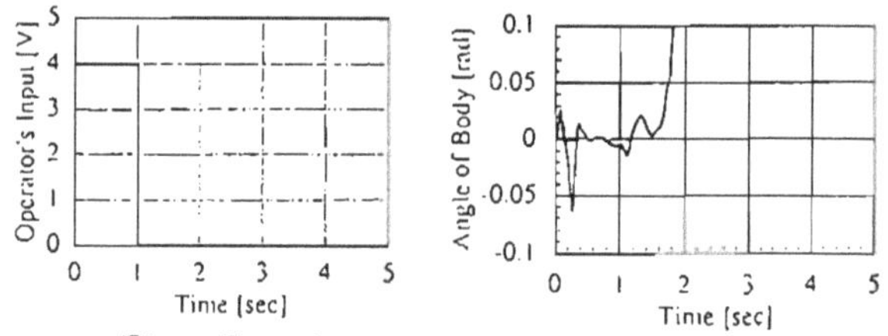

Fig. 5 Experimental Result with No Control

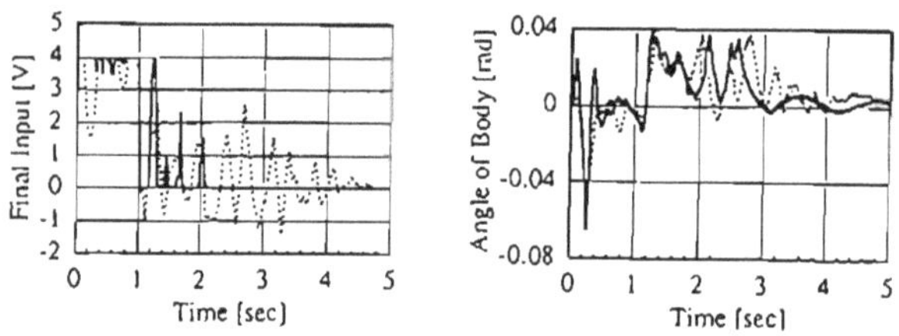

Fig.6 Experimental Result

5. Summary

A new approach on intelligent control of man-machine system is proposed in this report, in which the control system judges driver's operation autonomic and adaptively using the neural network. The forklift vehicle is taken as a typical example of man-machine system. Validity and usefulness of the proposed control system are verified experimentally.

Reference

(1) Fu, K. S., IEEE Transactions on Automatic Control, 16-1, (1971), 70-72.

(2) Saridis, G. N., IEEE Transactions on Automatic Control, 28-5, (1983), 311-321.

(3) Fukuda , Sibata, Transactions of JSME, 59-564, C, (1993), 2282-2289.

(4) Yoshida, Kidoushi, Proceedings of JSME, 940-10(1), (1994), 51-53.

(5) Yoshida, Proceedings of JSME, 960-15, (1996), 25-34.

Dynamics of Vehicle Suspensions with Valves Providing Frequency-Dependent Damping

G. Popov[1]

1 Introduction

Vibration isolation of mechanical systems, vehicles in particular, is mostly carried out by means of hydraulic dampers incorporating different type hydraulic valves. The valves have an important property to change their opening and, consequently, the damping provided either by themselves or under an external control and, therefore, they are widely used in passive, semi-active, and active suspension systems.

Sharp and Hassan [1], Hennecke and Zieglmeier [2], Wilkinson and Crolla [3], and many others studied the performance of active vehicle suspensions employing valves for automatic damping control. It has been established that active type suspensions are the best possible means for vibration isolation and vehicle handling and control. This is achieved, however, at the price of reduced reliability and high cost.

Horton and Crolla [4], Lizell [5], and Miller [6] investigated the switching ("on/off") semi-active vibration control strategy, which provides a "soft" or a "firm" damping depending on the current damping requirement. This strategy has been found very effective in improvement in ride and tire to ground contact. Unfortunately, this requires very fast valves capable to switch in 0.01 to 0.15s or so.

Another semi-active control strategy, based on a continuously operating valve, was analysed by Parker and Lau [7] and by Asami and Sekiguchi [8], who used different type valves to create a variable damping. The main drawback of this approach is its complexity, which could require an electronic processor and a servomotor to control the valve.

Passive suspensions with valves have not received too much attention in

[1]GEPOL Mechanical Engineering Consulting, 8340 Galeries d'Anjou Blvd., Suite 202, Anjou, Montreal, Quebec, Canada H1J 2K7

recent years, perhaps, for the reason that they were associated with classical dashpots, where an unchangable damping orifice was simply substituted by a valve to provide a softer damping. In a work by Popov and Sankar [9], however, it has been shown that a typical spool valve, operated by the pressure difference in the cylinder chambers, can provide an effective control over the first and second resonance transmissibilities for both sprung and unsprung masses, if the valve is correctly designed.

Lin and Willumeit [10] studied the dynamics of an adaptive damper with a vibrator-controlled valve. The valve and vibrator were designed to create a frequency-dependent damping, i.e. a large damping at system resonance and a small damping at higher frequencies. They demonstrated a strong potential of the frequency-dependent damping in vibration isolation of vehicles. The term "frequency-dependent damping" will here be referred to the damping coefficient rather than to the damping force.

In this article, the idea of using the vibrational motion of the valve to create a frequency-dependent damping is further developed and applied to a quarter-car model of vehicle. A comprehensive dynamic model of valve is presented; the equations of motion of the system are solved by the forth-order Runge-Kutta routine; and analysis of valve dynamics, optimization of valve parameters, and post-optimal parametric sensitivity analysis are presented.

2 Model of the System

The diagram of the damper is presented in Figure 1a. The actuator cylinder (absolute displacement X_2) is attached to the wheel-axle structure of mass m_2 and the piston rod is attached to the car body (X_1, m_1). The slide of the valve is positioned inside the piston rod together with an additional mass m_3 which is needed to reduce a high natural frequency of the valve. In a neutral position, the slide forms two gates of equal hydraulic resistance for the valve flow Q_v. A vibration of the sprung mass is transmitted to the valve which vibrates with the frequency and amplitude defined by its combined mass m_3, spring stiffness K_3, and the damping created by the throttle d_{th} and the friction between the slide and the sleeve. The amplitude of valve vibration is also a function of the sprung mass frequency. When the amplitude of valve vibration is large, the opening of the one of two in series gates is small, or even zero, and the damping delivered to the system is high. It happens at the frequency near by the sprung mass resonance and as the frequency increases, a progressively smaller damping is created. Figure 1b shows an alternative design of the valve gates with suppressed jet forces and Figure 1c presents the force diagram, where the subscripts j, s, i and d denote jet, valve spring, inertia, and valve damping, respectively.

The equations of motion for the system including sprung and unsprung masses and the valve are:

$$\begin{cases} m_1\ddot{X}_1 &= A_a\Delta P - C_{12}(\dot{X}_1 - \dot{X}_2) - K_{12}(X_1 - X_2) - C_{13}(\dot{X}_1 - \dot{X}_3) \\ & \quad -K_{13}(X_1 - X_3) \\ m_2\ddot{X}_2 &= -A_a\Delta P + C_{12}(\dot{X}_1 - \dot{X}_2) + K_{12}(X_1 - X_2) - C_{20}(\dot{X}_2 - \dot{X}_0) \\ & \quad -K_{20}(X_2 - X_0) \\ m_3\ddot{X}_3 &= F_j + F_i + F_d + F_s \end{cases}$$

$$\begin{cases} F_j = A_j|\Delta P|(X_3 - X_1)\frac{\alpha}{\beta} & \text{for } |X_3 - X_1| < X_{v0} \\ F_j = 0 & \text{for } |X_3 - X_1| \geq X_{v0} \end{cases}$$

$$\begin{cases} F_i &= A_i\frac{\alpha}{\alpha+S\sqrt{\beta}}(\ddot{X}_1 - \ddot{X}_2) \\ & \quad +A_i\frac{S(\alpha+2\beta)}{\sqrt{\beta}(\alpha+S\sqrt{\beta})^2}(\dot{X}_2 - \dot{X}_1)(\dot{X}_3 - \dot{X}_1)(X_3 - X_1) \\ & \quad \text{for } |X_3 - X_1| < X_{v0} \\ F_i &= 0 \quad \text{for } |X_3 - X_1| \geq X_{v0} \end{cases}$$

$$\begin{aligned} F_d &= A_{d1}\Delta P - A_{d2}(\dot{X}_3 - \dot{X}_1) \\ F_s &= -K_3(X_3 - X_1) \end{aligned}$$

$$\begin{cases} \Delta P = A_{p1}\frac{\beta}{(\alpha+S\sqrt{\beta})^2}(\dot{X}_2 - \dot{X}_1)|\dot{X}_2 - \dot{X}_1| & \text{for } |X_3 - X_1| < X_{v0} \\ \Delta P = A_{p2}(\dot{X}_2 - \dot{X}_1)|\dot{X}_2 - \dot{X}_1| & \text{for } |X_3 - X_1| < X_{v0} \end{cases}$$

where $\alpha = X_{v0}^2 - (X_3 - X_1)^2$, $\beta = X_{v0}^2 + (X_3 - X_1)^2$.

In the above equations, X_{v0} denotes the initial valve opening and all A's coefficients are constant and defined by the valve parameters. All displacements, velocities, accelerations, and forces are normalized with respect to the excitation amplitude X_0, $X_0\omega_1$, $X_0\omega_1^2$, and $m_3X_0\omega_1^2$, respectively.

3 Results and Discussion

The solutions of the equations formulated above were carried out for a quarter-car model with the following parameters: $m_1 = 375$ kg, $m_2 = 10$ kg, $K_1 = 40$ kN/m, $K_2 = 140$ kN/m, $C_1 = 1$kg/s, $C_2 = 100$ kg/s (tire) for a harmonic excitation of 0.05m in amplitude.

Analysis of the system dynamics reveals that a strongly nonlinear liquid jet force in combination with a relatively small stiffness of the valve spring may cause an asymmetrical vibration of the sprung and unsprung masses at frequencies above the first resonance. This property of suspension is not necessarily harmful; on the contrary, it can be used to obtain a different dynamic stiffness of the suspension for the strokes of compression and rebound.

The optimization study included the initial valve opening ϵ, the size of the bypass orifice d_0, and the size of the throttle d_{th} and was carried out for a valve of $d_v = 10$mm, $m_3 = 1$ kg, $K_3 = 10^3$ N/m, and fluid viscosity $\nu = 3 \cdot 10^5$ m^2/s. An optimal solution was found to be: $\epsilon = 0.089$, $d_0 = 3.7$mm, and $d_{th} = 1.058$mm which gives the peak displacement transmissibility of the sprung mass at both resonances of $T_1 = 1.758$. This does not necessarily represents the global minimum, a better solution, perhaps, can be found by involving more parameters in the search.

The post-optimal sensitivity study reveals the following. An increase of ϵ and d_0 from their optimum values can significantly reduce the second resonance peak of the sprung mass with only a small loss of performance at the first resonance, while a similar variation of the internal valve damping, d_{th}, has directly opposite effect. Variations of other parameters, such as valve mass, m_3, valve spring stiffness, K_3, valve diameter, d_v, and viscosity of fluid, ν, have also a strong influence on the system response. This can be seen from Figure 2, where the responses obtained with varied parameters are compared with the optimal response for the sprung mass. Here, the first resonance transmissibility is kept at 1.9. In Figure 3, the similar responses of the unsprung mass are compared with one that corresponds to the optimality condition of the sprung mass (curve 1).

4 Conclusion

This study demonstrates a strong potential of valves for passive suspension systems in reduction of vehicle vibrations by creating a frequency-dependent damping. The use of valves leaves a necessary margin of flexibility for practical design, while achieving different goals in suppression of undesired resonance peaks of both sprung and unsprung masses.

5 Acknowledgements

The author is deeply grateful to the Organizing Committee of the ISIFSM'98 Symposium, in particular, to Professor Arde Guran for encouraging me to present this paper and to Mrs. Linda Espeut for typing it, whose sincere efforts made this work be published.

6 References

[1.] Sharp, R.S. and Hassan, S.A., The Relative Performance Capabilities of Passive, Active and Semi-Active Car Suspension Systems, *Proc. Inst. Mech. Engrs.* 200, No. D3, 1986.

[2.] Hennecke, D. and Zieglmeier, F.J., Frequency Dependent Variable Suspension Damping – Theoretical Background and Practical Success, *Proc. Inst. Mech. Engrs.* C431/88, 101-111, 1988.

[3.] Wilkinson, P.A. and Crolla, D.A., Synthesis and Analysis of a Passive Alternative to Active Suspension, *Proc. Inst. Mech. Engrs.* C466/044/93, 169-179, 1993.

[4.] Horton, D.N.L. and Crolla, D.A., Theoretical Analysis of Semi-Active Suspension Fitted to an Off-Road Vehicle, *Vehicle System Dynamics* 15, 1986.

[5.] Lizell, M., Semi-Active Damping, *Proc. IMechE.* C429/88, 83-91, 1988.

[6.] Miller, L.R., The Effect of Hardware Limitations on an On/Off Semi-Active Suspension, *Proc. IMechE.* C442/88, 199-206, 1988.

[7.] Parker, G.A. and Lau, K.S., A Novel Valve for Semi-Active Vehicle Suspension Systems, *Proc. IMechE.* C427/88, 69-74, 1988.

[8.] Asami, T. and Sekiguchi, H., Study on an Oil Damper with a Variable Damping Mechanism, *JSME, Series III*, 31, No. 2, 370-379, 1988.

[9.] Popov, G. and Sankar, S., Performance of a Passive Vehicle Suspension with a Valve Damper, *Proc. CSME Forum 1994*, McGill University, Montreal, Canada, vol. II, 861-873, 1994.

[10.] Lin, Y. and Willumeit, H.P., The Dynamics of the Vibrator-Controlled Adaptive Damper and its Potential in Vibration Isolation, *Vehicle System Dynamics* 21, 353-369, 1991.

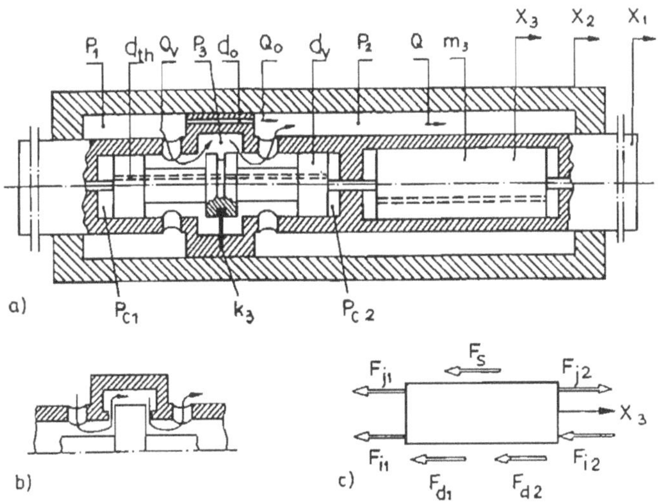

FIGURE 1. Schematics of the damper.

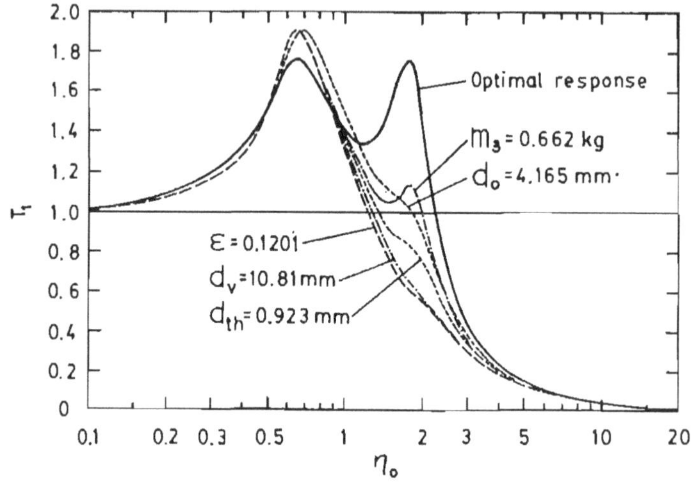

FIGURE 2. Optimal and near-optimal responses of the sprung mass.

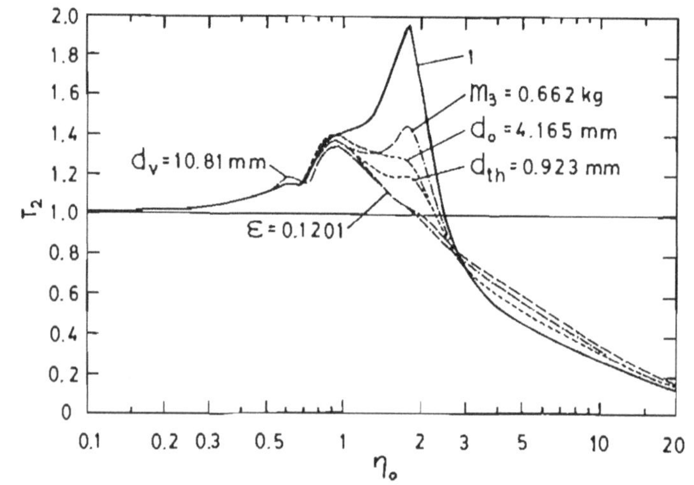

FIGURE 3. Unsprung mass responses corresponding to the optimal and near-optimal responses of the sprung mass.

Bounce Phenomena in Electrical Switching Devices

K. Suzuki[1]
Y. Itoh[2]

ABSTRACT A bounce phenomenon, occurs in electrical switching devices in the case of switching (opening and closing)a point of contact, which are used for control of a motor a lamp and so on. In this research, we analyzed a view point of mechanical bounce. Moreover, compared an analytical result with an experimental result. As a result, the bounce phenomenon can be explained theoretically, and the analytical result is in qualitative agreement with the experimental one.

1 Introduction

Electrical switching devices are widely used for controls of motors, lamps and so on. Then, an electrical contact bounce, so-called a chattering phenomena, usually occurs in electrical switching devices in case of switching, i.e. opening and closing a point of contact.

Previous works [1] have been done mainly from a view point of electrical bounce phenomenon. Therefore, there are few reports which have considered an electrical bounce as a mechanical bounce. In this paper, we analyzed from a view point of mechanical bounce phenomena. Thus, governing equations for the traveling core, the fixed core, the traveling contact and the fixed contact are obtained. Two kinds of collision, the transitional characteristic of the electromagnetic force, are also considered. Moreover, experiments on a transient phenomenon at the time of operating electrical switching device are conducted. In particular, the mechanical collisions between a traveling and a fixed contacts, and a traveling and a fixed cores are observed.

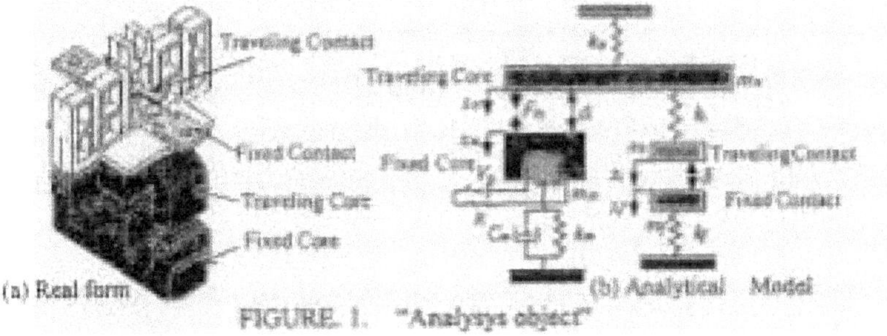

FIGURE. 1. "Analysys object"

2 Theoretical Analysis

2.1 Analytical Model

Figure 1 shows the analytical model of the electrical switching device. The electrical swiching device mainly consists of a traveling contact, a fixed contact, a traveling core, a fixed core.

In this paper, theoretical discussions are conducted for the case where the DC

[1] Fuji Electric Corporate R&D, Ltd., 1 Fuji-machi, Hino, Tokyo, 191-8502, Japan

[2] Kanagawa Industrial Technology Research, 705-1 Shimo-Imaizumi, Ebina, 243-0435, Japan

and AC voltage are applied in the electrical circuit.

2.2 Governing Equation of Electrical Switching Device

The governing equations of the traveling core, the fixed core, the traveling contact and the fixed contact are obtained in the case there exists no mechanical collisions, as follows;

$$m_m \ddot{x}_m = -c_m \dot{x}_m - k_m x_m - F_m \tag{1}$$

$$m_a \ddot{x}_a = -k_a x_a - k_t (x_a - x_t) + F_m \tag{2}$$

$$m_t \ddot{x}_t = -k_t (x_t - x_a) \tag{3}$$

$$m_f \ddot{x}_f = -k_f x_f \tag{4}$$

The electromagnetic force F_m in Eqs.(1) and (2) is given by

$$F_m = \frac{B_m^2 S_m}{2\mu_0} \tag{5}$$

where B_m, S_m and μ_0 are the magnetic flux density, the cross sectional area, and the permeability, respectively. In the case of transient motion, the governing equation of the magnetic flux density B_m is obtained, as follows;

$$\dot{B}_m = -\frac{\Re}{N^2} R B_m + \frac{V}{N S_m} \tag{6}$$

$$\Re = \frac{2}{\mu_0 S_m}(\Delta + x_m - x_a) + \int \frac{dl}{\mu_0 \mu_r S_m} \tag{7}$$

where $\Re$ and μ_r are the reluctance and the relative permeability.
When the collisions between the traveling core and the fixed core, the traveling contact and the fixed contact occurs, we apply the law of conservation of energy and the equation of coefficient of restitution [2]. The collision between the traveling core and the fixed core is expressed, as follows;

$$m_m \dot{x}_m + m_f \dot{x}_f = m_t \dot{x}_a' + m_a \dot{x}_a' \tag{8}$$

$$m_m \dot{x}_m' - m_a \dot{x}_a' = -e_a (m_m \dot{x}_m - m_a \dot{x}_a) \tag{9}$$

and the same way, the collision between the traveling contact and the fixed contact is expressed, as follows;

$$m_t \dot{x}_t + m_f \dot{x}_f = m_t \dot{x}_t' + m_f \dot{x}_f' \tag{10}$$

$$m_t \dot{x}_t' - m_f \dot{x}_f' = -e_c (m_t \dot{x}_t - m_f \dot{x}_f) \tag{11}$$

where the prime (') denotes the condition after the collision occurs.

It is noted from Eqs. (2) and (3) that the motions of the traveling core and the traveling contact are static coupling. From Eqs.(5) through (7) that the electromagnetic force is the functions of the time and the displacement between the fixed core and the traveling core. Moreover, from Eqs. (8) through (11) that the motions of the fixed core and the traveling core, the traveling and fixed contacts are coupled through the collisions.

In this paper, Eqs. (1) through (11) are solved numerically, by a general-purpose software package, in the case where the DC and AC voltage is applied. Thus, V is, respectively, given by V_0 and $V_0 \sin(2\pi ft + \varphi)$, where f is the frequency and the φ is the phase difference.

3 Discussion

3.1 Analytical Results

Figure 2 shows the transient motion of the electrical switching device for closing for the case where the DC and AC voltage are applied to the electrical circuit. All the variables are shown in non dimensional forms. We use δ, m_t, T_0, V_0, B_0 as the representative displacement, mass, time, voltage, magnetic flux density, respectively. T_0 and B_0 are given by

$$\sqrt{m_t / k_t} \text{ and } (V_0 T_0)/(NS_m) \qquad \text{N: Number of turns}$$

From Fig.2, for both cases, it is clarified that the mechanism of the mechanical bounce consists of four typical regions. Thus, in Region I, when the electrical circuit is closed, the electromagnet is gradually excited, and then the traveling core is moved toward the fixed core by the electromagnet force. The traveling contact also begins to move toward the fixed contact in accordance with the motion of the traveling core. In region II, several inelastic impacts not only between both contacts but also the traveling core and the fixed core are observed. In region III, both contacts seems to be in contact condition and to oscillate together, and then the traveling core and the fixed core also are in contact condition. Finally, in region IV, the oscillations decay to zero.

Moreover, the time lag in case of the DC voltage input, i.e. the time of first impact between the traveling core and the fixed core, is much smaller as compared with that of the AC voltage input, since the electromagnetic force on the actuator in case of the DC voltage input is much larger as compared with that of the AC voltage input.

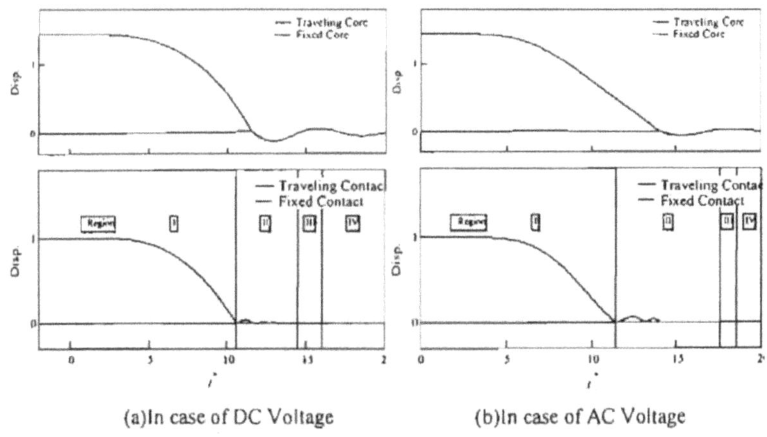

(a)In case of DC Voltage (b)In case of AC Voltage

FIGURE 2. " Transient Motion of Electrical Switching Device"

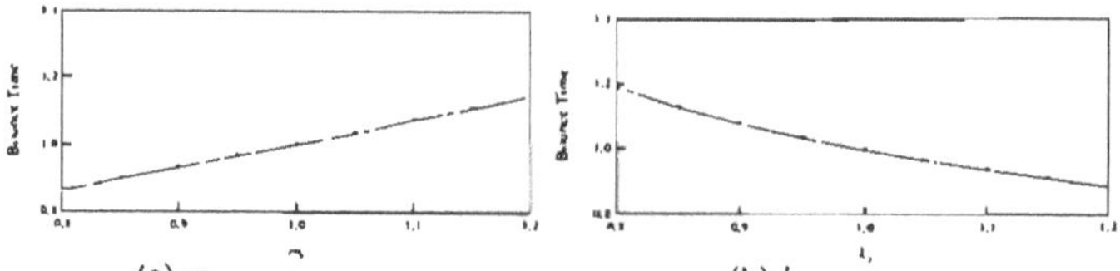

(a) m_t (b) k_t

FIGURE 3. " Bounce Time "

Furthermore, the bounce time of both contacts, defined as the time between the first and last impacts, is examined. Figure 3 shows the numerical results, as examples of the DC input, for the case where the mass m_t and the spring constant k_t of the traveling contact change. (In Fig.3, the values of m_t, k_t and the bounce time for the present device are set 1.0.) From Fig.3, the bounce time will decrease as m_t decreases or k_t increases as compared with the present device.

3.2 Experimental Results

Experiments are conducted not only to observe the mechanical contact bounce but also confirm the analytical results. In the experiments, the velocities of the traveling contact are measured by the laser vibration meter. The experimental results are shown in Fig.4, then the analytical results are also shown. The same tendencies are obtained for both cases although there exists a few differences between the theoretical and the experimental results. Although the graph of a core has not inserted, it is the same result.

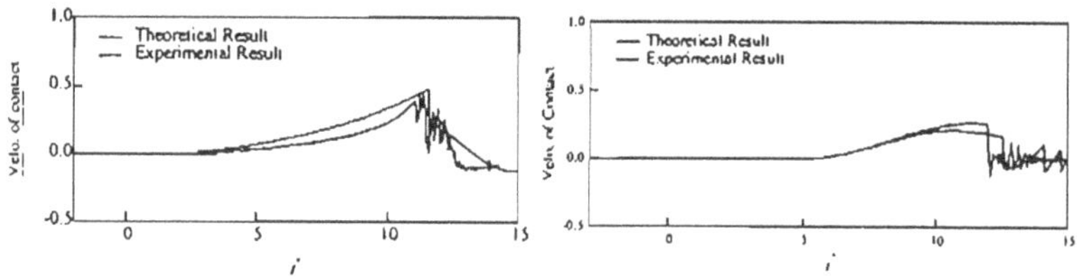

(a)In case of DC Voltage (b) In case of AC Voltage

FIGURE 4. " Experimental Results (velo. of traveling contact) "

4 Conclusion

In this paper, the electrical contact bounce is examined from a view of the mechanical collision. As a result, the characteristics of the mechanical contact bounce are clarified theoretically, for the case where the DC and AC voltage are applied to the electrical circuit to excite the electromagnet. Furthermore, the bounce phenomenon is also observed experimentally, and the experimental results are in qualitative agreement with the analytical ones.

5 References

[1]McBride, J. W. and Sharkh, S., Electrical Contact Phenomena During Impact, IEEE 37th Holm Conference on Electrical Contacts, 91CH 3066-8, (1991).

[2]Zhang, L. X. and *et. al.*, Basic Phenomena of Contact Bounce in the Electrical Switching System, Advanced Computational and Design Techniques in Applied Electromagnetic Systems, (1995), Elsevier Science B.V.

Impact of Structures and Rigid Bodies Undergoing Large Rotations for Vehicle Applications

J. A. C. Ambrósio[1]

ABSTRACT A formulation describing nonlinear deformations of flexible multibody systems in the presence of impact is presented, discussed and applied to a complex scenario of a vehicle rollover with occupant included. A Lagrangian formulation is used to represent the nonlinear geometric and material deformations of flexible multibody components. Though only invariant diagonal mass matrices are used for computational advantage, the coupling between the deformation and the rigid body motion is still described. A vehicle occupant represented by a twelve segment biomechanical model where contact surfaces modeled as ellipsoids are also included in the model. A continuous impact contact force model is used to describe the contact between rigid body segments and other system components. This not only insures the description of the local energy dissipation effects but also guarantees numerical stability during the integration of the equations of motion.

1 Introduction

Multibody systems are generally complex arrangements of structural and mechanical subsystems with different design purposes and mechanical behavior. Depending on the type of applications, operating speeds, external or internal loading of the components, the multibody system may experience small or large deformations that lead to a change of the system performance. The structures, on the other hand, may behave as multibody systems due to their large rotations or because they develop well defined mechanisms of deformation, as in crashworthiness applications.

Finite element based methodologies emerged to model the flexible effects of multibody systems efficiently [1-2]. Initially the description of the flexible effects have been superimposed on the gross rigid body motion with the structural deformations being calculated from the inertial loading of the mechanical components. However, the coupling between the gross motion and the system elastodynamics plays a important role when large or moderate angular velocities of the flexible system components are present. This problem

[1] IDMEC Polo Instituto Superior Técnico, Av. Rovisco Pais, P-1096 Lisboa, Portugal

has been addressed using finite element methods where the mass matrix includes the inertia coupling between the effects mentioned above [3-8].

Based on an updated Lagrangian formulation the equations of motion for a flexible body experiencing general geometric and material nonlinear deformations have been obtained [9]. This is an integrated methodology which is able to capture the full nonlinear structural behavior of the system components and maintain the body mass matrix constant while the inertial coupling is preserved. This methodology has been applied to a wide range of vehicle crashworthiness problems [10-12].

In a general application it is not always necessary to describe the distributed flexibility of the system. The deformations may be either localized or described using other models which include their effect in the contact model. Such is the case of the continuous impact/contact model proposed by Lankarani and Nikravesh [13]. This model is used here to describe the contact between the rigid components of the vehicle, occupant, ground and other impact surfaces during a rollover event with a vehicle occupant integrated in the system.

2 Rigid Multibody Dynamics

A multibody system is a collection of rigid and flexible bodies joined together by kinematic joints and force elements as depicted in figure 1. The presence of the kinematic joints is the main reason why it is normally necessary to use an extended set of coordinates to describe the system. These are called the dependent coordinates and define uniquely the position and orientation of the body in the system.

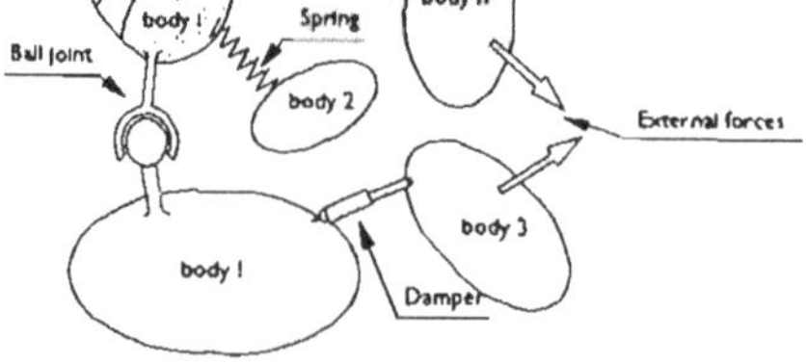

Figure 1. Schematic representation of a multibody system

For the ith body of the system $\mathbf{q}_i$ denotes a vector containing the translational coordinates $\mathbf{r}_i$, and a set of rotational coordinates $\mathbf{p}_i$. A vector of velocities for a rigid body i is defined as v_i contains a 3-vector of translational velocities $\dot{\mathbf{r}}_i$ and a 3-vector of angular velocities ω_i. The vector of accelerations for the body is denoted by $\dot{\mathrm{v}}_i$, the time derivative of v_i. For a multibody system containing *nb* bodies, the vectors of coordinates, velocities, and accelerations are $\mathbf{q}$, $\mathbf{v}$ and $\dot{\mathbf{v}}$ that contain the elements of $\mathbf{q}_i$, $\mathbf{v}_i$ and $\dot{\mathbf{v}}_i$, respectively, for i=1,...,nb.

The constraint equations describing the relative motion between contiguous bodies arise from the kinematic joints. The kinematic constraints are described by *mr* independent equations as

$$\Phi(\mathbf{q}) = \mathbf{0} \tag{1}$$

The first and second derivatives of the constraints yield the kinematic velocity and acceleration equations, respectively

$$\dot{\Phi}(\mathbf{q},t) = \mathbf{0} \quad \equiv \quad \mathbf{D}\,\mathbf{v} = \mathbf{\nu} \tag{2}$$

$$\ddot{\Phi}(\mathbf{q},\dot{\mathbf{q}},t) = \mathbf{0} \quad \equiv \quad \mathbf{D}\,\dot{\mathbf{v}} = \mathbf{\gamma} \tag{3}$$

where $\mathbf{D}$ is the Jacobian matrix of the constraints. The equations of motion for the system of rigid bodies are written [14]

$$\mathbf{M}\ddot{\mathbf{q}} + \mathbf{D}^T\mathbf{\lambda} = \mathbf{g} \tag{4}$$

where $\mathbf{M}$ is the inertia matrix, λ is a vector of unknown Lagrange multipliers, and $\mathbf{g} = \mathbf{g}(\mathbf{q},\mathbf{v})$ contains the forces, moments and gyroscopic terms.

Equations (3) and (4) are a system of differential and algebraic equations that must be solved together to obtain the accelerations. In the numerical solution of the system equations of motion a predictor-corrector, variable order algorithm is used [15] together with a constraint violation stabilization scheme [16] or with an Augmented Lagrangian methodology [17]. This algorithm also includes error estimation and step-size control which is particularly suitable for impact simulations where sudden forces may appear as a result of contact between different parts of the impacting structures.

3 Distributed Flexibility in Multibody Systems

The most general motion of a flexible body involves a continuous change of its shape while undergoing large rotations, as shown in figure 2. The equations of motion of the flexible body are obtained using the principle of virtual works followed by an updated Lagrangian formulation [9]. Let the finite element method be used to represent the equations of motion of the flexible body. Referring to figure 2, the assembly of all finite elements used in the discretization of the flexible body results in the equation of motion for a single flexible body written as:

$$\begin{bmatrix} \mathbf{M}_{rr} & \mathbf{M}_{rf} & \mathbf{M}_{rf} \\ \mathbf{M}_{\phi r} & \mathbf{M}_{\phi\phi} & \mathbf{M}_{\phi f} \\ \mathbf{M}_{fr} & \mathbf{M}_{f\phi} & \mathbf{M}_{ff} \end{bmatrix} \begin{bmatrix} \ddot{\mathbf{r}} \\ \dot{\omega}' \\ \ddot{\mathbf{u}}' \end{bmatrix} = \begin{bmatrix} \mathbf{g}_r \\ \mathbf{g}'_\phi \\ \mathbf{g}'_f \end{bmatrix} - \begin{bmatrix} \mathbf{s}_r \\ \mathbf{s}'_\phi \\ \mathbf{s}'_f \end{bmatrix} - \begin{bmatrix} 0 \\ 0 \\ {}^t_{t'}\mathbf{F} \end{bmatrix} - \begin{bmatrix} 0 & 0 & 0 \\ 0 & 0 & 0 \\ 0 & 0 & {}^t_{t'}\mathbf{K}_L + {}^t_{t'}\mathbf{K}_{NL} \end{bmatrix} \begin{bmatrix} 0 \\ 0 \\ \mathbf{u}' \end{bmatrix} \tag{5}$$

where $\ddot{\mathbf{r}}$ and $\dot{\omega}'$ are, respectively, the translational and angular accelerations of the body-fixed reference frame and $\ddot{\mathbf{u}}'$ denotes the nodal accelerations measured in the body-fixed coordinate system. Note that the local coordinate frame $\xi\eta\zeta$ attached to a point of the flexible body is used to represent the body gross motion and to describe the deformation of the body. Vector $\mathbf{u}'$ denotes the increments of displacements from the previous equilibrium configuration to the current configuration measured in the body-fixed coordinate system.

Equation (5) describes thoroughly the motion of the flexible body. However, this equation is highly nonlinear. The sources of this nonlinear behavior are: a configuration dependent mass matrix; changing external applied forces; gyroscopic and centrifugal forces; nonconstant stiffness matrices. The configuration dependent mass matrix for the flexible body results from the assembly of the individual contributions of each finite element.

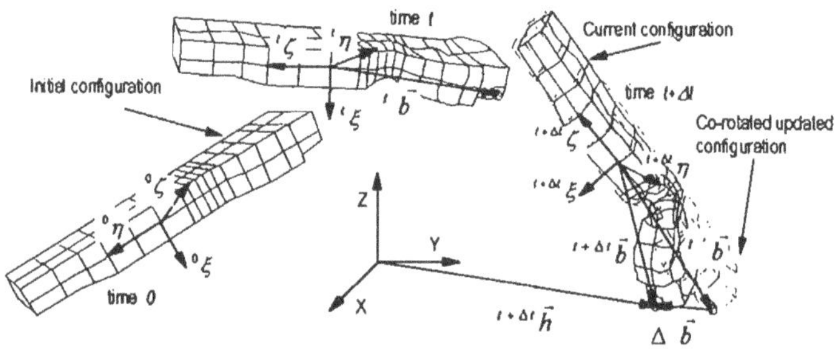

Figure 2. General motion of a flexible body

The right-hand side of equation (5) contains a time varying vector of gyroscopic and centrifugal forces which are evaluated for each element. The vector of the external generalized applied forces for each element is denoted by $\mathbf{g}_f$. Matrices ${}^{\,t}_{t'}\mathbf{K}_L + {}^{\,t}_{t'}\mathbf{K}_{NL}$ are the linear and nonlinear stiffness matrices, respectively, and $\mathbf{f}$ is the vector of equivalent nodal forces due to the actual state of stress. For a nonconstant constitutive tensor ${}_{t'}\mathbf{C}$ both stiffness matrices are nonlinear, as in the case of a multibody system experiencing elasto-plastic deformations of one or more of its components.

During the integration of the system equations of motion the mass matrix is inverted every time step in order to solve the equations explicitly for the accelerations. A simpler form of the equations of motion for a flexible body is obtained using a lumped mass formulation and substituting the local nodal accelerations $\ddot{\mathbf{u}}'$ by the nodal accelerations relative to the inertial frame $\ddot{\mathbf{q}}_f$.

Let the vectors of nodal accelerations be partitioned into

$$\ddot{\mathbf{u}}' = \left[\ddot{\delta}'^T \quad \ddot{\theta}'^T\right]^T \tag{6}$$

The relation between relative and absolute nodal accelerations for a node k is

$$\ddot{\mathbf{q}}'_{f_k} \equiv \begin{bmatrix} \ddot{\mathbf{d}}' \\ \ddot{\alpha}' \end{bmatrix}_k = \ddot{\mathbf{u}}'_k + \begin{bmatrix} \mathbf{A}^T & -\left(\tilde{\mathbf{x}}_k + \tilde{\delta}_k\right)' \\ \mathbf{0} & \mathbf{I} \end{bmatrix} \begin{bmatrix} \ddot{\mathbf{r}} \\ \dot{\tilde{\omega}}' \end{bmatrix} + \begin{bmatrix} \tilde{\omega}'\tilde{\omega}'(\mathbf{x}_k + \delta_k)' + 2\tilde{\omega}'\dot{\delta}'_k \\ \tilde{\omega}'\dot{\theta}'_k \end{bmatrix} \tag{8}$$

where $\mathbf{x}_k$ is the vector containing the node position in the reference configuration. Upon assembling equations (8) for all n nodes and substituting them into equation (5), one obtains

$$\begin{aligned} &\sum_{k=1}^{n} m_k \ddot{\mathbf{d}}'_k = \mathbf{g}_r \\ &\sum_{k=1}^{n} \left[m(\tilde{\mathbf{x}}' + \tilde{\omega}')\ddot{\mathbf{d}}' \right]_k = \mathbf{g}'_\theta \\ &\mathbf{M}_{ff}\ddot{\mathbf{q}}'_f = \mathbf{g}'_f - \mathbf{f} - \left({}^t_t\mathbf{K}_L + {}^t_t\mathbf{K}_{NL} \right)\mathbf{u}' \end{aligned} \tag{9}$$

These equations describe the motion for the center of mass of a system of particles. The last equation represents the motion for the nodes of the flexible body, expressed in the body-fixed coordinate system. Due to the use of the lumped mass formulation the mass matrix $\mathbf{M}_{ff}$ is diagonal

$$\mathbf{M}_{ff} = \mathrm{Diag}\,(\mathrm{m}_1\,\mathbf{I}\,,\mathbf{0}\,,\cdots,\mathrm{m}_k\,\mathbf{I}\,,\mathbf{0}\,,\cdots,\mathrm{m}_n\,\mathbf{I}\,,\mathbf{0}) \tag{10}$$

where m_k is the mass of node k, and **I** and 0 are 3×3 identity and zero matrices associated with translational and rotational degrees of freedom, respectively.

Sometimes it is useful to locate the origin of the body-fixed coordinate system in a point of the flexible body different from its center of mass. Assume that the flexible body has a rigid part and a flexible part with the body-fixed coordinate frame being attached to the rigid part center of mass, as shown in figure 3. The flexible part is attached to the rigid part by the nodes, belonging to the boundary ψ, that share the body-fixed coordinate frame with the rigid part. For a rigid body, the Newton-Euler equations of motion are

$$\begin{aligned} &m\ddot{\mathbf{r}} = \mathbf{f} \\ &\mathbf{J}'\dot{\omega}' = \mathbf{n}' - \tilde{\omega}'\mathbf{J}'\omega' \end{aligned} \tag{11}$$

where $\mathbf{f}$ and $\mathbf{n}'$ are the external forces applied over the center of mass of the rigid part and moments applied over the body and $\mathbf{J}'$ is the inertia tensor. Equations (11) can be used instead of equation (9) provided that proper kinematic constraints are introduced between the flexible and rigid parts of the body. These kinematic constraints enforce null local displacements, velocities and accelerations for the nodes in the boundary ψ

$$\mathbf{u}' = \dot{\mathbf{u}}' = \ddot{\mathbf{u}}' = \mathbf{0} \tag{12}$$

These constraint equations can be applied to equation (9c) for each one of the boundary nodes using the Lagrange multiplier technique. At a later step these multipliers are eliminated from the equations of motion resulting in the dynamic equations [9]

$$\begin{bmatrix} m\mathbf{I}+\overline{\mathbf{A}}\underline{\mathbf{M}}^{*}\overline{\mathbf{A}}^{T} & -\overline{\mathbf{A}}\underline{\mathbf{M}}^{*}\mathbf{S} & \mathbf{0} \\ -\left(\overline{\mathbf{A}}\underline{\mathbf{M}}^{*}\mathbf{S}\right)^{T} & \mathbf{J}'+\mathbf{S}^{T}\underline{\mathbf{M}}^{*}\mathbf{S} & \mathbf{0} \\ \mathbf{0} & \mathbf{0} & \mathbf{M}_{ff} \end{bmatrix} \begin{bmatrix} \ddot{\mathbf{r}} \\ \dot{\boldsymbol{\omega}}' \\ \ddot{\mathbf{q}}'_{f} \end{bmatrix} = \begin{bmatrix} \mathbf{f}+\overline{\mathbf{A}}\mathbf{C}'_{\delta} \\ \mathbf{n}'-\tilde{\boldsymbol{\omega}}'\mathbf{J}'\boldsymbol{\omega}'-\mathbf{S}^{T}\mathbf{C}'_{\delta}-\overline{\mathbf{I}}^{T}\mathbf{C}'_{\theta} \\ \mathbf{g}'_{f}-\mathbf{f}-\left({}^{t}_{t}\mathbf{K}_{L}+{}^{t}_{t}\mathbf{K}_{NL}\right)\mathbf{u}' \end{bmatrix} \tag{13}$$

where $\underline{\mathbf{M}}^{*}$ is a diagonal mass matrix containing the mass of the $\underline{n}$ boundary nodes. Matrices $\overline{\mathbf{A}}^{T}$, $\mathbf{S}$ and $\overline{\mathbf{I}}$ are made from (3x3) matrices as:

$$\overline{\mathbf{A}}=\begin{bmatrix}\mathbf{A} & \cdots & \mathbf{A}\end{bmatrix}\ ;\ \mathbf{S}=\begin{bmatrix}\left(\tilde{\mathbf{x}}'_{1}+\tilde{\boldsymbol{\delta}}'_{1}\right)^{T} & \cdots & \left(\tilde{\mathbf{x}}'_{\underline{n}}+\tilde{\boldsymbol{\delta}}'_{\underline{n}}\right)^{T}\end{bmatrix}^{T}\ ;\ \overline{\mathbf{I}}=\begin{bmatrix}\mathbf{I} & \cdots & \mathbf{I}\end{bmatrix}^{T}$$

Vectors $\mathbf{C}'_{\delta}$ and $\mathbf{C}'_{\theta}$ represent, respectively, the reaction force and moment of the flexible part of the body over the rigid part. These reaction forces and moments are written as

$$\begin{aligned} \mathbf{C}'_{\delta} &= \mathbf{g}'_{\underline{\delta}}-\mathbf{f}_{\underline{\delta}}-\left({}^{t}_{t}\mathbf{K}_{L}+{}^{t}_{t}\mathbf{K}_{NL}\right)_{\underline{\delta}\delta}\boldsymbol{\delta}'+\left({}^{t}_{t}\mathbf{K}_{L}+{}^{t}_{t}\mathbf{K}_{NL}\right)_{\underline{\delta}\theta}\boldsymbol{\theta}' \\ \mathbf{C}'_{\delta} &= \mathbf{g}'_{\underline{\theta}}-{}^{t}_{t}\mathbf{F}_{\underline{\theta}}-\left({}^{t}_{t}\mathbf{K}_{L}+{}^{t}_{t}\mathbf{K}_{NL}\right)_{\underline{\theta}\delta}\boldsymbol{\delta}'+\left({}^{t}_{t}\mathbf{K}_{L}+{}^{t}_{t}\mathbf{K}_{NL}\right)_{\underline{\theta}\theta}\boldsymbol{\theta}' \end{aligned} \tag{14}$$

In these equations, the subscripts δ and θ refer to the partition of the vectors and matrices with respect to the translational and rotational nodal degrees of freedom. The underlined subscripts are referred to the boundary nodal displacements between the rigid and flexible parts.

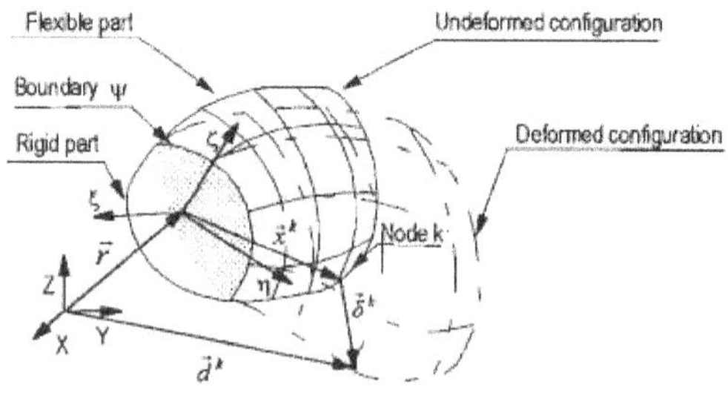

Figure 3. Flexible body with a rigid part and a deformable part

Equation (13) is the finite element equation of motion for a flexible body. When finite elements with rotational degrees of freedom are used to describe the flexible body, some zero elements appear in the diagonal of the mass submatrix $\mathbf{M}_{ff}$. Rotational inertias obtained by lumping the off-diagonal terms of the consistent mass matrix $\mathbf{M}_{ff}$ are used to replace the zero coefficients.

4 Continuous Impact Force Including Local Deformations

Even when the model accounts for the flexibility of the system components the local deformations are responsible for some level of energy dissipation. A model for the contact force must use the material and geometric properties of the surfaces in contact and contribute to a stable integration of the equations of motion. This is achieved with a continuous contact force model [13].

Let the contact force between two bodies of the system or between a component and an external object be a function of the pseudo-penetration and pseudo-velocity of penetration given by

$$\mathbf{f}_{s,i} = \left(K\delta^n + D\dot{\delta}\right)\mathbf{u} \tag{15}$$

where D is a damping coefficient and K is a generalized stiffness coefficient which depends on the geometry of the surfaces in contact and their material properties. The pseudo-penetration velocity $\dot{\delta}$ represents the actual relative velocity between the two contacting surfaces in the point of contact.

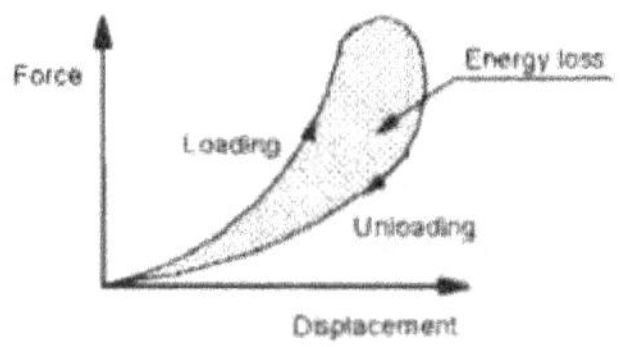

Figure 4. Force displacement relationship accounting for histeresis

The damping coefficient D introduces the hysteresis form of the energy dissipation leading to the energy loss pictured by figure 4. This coefficient can be shown to be a function of the impact velocity $\dot{\delta}^{(-)}$, relative stiffness of the contacting surfaces and restitution coefficient e, and it is written as

$$D = \frac{3K\left(1-e^2\right)}{4\dot{\delta}^{(-)}}\delta^n \tag{16}$$

where the restitution coefficient e reflects the type of impact. The nonlinear contact force is obtained by substituting equation (16) into equation (15)

$$\mathbf{f}_{s,i} = K\ \delta^n\left[1 + \frac{3\left(1-e^2\right)}{4}\frac{\dot{\delta}}{\dot{\delta}^{(-)}}\right]\mathbf{u} \tag{17}$$

This equation represents impact conditions for contacting velocities lower than the propagation speed of elastic waves, i.e., $\dot{\delta}^{(-)} \leq 10^{-5}\sqrt{E/\rho}$.

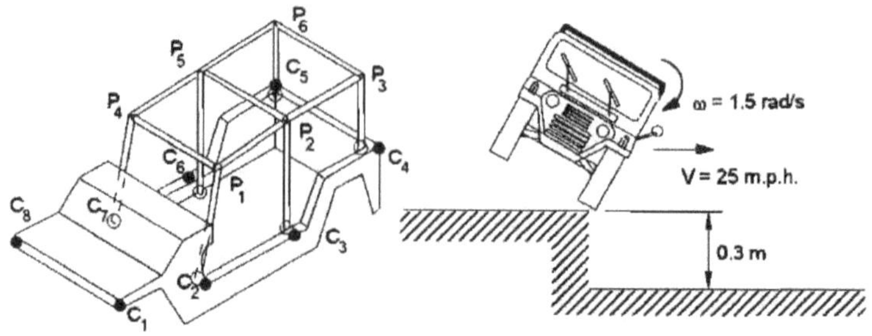

Figure 5 Vehicle model and initial conditions

5 Application to Vehicle Impact With Rollover

5.1 *Vehicle Model*

An all terrain vehicle with a deformable rollbar cage is used in the simulation of a complex impact scenario represented by the rollover described in figure 5. The vehicle is made of 10 rigid bodies and a deformable chassis. All suspension elements, including wheels and tires are introduced in the model. The contact between the tires and the ground is modeled by a comprehensive pneumatic tire model introduced by Gim [19].

The rollbar cage is modeled by nonlinear beam elements that are attached to the main chassis. In this model the chassis is the rigid part of the partially flexible body. The vehicle model has a total of 51 degrees of freedom.

5.2 *Biomechanical Model With Restraint Systems*

Biomechanical models of the human body for crashworthiness applications, with the exception of some aerospace applications, are commonly constructed with rigid bodies only. Local flexibility effects of the biomechanical segments are described by the contact and impact force model presented before. A three-dimensional, whole body response, biomechanical model of the human body using twelve rigid bodies interconnected by eleven kinematic joints is presented in figure 6. There, the potential contact surfaces are also represented. Joint resistance torques that characterize passive muscle action are modeled by viscous torsional dampers located in each kinematic joint. When two contiguous bodies interconnected by a joint reach physically unacceptable positions, penalty torques are applied to each body to limit such relative motionUsing a proper database, models with different sizes and masses can be easily obtained for different application cases.

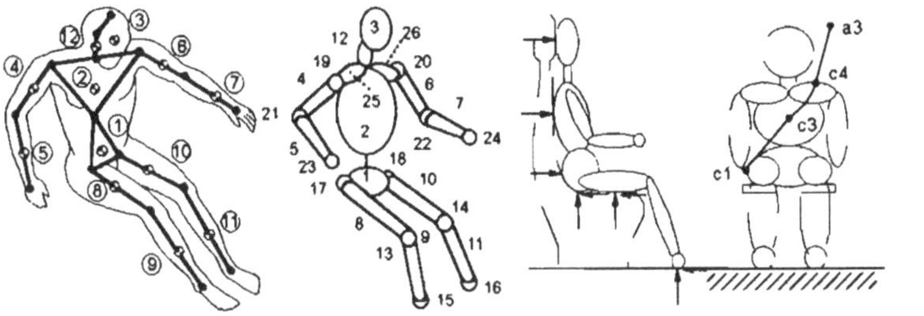

a) Bodies and joints b) Contact surfaces c) Seat contacts and restraint systems

Figure 6. Three-dimensional biomechanical model

5.3 *Vehicle With Occupant Integrated Simulation*

A fully coupled vehicle-occupant model and the interaction between the compartment and occupant is described in the framework of a multibody dynamics formulation in order to illustrate the issues described before during complex contact scenarios [18]. The results of the application of this complex model are presented in figure 7 for a rollover simulation.

Figure 7. Occupant and vehicle integrated models in a rollover simulation.

This type of models describe not only the structural deformations at the vehicle level but also the potential for occupant injury in terms of an injury criteria. A permanent deformation of 140 mm was observed in the rollbar cage at node P_1, which correlates very well with experimental results [20]. This result is presented in figure 8 where the simulation of the rollover with other methodologies is also pictured. Moreover, the vehicle rollover motion is similar to that monitored during the actual experiment.

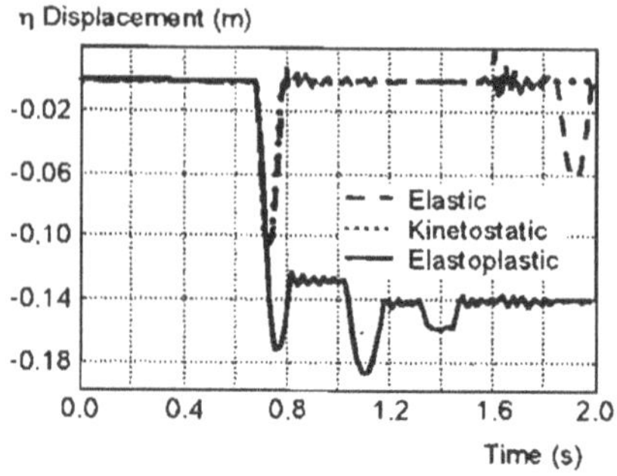

Figure 8 Deformation of node P_1 of the rollbar cage in the local Y direction

In the rollover it was found that an extremely high contact forces between occupant and vehicle developed. It is clear from the forces that develop at the seat belts level, described in figure 9, that the occupant has to be heavily constrained to the seat. However, contact forces between the occupant's head and the internal structure of the vehicle are still very high, corresponding to an HIC of 5400 which means that the potential for survival is null [18].

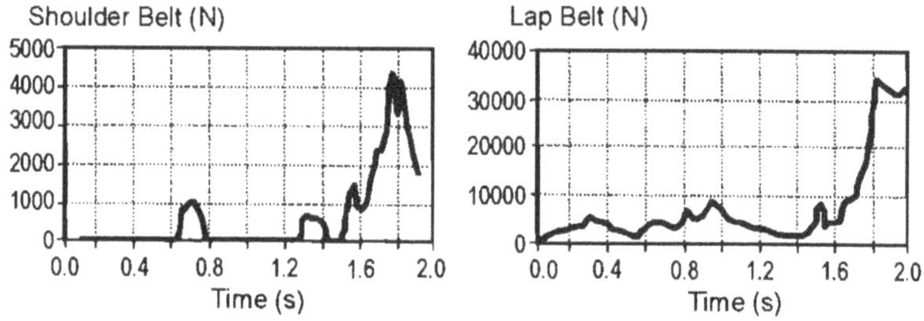

Figure 9 Forces in the seat belts

6 Conclusions

Different aspects of the impact involving complex multibody systems have been addressed in this work. Formulations allowing for the combined multibody and nonlinear finite element description of the flexible parts have been presented. Based on a general description of the continuum mechanics equations and using an updated Lagrangian formulation, the method proposed has no restrictions on the type of deformations appearing in a multibody model. When the contact develops between rigid parts of the multibody a continuous contact impact model has been applied. The simulation of a vehicle and integrated occupant has been presented in order to show that the complex interactions can be efficiently modeled. The results show that most valuable information about the system deformation and occupant injury potential can be extracted from these integrated models.

Further research efforts are still required in order to improve the efficiency of multibody dynamics as applied to flexible systems. These are: adequate modeling of nonlinear flexible bodies with formulations using only structural stiffness and mass terms obtained in standard finite element formulations; to allow a greater interaction with commercially available codes with the possibility to introduce more complex crushable elements; introduction of piecewise analysis techniques based on momentum balance methods for an efficient evaluation of velocity jumps during impact; development of new methodologies for constraint addition and deletion capabilities to deal in a more general manner with the contact problem, and finally, the exploitation of more efficient integration algorithms.

7 Acknowledgements

The authors gratefully acknowledge the support of JNICT - Junta Nacional de Investigação Científica e Tecnológica through Project PBICT/P/CEG/2348/95 and AGARD - Advisory Group for Aerospace Research and Development through project P124.

8 References

[1] Erdman, A.G., Sandor, G.N. (1972) Kineto-elastodynamics – a review of the state of the art and trends. Mech. Mach. Theory **7**, 19-33

[2] Thompson, B.S., Sung, C.K. (1986) Survey of finite element techniques for mechanism design. Mech. Mach. Theory **21**(4), 351-359

[3] Song, J.O., Haug, E.J. (1980) Dynamic analysis of planar flexible mechanisms. Comp. Meth. In Appl. Mech. And Engng. **24**, 359-381

[4] Shabana, A. , Wehage, R. (1983) Variable degree of freedom component mode analysis of inertia variant flexible mechanical systems. ASME J. of Mechanisms, Transmissions and Automation in Design **105**, 371-378

[5] Pereira, M.S., Proença, P.L. (1991) Dynamic analysis of spatial flexible multibody systems using joint coordinates. Int. J. Num. Meth. in Engng. **32**, 1799-1812.

[6] Kane, T.R., Ryan, R.R., Banerjee, A.K. (1987) Comprehensive theory for the dynamics of a general beam attached to a moving rigid base. J. of Guidance, Control and dynamics **10**(2), 139-151

[7] Cardona, A., Geradin, M. (1988) A beam finite element nonlinear theory with finite rotations. Int. J. Num. Meth. In Engng. **26**, 2403-2438

[8] Geradin, M., Cardona, A., Doan, D.B., Duysens, J. (1994) Finite element modeling concepts in multibody dynamics. In: M. Pereira, J. Ambrósio (eds.) Computer aided analysis of rigid and flexible mechanical systems. NATO ASI Series E. Vol. 268, 233-284. Dordrecht: Kluwer Academic Publishers

[9] Ambrósio, J.A.C., Nikravesh, P.E. (1992) Elastic-plastic deformations in multibody dynamics. Nonlinear Dynamics **3**, 85-104

[10] Ambrósio, J.A.C., Nikravesh, P.E., Pereira, M.S. (1990) Crashworthiness analysis of a truck. J. Math. Comp. Modeling **14**, 959-964

[11] Ambrósio, J.A.C. Silva, M.P.T., and Gonçalves, J.P. (1996) Development of energy absorbing devices using a kinetostatic multibody dynamics methodology. Int. J. Crashworthiness. **1**(2), 127-143

[12] Pereira, M.S., Ambrósio, J.A.C., Dias, J.P. (1997) Crashworthiness analysis and design using rigid-flexible multibody dynamics with application to train vehicles. Int. J. Num. Meth. in Engng. **40**, 655-687

[13] Lankarani, H.M., Nikravesh, P.E. (1994) Continuous contact force models for impact analysis in multibody systems. Nonlinear Dynamics **5**(2), 193-207

[14] Nikravesh, P.E. (1988) Computer aided analysis of mechanical systems. Englewood Cliffs: Prentice-Hall.

[15] Shampine, L.F., Gordon, M.K. (1975) Computer solution of ordinary differential equations: the initial value problem. San Francisco: V.H.Freeman and Co.

[16] Baumgarte, J. (1972) Stabilization of constraints and integrals of motion. Computer Methods in Applied Mechanics Engineering. **1**, 1-16

[17] García de Jalón, J., Bayo, E. (1993) Kinematic and dynamic simulation of multibody systems - the real time challenge. New York: Springer-Verlag

[18] Silva, M.P.T, Ambrósio, J.A.C., Pereira, M.S. (1997) A multibody approach to the vehicle and occupant integrated simulation. Int. J. of Crashworthiness **2**(1), 73-90

[19] Gim, G. (1988) Vehicle dynamic simulation with a comprehensive model for pneumatic tyres. Ph.D. Dissertation. University of Arizona

[20] Ambrósio, J.A.C., Pereira, M.S. (1997) Multibody dynamic tools for crashworthiness and impact. In: J.A.C. Ambrósio, M.S. Pereira, F.P Silva (eds.) Crashworthiness of transportation systems: structural impact and occupant protection. NATO ASI Series E. Vol. 332, 475-521. Dordrecht: Kluwer Academic Publishers

SESSION IV–(A)

Modeling, Computations

The Role of Tribology in the Modeling of Dynamic Contact Problems ... and Vice Versa

A. Soom[1]

Abstract

In many contact problems friction is included as boundary condition or a force that is modeled, as a function of sliding velocity or other parameters. Many of these models such as the coefficient of friction, (the simplest one), contain many hidden variables and assumptions. Tribology research over the past fifty or more years has advanced our understanding of dry and lubricated contacts mostly from static or quasistatic points of view. These quasi-static or steady sliding problems can offer useful points of departure when studying and modeling dynamic or unsteady contact phenomena. It often allows a separation of constitutive interface behavior from the dynamic response of the overall mechanical system. In this way, some of the hidden assumptions can be revealed so that dynamicists don't seek nonexistent answers from tribologists and tribologists don't too readily ascribe unexplained results on system dynamics. In this address, a few examples of dry and lubricated contact problems will be presented to illustrate these ideas.

[1]Department of Mechanical and Aerospace Engineering, State University of New York at Buffalo, Buffalo, NY 14260

Finite Element Treatment of Tribological Problems

N. Strömberg[1]

ABSTRACT The present paper concerns the numerical treatment of contact, friction and wear between a thermoelastic body and a rigid foundation. The governing equations of thermoelasticity, Signorini contact, Coulomb friction and Archard wear are put together to a system of equations which is solved by using a Newton method.

1 Governing Equations

Let us consider a discretized thermoelastic body for which a potential contact surface is unilaterally constrained by a rigid support. It is assumed that no heat can transfer through the support. The tribological laws are given by Signorini's contact conditions, Coulomb's law of friction and Archard's law of wear. The total number of kinematic freedoms is η_u, the total number of nodal temperatures is η_T and the number of potential contact nodes is η_c. The displacement vector is represented by $\mathbf{u} \in \Re^{\eta_u}$ and the nodal temperatures by $\mathbf{T} \in \Re^{\eta_T}$.

For each contact node, $i = 1, \ldots, \eta_c$, three unit directions are defined: one normal direction and two tangential directions with respect to the support. Using these unit directions, it is possible to express the normal displacement and the tangential displacements at each contact node in the following way:

$$u_{in} = \mathbf{C}_{in}\mathbf{u} \in \Re, \qquad u_{it} = \mathbf{C}_{it}\mathbf{u} \in \Re, \qquad u_{io} = \mathbf{C}_{io}\mathbf{u} \in \Re,$$

where subscripts n, t and o stand for normal, tangential and orthogonal direction, respectively, and $\mathbf{C}_{in}$, $\mathbf{C}_{it}$ and $\mathbf{C}_{io}$ are the i:th rows of the transformation matrices $\mathbf{C}_n \in \Re^{\eta_c \times \eta_u}$, $\mathbf{C}_t \in \Re^{\eta_c \times \eta_u}$ and $\mathbf{C}_o \in \Re^{\eta_c \times \eta_u}$.

The equilibrium of forces can be written as

$$\bar{\mathbf{K}}\mathbf{u} + \hat{\mathbf{K}}(\mathbf{T} - \mathbf{T}_0) = \mathbf{F} - \mathbf{C}_n^T\mathbf{P}_n - \mathbf{C}_t^T\mathbf{P}_t - \mathbf{C}_o^T\mathbf{P}_o, \tag{1}$$

[1]Division of Mechanics, Department of Mechanical Engineering, Linköping Institute of Technology, S-581 83 Linköping, Sweden, e-mail: nicst@ikp.liu.se

where $\bar{\mathbf{K}} \in \Re^{\eta_u \times \eta_u}$ is the elastic stiffness matrix, $\hat{\mathbf{K}} \in \Re^{\eta_u \times \eta_T}$ is the matrix corresponding to the thermal dilatation, $\mathbf{T}_0 \in \Re^{\eta_T}$ is the nodal reference temperatures, $\mathbf{F} \in \Re^{\eta_u}$ is the external nodal force vector, $\mathbf{P}_n \in \Re^{\eta_c}$ is the normal contact force vector, and $\mathbf{P}_t \in \Re^{\eta_c}$ and $\mathbf{P}_o \in \Re^{\eta_c}$ are the tangential contact forces.

Utilizing the standard heat conduction equation of uncoupled thermoelasticity, the balance of energy can be expressed by

$$\mathbf{M}\dot{\mathbf{T}} + \mathbf{O}\mathbf{T} = \mathbf{L}(\mathbf{u}, \mathbf{P}_n), \tag{2}$$

where $\mathbf{M} \in \Re^{\eta_T \times \eta_T}$ is the matrix corresponding to the heat capacity of the body, $\mathbf{O} \in \Re^{\eta_T \times \eta_T}$ is the matrix corresponding to the thermal conductivity and $\mathbf{L}(\mathbf{u}, \mathbf{P}_n) \in \Re^{\eta_T}$ is the dissipated heat due to friction and wear, where each component, $j = 1, \ldots, \eta_T$, is given by

$$L_j(\mathbf{u}, \mathbf{P}_n) = \sum_{i=1}^{\eta_c} N^j(\mathbf{x}_i)(\frac{k}{I_i}(P_{in})_+^2 + \mu(P_{in})_+)\sqrt{(\dot{u}_{it})^2 + (\dot{u}_{io})^2}. \tag{3}$$

Hear, $N^j(\mathbf{x}_i)$ represents the shape functions, $\mathbf{x}_i$ is the location of the i:th contact node and I_i is a weighting factor depending on the quadrature rule used in the space discretization of the contact surface. Furthermore, a superimposed dot stands for time derivative, $(x)_+ = (x + |x|)/2$ and, finally, μ and k are the coefficients of friction and wear, respectively.

In order to account for wear in the normal contact law, Signorini's unilaterally contact conditions are extended in the following manner (for $i = 1, \ldots, \eta_c$):

$$P_{in} \in \Re_+ : \quad (u_{in} - w_i - g_i)(P_{in}' - P_{in}) \leq 0 \quad \forall P_{in}' \in \Re_+, \tag{4}$$

where w_i is the nodal value of the internal state variable measuring the wear gap, g_i is the initial gap between the contact node and the support, $\Re_+$ is the non-negative orthant of $\Re$.

For each contact node, $i = 1, \ldots, \eta_c$, Coulomb's law of friction is given by the following dissipation principle: find $(P_{it}, P_{io}) \in \mathcal{F}(P_{in})$ such that

$$\dot{u}_{it}(P_{it}' - P_{it}) + \dot{u}_{io}(P_{io}' - P_{io}) \leq 0 \quad \forall (P_{it}', P_{io}') \in \mathcal{F}(P_{in}), \tag{5}$$

where

$$\mathcal{F}(P_{in}) = \{(P_{it}, P_{io}) : \sqrt{(P_{it})^2 + (P_{io})^2} \leq \mu(P_{in})_+\}$$

is Coulomb's set of admissible tangential contact forces.

Finally, the evolution of the internal state variable w_i is governed by Archard's law of wear. That is

$$\dot{w}_i = \frac{k}{I_i}(P_{in})_+\sqrt{(\dot{u}_{it})^2 + (\dot{u}_{io})^2}, \qquad \text{for } i = 1, \ldots, \eta_c. \tag{6}$$

The tribological problem is now stated in equations (1)-(6). That is, for a given load history $\mathbf{F}(t)$, $t \in [0,T]$, on a time interval $[0,T]$, the following quasi-static problem is to be solved: find $t \mapsto \mathbf{u}(t)$, $t \mapsto \mathrm{T}(t)$, $t \mapsto \mathbf{P}_n(t)$, $t \mapsto \mathbf{P}_t(t)$ and $t \mapsto \mathbf{P}_o(t)$ such that the functions $\mathbf{u}(t)$, $\mathrm{T}(t)$, $\mathbf{P}_n(t)$, $\mathbf{P}_t(t)$ and $\mathbf{P}_o(t)$ satisfy conditions (1)-(6). A more detailed derivation of this problem can be found in [1] as well as the derivation of the numerical method presented below.

2 The Newton Method

The tribological problem defined above is treated by first approximating the time rates appearing in (2), (3), (5) and (6) at a time t_{k+1} by using a backward Euler discretization, then defining $(\bar{u}_{it}, \bar{u}_{io}) = (u_{it}(t_{k+1}) - u_{it}(t_k), u_{io}(t_{k+1}) - u_{io}(t_k))$, rewriting (4) and (5) using an approach of projections and, finally, putting together (1) and (2).

In such manner, letting $\mathbf{y} = (\mathbf{u}, \mathbf{P}_n, \mathbf{P}_t, \mathbf{P}_o)$, the following system of equations is derived at a time t_{k+1}: $\mathbf{H}(\mathbf{y}) =$

$$\left\{ \begin{array}{c} \bar{\mathbf{K}}\mathbf{u} - \tilde{\mathbf{K}}(\mathbf{Q} + \bar{\mathbf{L}}(\mathbf{u}, \mathbf{P}_n)) + \mathbf{C}_n^T\mathbf{P}_n + \mathbf{C}_t^T\mathbf{P}_t + \mathbf{C}_o^T\mathbf{P}_o + \hat{\mathbf{K}}\mathrm{T}_0 - \mathbf{F} \\ (-P_{in} + \Pi_{in}(\mathbf{y}))_{i=1}^{i=\eta_c} \\ \left(-\begin{pmatrix} P_{it} \\ P_{io} \end{pmatrix} + (\mathbf{\Pi}_{ito}(\mathbf{y}))^T\right)_{i=1}^{i=\eta_c} \end{array} \right\} = 0,$$

where $\tilde{\mathbf{K}} = \hat{\mathbf{K}}(\mathbf{M} + (t_{k+1} - t_k)\mathbf{O})^{-1}$, $\mathbf{Q} = \mathbf{M}\mathrm{T}(t_k)$ and

$$\bar{L}_j(\mathbf{u}, \mathbf{P}_n) = \sum_{i=1}^{\eta_c} N^j(\mathbf{x}_i)(\frac{k}{I_i}(P_{in})_+^2 + \mu(P_{in})_+)\sqrt{(\bar{u}_{it})^2 + (\bar{u}_{io})^2}$$

is the time discretization of (3). Furthermore, for any $r > 0$,

$$\Pi_{in}(\mathbf{y}) = (P_{in} + r(u_{in} - w_i(t_k) - \frac{k}{I_i}(P_{in})_+\sqrt{(\bar{u}_{it})^2 + (\bar{u}_{io})^2} - g_i))_+$$

is the projection of $P_{in} + r(u_{in} - w_i - g_i)$ onto $\Re_+$ with the implicit approximation of (6) inserted, and

$$\mathbf{\Pi}_{ito}(\mathbf{y}) = \left\{ \begin{array}{ll} (P_{it}(r), P_{io}(r)) & \text{if } (P_{it}(r), P_{io}(r)) \in \mathcal{F}(P_{in}) \\ \dfrac{\mu(P_{in})_+(P_{it}(r), P_{io}(r))}{\sqrt{(P_{it}(r))^2 + (P_{io}(r))^2}} & \text{otherwise} \end{array} \right.$$

is the projection of $(P_{it}(r) = P_{it} + r\bar{u}_{it}, P_{io}(r) = P_{io} + r\bar{u}_{io})$ onto $\mathcal{F}(P_{in})$. Notice that the two projections above are derived from the variational inequalities, representing the laws of Signorini and Coulomb, given in (4) and (5).

The system of equations given above, i.e. $\mathbf{H}(\mathbf{y}) = \mathbf{0}$, is solved using the following Newton algorithm:

Algorithm: Let β, γ and ϵ be given scalars with $\beta \in (0,1)$, $\gamma \in (0,1/2)$ and ϵ small. Repeat the following steps for each time increment $k+1$:

0: Let $\mathbf{y}^0$ be given from the previous time step k and set $l = 0$.

1: Find a search direction $\mathbf{z}$ such that

$$\mathbf{H}(\mathbf{y}^l) + \mathbf{H}'(\mathbf{y}^l;\mathbf{z}) = \mathbf{0}, \tag{7}$$

where $\mathbf{H}'(\mathbf{y}^l;\mathbf{z})$ is the directional derivative of $\mathbf{H}(\mathbf{y}^l)$.

2: Let $\alpha^l = \beta^{m_l}$, where m_l is the first non-negative integer m for which the following decrease criterion holds:

$$g(\mathbf{y}^l + \beta^m \mathbf{z}) \leq (1 - 2\gamma\beta^m) g(\mathbf{y}^l), \qquad g(\mathbf{y}) = \frac{1}{2}\mathbf{H}^T(\mathbf{y})\mathbf{H}(\mathbf{y}).$$

3: Set $\mathbf{y}^{l+1} = \mathbf{y}^l + \alpha^l \mathbf{z}$.

4: If $g(\mathbf{y}^{l+1}) \leq \epsilon$, then terminate with $\mathbf{y}^{l+1}$ as an approximate zero of $\mathbf{H}(\mathbf{y})$. Otherwise, replace l by $l+1$ and return to step 1.

After convergence in each time step is achieved, the variables $\mathbf{T}$, u_{il}, u_{io} and w_i are updated.

The essential task of the algorithm is to find the search direction. That is solving (7) in step 1 of the algorithm, which generally is a system of non-linear equations. This is handle by introducing an appropriate substitution of $\mathbf{H}'(\mathbf{y}^l;\mathbf{z})$ such that (7) results in a system of linear equations which still yields an effective search direction, *see* [1].

The numerical method presented above has proven to be robust and efficient; it has been utilized to solve a number of tribological problems of didactic character [1]. For instance, the evolution of contact stresses and wear gaps have been determined for a number of fretting problems. It has been found that the contact stresses strongly depend on the removal of material due to wear. It has also been found that the frictional heat generation strongly depends on the wear rate. This and results from other similar calculations will be presented at the conference.

3 References

[1] Strömberg, N., *Thermomechanical Modelling of Tribological Problems*, Linköping Studies in Science and Technology. Dissertations No. 497, Linköping, Sweden, 1997

Efficient Computation of General Hertzian Traction Integrals

M. Törmänen[1]

This paper presents an efficient numerical procedure for computing the general relationship between traction and motion at a rolling and spinning Hertzian point contact. The contact model is based on the dry friction concept which also displays the behaviour of contacts where solidified oil films are present.

1 Introduction

Many mechanical systems contain power-transferring point contacts. One category is mechanical traction drive variators where these contacts are found between bodies that are pressed together and moved against one another. Thereby, the power is transferred from one body to the other by traction forces. In this treatise, the relative motion is considered as a combination of rotational and translational velocity, and the contact shape as a plane ellipse. This view agrees well with Hertz' contact theory. Therefore, these contacts will here be referred to as *rolling and spinning Hertzian point contacts.*

The aim of this work was to establish an efficient numerical procedure for computation of the tractive forces at a Hertzian point contact with arbitrary rolling and spinning relative motion. This is achieved with good accuracy and short computation time. The analysis is based on the contact model of Mägi, [1], which also covers contacts where isothermal solidifying lubricants are present. To ensure that the procedure performs well for different cases of relative motion, various numerical methods are considered. This includes also robust methods for reference value calculation and numerical noise detection. With this procedure it is possible to analyse systems undergoing transient geometry changes and shifting operating conditions. A more thorough description will be found in [2].

2 Theoretical Background

The relative motion that causes the tractive effort between two mutually compressed bodies is uniquely defined by one single point in the common tangential plane, see fig. 1. This point is the instantaneous centre of relative rotation and is here referred to as the *speed pole*. In the same manner the resulting traction can be defined with one single force at a specific location. This is the fundamental form of the pole concept of Mägi, [1].

1. Department of Machine and Vehicle Design, Chalmers University of Technology, S-412 96 Göteborg, Sweden. E-mail: mitn@mvd.chalmers.se

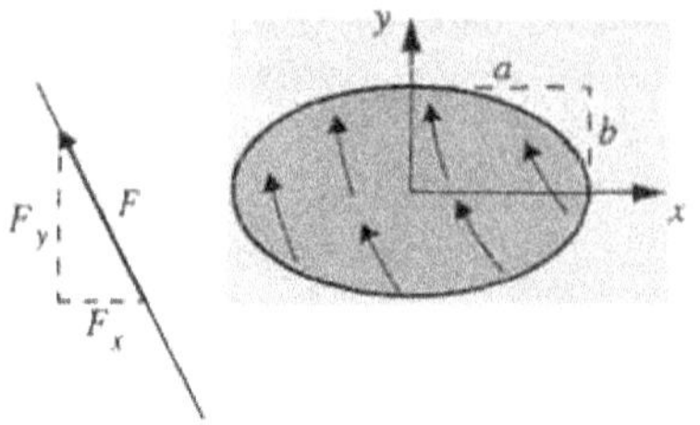

FIGURE 1. Traction-motion interaction for a Hertzian point contact depicted in the common tangential plane. The figure illustrates the sliding relative motion about the speed pole *sp* and the corresponding line of action of the tractive force *F*.

The constitutive relationships between the resulting traction and the relative motion are described by the following three integrals found in ref. [1]:

$$\theta_x = \frac{F_x}{\mu N} = \frac{3}{2\pi}\iint_D \sqrt{1-\xi^2-\eta^2}\,\frac{\eta/\alpha-\varepsilon_y}{\sqrt{(\varepsilon_x-\alpha\xi)^2+(\varepsilon_y-\eta/\alpha)^2}}\,d\xi d\eta\,, \tag{1}$$

$$\theta_y = \frac{F_y}{\mu N} = \frac{3}{2\pi}\iint_D \sqrt{1-\xi^2-\eta^2}\,\frac{\varepsilon_x-\alpha\xi}{\sqrt{(\varepsilon_x-\alpha\xi)^2+(\varepsilon_y-\eta/\alpha)^2}}\,d\xi d\eta\,, \tag{2}$$

$$\theta_\varphi = \frac{M}{\mu N\sqrt{ab}} = \frac{3}{2\pi}\iint_D \sqrt{1-\xi^2-\eta^2}\,\sqrt{(\varepsilon_x-\alpha\xi)^2+(\varepsilon_y-\eta/\alpha)^2}\,d\xi d\eta\,, \tag{3}$$

$$D = \{(\xi,\eta):\ (\xi^2+\eta^2\le 1)\}\,. \tag{4}$$

These non-dimensional integrals depend, here, only on the non-dimensional speed pole location $(\varepsilon_x, \varepsilon_y)$ and the non-dimensional contact ovality (α). Equation (3) yields the perpendicular distance to the line of action from the speed pole.

3 Problem Statement

Already today, several efficient numerical procedures for calculation of double integrals are available. However, most of these procedures use methods that do not work well when the integrands to the double integrals considered are affected of singularities or discontinuities. This kind of problem occurs for the traction integrals when the speed pole is located within the contact. It results in decreased performance, i.e. slower computation and larger errors. Fig. 2 left displays the error signal from a commercially available two-dimensional integration routine, [3], when it is applied on the traction integrals when the contact ovality α and the relative accuracy requested is 1.5 and 10^{-6}, respectively. As it is shown in fig. 2 left, it is very obvious that the integrals become troublesome to evaluate when the speed pole is located within the contact, e.g., point *P* in fig. 2 right. In fact, the contact boundary surrounds the distribution of the errors and the accuracy is here as low as 10^{-3}.

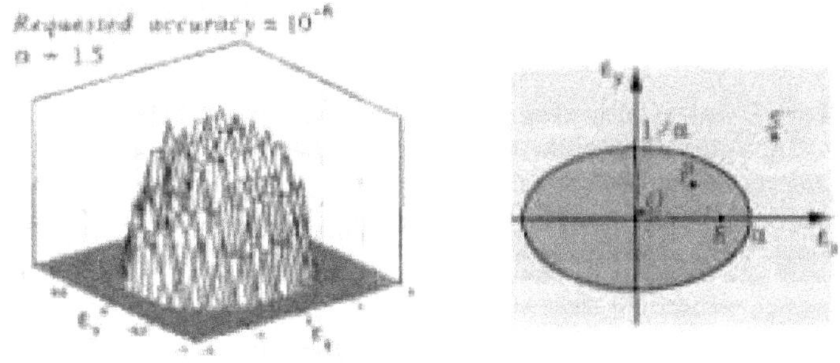

FIGURE 2. Left: Mesh of the error signal from the two-dimensional integration routine *D01DAF* when evaluating the traction integrals for various speed pole locations. Right: Illustration of different speed pole locations where different numerical methods are used.

In contrary, when the speed pole is located outside the contact, point S, the integrals are troubleless to evaluate for a requested accuracy up to 10^{-8}

4 New Efficient Numerical Procedure

4.1 Method A: Coordinate Transformation

By using anti-symmetrical polar coordinates, see ref. [2], about the speed pole when the speed pole is located within the contact, point P in fig. 2 right, with exception of the contact origin, the integrals can be evaluated nicely with some commercially available integration routine. This method works very well within the whole contact since the integrand functions do not have any discontinuities and provides an error signal that is zero for a requested accuracy up to 10^{-8}. For the case when the speed pole is located in the origin there exist analytical solutions, see ref. [1]. This method is very good to use in analyses where the constitutive relationships are not computed repetitiously. However, when implementing this method to programs where these relationships must be evaluated multiple times in each time step, e.g. Multi-body Dynamics software, fast computation is as important as high accuracy.

4.2 Method B: Two-dimensional Taylor Expansion

When the speed pole is located close to the origin, point Q in fig. 2 right, efficient solutions can be established by using two-dimensional series expansion where the coefficients are analytically calculated, [2]. The expansions to θ_x and θ_y are of first order, while the solution to θ_φ is of second order. The later is due to the relationships $\partial\theta_\varphi/\partial\varepsilon_y = \theta_x$ and $\partial\theta_\varphi/\partial\varepsilon_x = -\theta_y$

4.3 *Method C: Reduction of Integration Dimensions*

When the speed pole is located at the semi-axes or the extensions to those, point R in fig. 2 right, a method has been developed by Mägi, [1], which reduces the two-dimensional integrals to one-dimensional integrals by using complete elliptical integrals of first and second order. These elliptical integrals are evaluated with geometric series and enables very fast and accurate evaluations of the remaining integrals. These numerical procedures is shown in ref. [4].

4.4 *Method D: Maclaurin Expansion*

When the speed pole is located within the contact, point P in fig. 2 right, very efficient solutions to the integrals is found with Maclaurin expansion about either one of the semi-axes to the contact ellipse, [2]. The coefficients is computed in the same manner as the double-integrals in method C. Within the whole contact the max. error can be as low as 10^{-8}. It is shown that the accuracy decrease rapidly when the speed pole moves outside the contact area. However, there exist areas near the semi-axes or the extensions to those where this method is still useful.

4.5 *General Procedure*

For a given accuracy an efficient numerical procedure is then built by using the three efficient methods *B,C, and D* to provide solutions in those regions where they are the most efficient and letting the remaining regions to some commercial method. Thereby, the region will dynamic and will be larger for lower accuracies. It is shown, [2], that these efficient methods cover the usual speed pole locations.

5 Conclusion

The general numerical procedure takes advantage of the most efficient methods now available to evaluate the traction motion relationships for an accuracy up to 10^{-8}.

6 References

[1] Mägi, M., On Friction at Elliptic Hertzian Contact Areas with Special Reference to Variable Drive Application, Lic. thesis, Göteborg, Sweden, 1964.

[2] Törmänen, M., Design of A Powerful Tool for Analysis of Mechanical Systems Containing Rolling and Spinning Hertzian Point Contacts, Lic. thesis, Göteborg, Sweden, 1998. (To be published)

[3] NAG Fortran Library Manual, Mark 15, Numerical Algorithms Group Ltd, 1991.

[4] Mägi, M., Versatile numerical procedures - based on speed and force pole offsets - for prediction of efficiency of rolling contact type CVTs, VDI-Berichte Nr. 1207, 1995.

On Modeling Rapid Granular Flows

G. Ahmadi[1]

ABSTRACT In this paper the recent developments concerning modeling of rapid flows of granular material is presented. Particular emphasis is given to thermodynamically-based continuum models. It is shown that these models include a transport equation for the particulate fluctuation energy, and in more advanced version, for evolution of components of the kinetic stress tensor. The implication of generalized (extended) thermodynamical principles in modeling particulate flows in a state of rapid motion are described. The special cases of rapid simple shear flows of granular materials is discussed. Comparisons of the model predictions for granular stresses and normal stress differences with the available simulation results are presented. The effects of inter-granular friction, and interstitial fluids on flows of particulate media is also discussed.

1 Introduction

Last two decades have seen a rather exciting development in modeling rapid granular flows. In one approach the granules are treated as molecules of a dense gas and the techniques of the kinetic theory of gases are used to derive the governing equations for the mean motion. In the other approach, similarities between the rapid flows of granular materials and the turbulent fluid flows is exploited and turbulence modeling techniques are used. In this approach, ensemble averaging is used to derive the basic balance laws for the mean motion. The closure techniques of hydrodynamic turbulence are then used for prescibing the needed constitutive equations. In this paper continuum-based models for rapid flows of granular material are described. The governing equations are discussed and the implication of second law of thermodynamics in modeling of particulate flows is described. Particular attention was given to the theory of thermodynamically consistent rate-dependent model for flow of granular materials as developed by Ahmadi and Abu-Zaid [1]. The special case of simple shear flow of granular materials in the grain inertia regime is also described.

[1]Department of Mechanical and Aeronautical Engineering, Clarkson University, Potsdam, NY 13699-5725

1.1 Basic Equations

The mass-weighted averaged balance laws for granular materials in grain inertia regime are:

Mass

$$\frac{\partial p}{\partial t} + (\rho v_i)_{,i} = 0 \ , \tag{1}$$

Linear Momentum

$$\rho \dot{v}_i = \tau_{ji,j} + \rho f_i \ , \tag{2}$$

Fluctuation Energy

$$\rho \dot{k} = \tau_{ij} v_{j,i} + K_{i,i} - \rho \epsilon \ . \tag{3}$$

In these equations, ρ is the mass density, v_i is the mass-weighted mean velocity vector, τ_{ij} is the sum of the collisional and kinetic stress tensors, f_i is the body force per unit mass, k is the fluctuation kinetic energy per unit mass, ϵ is the dissipation rate per unit mass, and K_i is the fluctuation energy flux vector. For incompressible grains with a density ρ_0, $\rho = \rho_0 \nu$, where ν is the solid volume fraction. Equations (1)-(3) are the statements of conservation of mass, balance of linear momentum, and balance of fluctuation energy for the mean granular motion in grain inertia regime.

According to the thermodynamical formulation of Ahmadi and Abu-Zaid [1], constitutive equations for the granular stress tensor and the fluctuation energy flux vector are given as

$$\begin{aligned} \tau_{ij} &= -p^T \delta_{ij} + 2\mu^T d_{ij}^D \\ &\quad + \rho k T^2 \left\{ \alpha \frac{\hat{D}}{Dt} d_{ij}^D + \zeta T d_{ik}^D d_{kl}^D d_{ij}^D + \beta \left[\frac{1}{3} d_{lk}^D d_{kl}^D \delta_{ij} - d_{ik}^D d_{kj}^D \right] \right\} \end{aligned} \tag{4}$$

$$K_i = \kappa^T \left[k_{,i} - \frac{k_T}{T}_{,i} \right] \ , \tag{5}$$

where p^T is the granular pressure, T is a time scale of granular fluctuation, μ^T is the granular viscosity, κ^T is the granular conductivity, and α, β, and ζ are material parameters which, in general, are functions of physical properties of particles. In Equation (5), d_{ij}^D is the deviatoric part of the mass-weighted mean deformation rate tensor d_{ij}, and $\frac{\hat{D}}{Dt}$ denotes the co-rotational (Jaumann) derivative. The expression for the granular pressure is given as

$$p^T = \frac{2}{3} \rho k [1 + 2(1 + r) \nu \chi] \ . \tag{6}$$

2. Simple Shear Flows

Consider a steady simple shear flow with the components of velocity vector given by

$$u(y) = \dot{\gamma} y, \qquad v = w = 0 \ , \tag{7}$$

where $\dot{\gamma}$ is the constant shearing rate. The normal and shear stresses become

$$\tau_{11} = -p^T - \rho k T^2 \left(\frac{\alpha}{2} + \frac{\beta}{12} \right) \dot{\gamma}^2 \ , \tag{8}$$

$$\tau_{22} = -p^T - \rho k T^2 \left(\frac{\alpha}{2} - \frac{\beta}{12} \right) \dot{\gamma}^2 \ , \tag{9}$$

$$\tau_{33} = -p^T + \rho k T^2 \frac{\beta}{6} \dot{\gamma}^2 \ , \tag{10}$$

and

$$\tau_{21} = \mu^T \left(1 + T^2 \frac{\zeta}{4\hat{C}^{\mu}} \dot{\gamma}^2 \right) \dot{\gamma} \ . \tag{11}$$

Equations (7)-(10) clearly show that the granular normal stresses are anisotropic.

The first normal stress difference as predicted by the equations (8)-(10) and that of Ahmadi and Abu-Zaid [1] along with the digital simulation data of Walton and Braun [2] for smooth nearly elastic particles are shown in Figure 1a. This figure shows that both model predictions for first normal stress difference are in good agreement with the simulation data. It is also observed that the first normal stress difference as predicted by Equations (8)-(10) increases more rapidly as the restitution coefficient decreases when compared with the model of Ahmadi and Abu-Zaid [1].

Figure 1b compares the predicted variations shear stress with the experimental data of Savage and Sayed [21]. The nondimensional shear stress $\bar{\tau}_{21}$ as given by Equation (11), for $\nu_m = 0.64356$ and $r = 0.9$ are plotted by solid lines in these figures. The predictions of the model of Lun et al. [4] are also shown by dashed lines for comparison. These figures show that the predictions of different models are in reasonable agreement with the experimental data. However, the model of [4] somewhat overestimate the experimental data.

3. Conclusions

Based on the presented results, the following conclusions may be drawn:

1. The rate-dependent model is capable of predicting the normal stress differences during rapid shearing of granular materials.

2. The predictions of the model for the first normal stress differences are in good agreement with the simulation data of Walton and Braun [2].

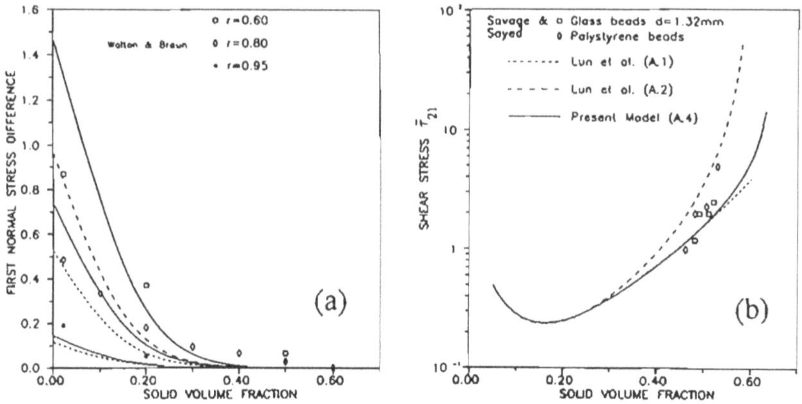

FIGURE 1. Comparison of model predictions with (a) simulation results of Walton and Braun [2], (b) with experimental data of Savage and Sayed [3] and model of Lun eet al. [4].

3. The model prediction for the granular shear stress is in good agreement with the experimental data of Savage and Sayed [4].

4. References

[1] Ahmadi, G. and Abu-Zaid, S., "A rate-dependent thermodynamical model for rapid granular flows", *J. Non-Newtonian Fluid Mech.*, Vol. 35, pp. 15-35, 1990.

[2] Walton, O.T. and Braun, R.L., "Stress calculations for assemblies of inelastic spheres in uniform shear", *Acta. Mech.*, Vol. 63, pp. 73-86, 1986.

[3] Savage, S.B. and Sayed, M., "Stresses developed by dry cohessionless granular materials sheared in an annular shear cell", *J. Fluid Mech.*, Vol. 142, pp. 391-430, 1984.

[4] Lun, C.K.K., Savage, S.B., Jeffrey, D.J. and Chepruniy, C., "Kinetic theories for granular flow: Inelastic particles in Couette flow and slightly inelastic particles in a general flow field", *J. Fluid Mech.*, Vol. 140, pp. 223-256, 1984.

Dynamic Normal Loads and Friction in Plowing Contacts

D. P. Hess[1]

ABSTRACT The influence of harmonic normal loads on sliding friction is investigated through analysis of contacts consisting of spherical sliders of hard materials on softer metal surfaces. Friction for such contacts is assumed to result from a plowing component and a shearing component. Calculations and experiments show that the relation between the coefficient of friction and normal load is highly nonlinear. In the presence of harmonic variations in normal load, this non-linearity causes a shift in the average coefficient of friction. For ideal lubricated contacts, the shearing component of friction is very small and for this case, it is shown that the maximum average reduction in the coefficient of friction is ten percent. When the shearing component is more significant, as with dry contacts, the shift is less. For example, when the interfacial shear strength is one-sixth the hardness of the softer material, the maximum average reduction in the coefficient of friction is five percent.

1 Introduction and Background

Contacts are often subjected to dynamic loads that may be generated from within the contact due to surface roughness or waviness, or externally from moving components or vibration. Understanding the effect of dynamic loads on friction is essential for accurate simulation of machine dynamics as well as for accurate interpretation of experimental data.

Several investigators [1-4] have studied the effect of dynamic loads and vibration on friction. This previous work focused primarily on the effect of dynamic loads and vibration on sliding elastic contacts in which the nonlinear contact stiffness accounts for a reduction in average friction for continuous contact. Also, earlier dynamic models [3,4] consider simple adhesion as the primary mechanism of friction.

Other researchers have investigated the plowing mechanism of friction for hard sliders of various shapes sliding on dry and lubricated metal surfaces [5,6]. This earlier work focused on understanding the plowing mechanism of friction through experiments and modeling. In this paper, the effect of dynamic loads on sliding plastic contacts is examined, and plowing and shear are assumed to be the primary mechanisms of friction.

[1]Department of Mechanical Engineering, University of South Florida, Tampa, FL 33620

2 Spherical Plowing Contact

Consider the spherical slider shown in Fig. 1. The contact forces acting on an elemental area dA are $dN = p\cos\beta dA$ and $dF = p\sin\beta\cos\gamma dA + s\sin\gamma dA$ where $dA = R^2\sin\beta\, d\beta\, d\gamma$, R is the radius of the sphere, and p and s are the normal pressure and shear stress acting on dA. The total contact forces, determined by integrating dN and dF over the front half of the sphere, are

$$N = \frac{\pi d^2 p}{8} \tag{1}$$

$$F = R^2 p\left[\sin^{-1}\left(\frac{d}{2R}\right) - \left(\frac{d}{2R}\right)\left(1-\left(\frac{d}{2R}\right)^2\right)^{1/2}\right] + 2R^2 s\left[1-\left(1-\left(\frac{d}{2R}\right)^2\right)^{1/2}\right] \tag{2}$$

where d is the track or groove width. Dividing Eq. (2) by Eq. (1) gives an expression for the coefficient of friction

$$\mu = \frac{2}{\pi}\left(\frac{2R}{d}\right)^2\left\{\sin^{-1}\left(\frac{d}{2R}\right) - \left(\frac{d}{2R}\right)\left(1-\left(\frac{d}{2R}\right)^2\right)^{1/2} + \frac{2s}{p}\left[1-\left(1-\left(\frac{d}{2R}\right)^2\right)^{1/2}\right]\right\} \tag{3}$$

The first two terms in Eq. (3) represent the plowing component of friction and the last term represents the friction due to shear. The normal pressure p is approximately equal to the hardness of the softer metal. The shear stress s of the interface approaches the shear strength of the softer metal for dry conditions. However, this shear stress becomes negligible for lubricated conditions since it approaches the shear strength of the lubricant film.

Tsukizoe and Sakamoto [6] performed experiments with various spherical sliders made of diamond and bearing steel sliding on softer low carbon steel, aluminum and copper surfaces. Their data revealed a nonlinear relation between the coefficient of friction and normal load that is predicted quite well through calculations using Eqs. (1) and (3).

Consider a dynamic load $W = W_o(1+\cos t)$ acting on the slider in Fig. 1. This normal load includes a steady component W_o and a simple harmonic component $W_o\cos t$. The instantaneous coefficient of friction μ, the steady coefficient of friction μ_o associated with W_o, and the average coefficient of friction μ_{av} have been computed for various sliders. Due to the inherent non-linearity, the time-varying component of the instantaneous coefficient of friction is not a simple harmonic. As a result, the average coefficient of friction μ_{av} is lower than the steady coefficient of friction μ_o. This shift is expressed in terms of the ratio of average to steady values, μ_{av}/μ_o.

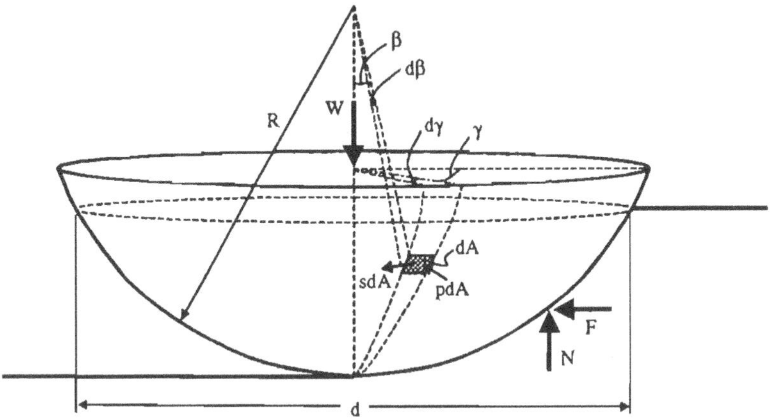

FIGURE 1. Spherical slider.

Calculations of μ_{av}/μ_o versus maximum $d/2R$ for two levels of interfacial shear strength s are presented in Fig. 2. This data is independent of load, slider size and parameter p. When the shear strength s is p/6, which approaches dry conditions, up to a 5% reduction in average coefficient of friction occurs when the maximum $d/2R$ is about 0.5, while up to an 11% increase occurs as the maximum $d/2R$ approaches unity. As the groove width and load decrease, the shift in average coefficient of friction, due to harmonic loading $W = W_o(1+\cos t)$, approaches zero. At the other extreme, when the shear strength s is zero, which approaches lubricated conditions, up to a 10% reduction in average coefficient of friction occurs for low loads, while up to an 11% increase occurs as the maximum $d/2R$ approaches unity.

3 Summary

Plastic contacts consisting of hard spherical sliders sliding on softer metal surfaces and subjected to dynamic loads have been modeled assuming the primary mechanisms of friction to be from plowing and shear. In the presence of harmonic normal loading, shifts in average friction have been found to occur due to the inherent nonlinear dependence of the coefficient of friction on normal load for spherical sliders. For harmonic loading with low to moderate penetration, i.e., for maximum groove width to slider diameter ratios d/2R up to about 0.85, reductions in coefficient of friction of up to about 5% have been found for dry contacts and up to 10% have been found for boundary lubricated contacts. For harmonic loading with high penetration, i.e., maximum d/2R greater than 0.9, increases in coefficient of friction of up to 11% have been found.

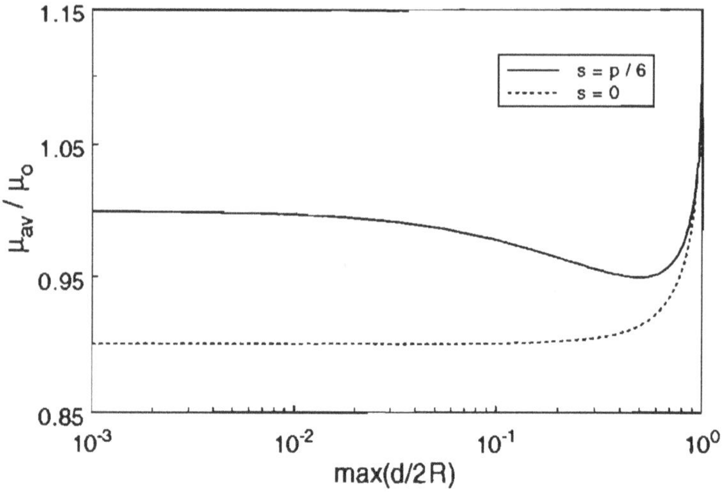

FIGURE 2 Average coefficient of friction versus maximum groove width for spherical sliders under harmonic loading.

4 Acknowledgment

The author gratefully acknowledges the support of the National Science Foundation under Grant No. CMS-9629217.

5 References

[1] Tolstoi, D. M., Significance of the Normal Degree of Freedom and Natural Normal Vibrations in Contact Friction, *Wear*, 10, 199-213, 1967.

[2] Godfrey, D., Vibration Reduces Metal to Metal Contact and Causes an Apparent Reduction in Friction, *ASLE Transactions*, 10, 183-192, 1967.

[3] Sakamoto, T., Normal Displacement & Dynamic Friction Characteristics in a Stick-Slip Process, *Tribology International*, 20, 25-31, 1987.

[4] Hess, D. P. and Soom, A., Normal Vibrations and Friction Under Harmonic Loads: Part I – Hertzian Contacts and Part II – Rough Planar Contacts," *ASME Journal of Tribology*, 113, 80-92, 1991.

[5] Bowden, F. P., Moore, A. J. W., and Tabor, D., The Ploughing and Adhesion of Sliding Metals, *Journal of Applied Physics*, 14, 80-91, 1943.

[6] Tzukizoe, T. and Sakamoto, T., Friction in Scratching without Metal Transfer, *Bulletin of the JSME*, 18, 65-72, 1975.

Identification of Friction Models From Experimental Data: A Case Study

D. M. McFarland[1]
A. V. Srinivasan[1]

ABSTRACT A standard system identification technique is used to fit mathematical laws of dynamic friction to experimental data. The conversion of the data, originally recorded graphically, to discrete time series and their subsequent processing is discussed. Identified friction models are compared on the basis of their accuracy in estimating energy dissipation per cycle of vibration.

1 Introduction

Despite widespread interest in friction and the attendant energy dissipation due to rubbing at interfaces of machine components, only a limited amount of experimental data on motion in the presence of friction is available in the open literature. Much of what is available is in forms not amenable to automatic processing. This paper describes the systematic analysis of test data taken using a single-degree-of-freedom, sliding friction rig during a program previously conducted by the second author. Using the technology of the day, these data were initially recorded as traces on photographic film. These traces have now been further processed to refine mathematical models of the friction that occurred at the rig's instrumented interface.

Following a brief discussion of how the traces were scanned and digitized, the calculation of friction forces from measured interfacial displacements and test rig accelerations is reviewed. Standard system identification techniques are then employed to compare several functional forms for the laws of friction describing the dominant phenomena at the sliding interface.

2 Problem Statement

As part of a research program conducted in the mid-1980s (reported fully in [1] and [2]), data were taken from a single-degree-of-freedom, sliding friction rig. The apparatus sat on a slip table and was driven by an electromechanical shaker

[1]Strategic Technologies Inc., East Glastonbury, Connecticut

at a controlled level of base acceleration, $\ddot{x}_b(t)$. The base acceleration, the acceleration $\ddot{x}(t)$ of the sliding mass and the relative motion

$$\varepsilon(t) = x(t) - x_b(t) \tag{1}$$

were measured directly. The friction force $Q(t)$ was due to the rubbing of inserts, which were made of titanium and coated with tungsten carbide.

The accelerations of the base and slider and the relative displacement were recorded photographically. Three such figures were available from the report [2], showing the response to a $5\,g$ base acceleration at an off-resonance driving frequency (135 Hz) and to $5\,g$ and $8\,g$ base accelerations at resonance (143 Hz).

The present effort uses the available data to

1. prepare discrete time histories suitable for use with modern system identification software,
2. calculate the time histories of the corresponding friction force Q,
3. use standard system ID software to estimate the parameters of some candidate mathematical models $Q_i(\dot{\varepsilon}, \operatorname{sgn}\dot{\varepsilon}, \dots)$, and
4. compare the models Q_i to Q by computing the energy dissipation per cycle predicted by each.

3 Analysis

The three available figures, comprising ten usable traces, were scanned into image files and processed using the program *NIH Image* on a Macintosh computer. Because of the quality of the images, the digitizing of the traces could not readily be automated, and a manual procedure of picking critical points on the curves was eventually adopted. The resulting files are portable, and were further processed in *Matlab* on a Unix machine.

The authors have developed a general procedure for the calculation of friction forces in three dimensions at the interfaces of multi-degree-of-freedom structures [3]. The present problem is the simplest application of this approach, having only a single interface and a single degree of freedom of relative motion. The calculation of Q from measured quantities is accordingly very straightforward. Directly from the equation of motion of the system we obtain

$$Q = -m\ddot{x} - c\dot{\varepsilon} - k\varepsilon. \tag{2}$$

Since the mass, stiffness and viscous damping coefficient of the experimental rig are known and ε and $\ddot{x}$ were measured directly, this form is very convenient here—only the relative velocity $\dot{\varepsilon}$ needs to be computed before Q can be found. This was done by taking finite differences of the measured ε (a smooth signal).

With ε and its derivatives known and Q computed at the same discrete times, the parameters of mathematical models $Q_i = Q_i(\dot{\varepsilon}, \operatorname{sgn}\dot{\varepsilon}, \dots)$ can be evaluated. This was done using the ARX routine from the *Matlab* System Identification Toolbox to estimate the coefficients of the model functions Q_i.

	α_1	α_2	α_3	α_4
Q_1	28 880	—	—	—
Q_2	—	219.8	—	—
Q_3	18 462	101.5	—	—
Q_4	17 945	106.5	1 472 400	—
Q_5	16 807	55.9	-11 639 000	−5.98

Table 1: Identified parameters of friction force models.

Model	Dissipation, N m	Error, %
Q	0.0127	—
Q_1	0.0130	2.8
Q_2	0.0102	−19.1
Q_3	0.0131	3.0
Q_4	0.0131	3.0
Q_5	0.0125	−1.3

Table 2: Average dissipation per cycle due to friction.

4 Results

Table 1 shows the results of fitting functions of the form

$$Q_i = \alpha_1 \dot{\varepsilon} + \alpha_2 \operatorname{sgn} \dot{\varepsilon} + \alpha_3 \varepsilon + \alpha_4 \ddot{\varepsilon} \tag{3}$$

to the friction force Q computed from measured data. A — in the table indicates that a model was assumed not to be a function of a particular parameter.

The original force Q and approximations computed using two of these models are plotted in Fig. 1. Clearly, the general model, Q_5, agrees much better with the real friction force than does the Coulomb model, Q_2.

Energy dissipation per cycle was computed through numerical integration of

$$\Delta E = \int_0^T Q \dot{\varepsilon} \, dt \tag{4}$$

where T is the period of vibration (forcing), with the results shown in Table 2. Of the models examined, Coulomb friction gives the poorest estimate of dissipation due to rubbing at the interface. However, since the Coulomb model underestimates the dissipation, designs based on it will be conservative, if perhaps suboptimal.

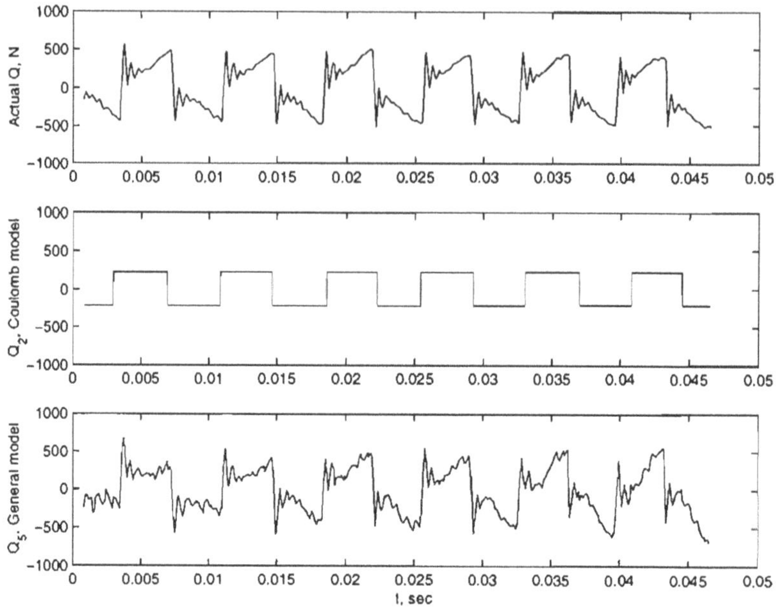

Figure 1: Actual friction force Q and two approximations using ID results.

5 Conclusions

Graphical data from a past experiment have been successfully digitized and used in the identification of mathematical models of laws of friction. A Coulomb model was found to significantly underestimate energy dissipation.

References

[1] A. V. Srinivasan, B. N. Cassenti, and D. G. Cutts. Study of characteristics of dry friction damping (annual report). Technical Report R84–956479–1, United Technologies Research Center, East Hartford, Connecticutt, March 1984.

[2] A. V. Srinivasan, B. N. Cassenti, and D. G. Cutts. Study of characteristics of dry friction damping (final report). Technical Report R85–956479–2, United Technologies Research Center, East Hartford, Connecticutt, April 1985.

[3] A. V. Srinivasan and D. M. McFarland. Characterization of laws of friction in the context of engine blade dynamics. *ASME Journal of Engineering for Power*. To appear.

A New Contact Surface Smoothing Procedure for the Implicit Finite Element Analysis of Frictional Contact

V. Padmanabhan[1]
T. A. Laursen[1]

ABSTRACT A new Hermite cubic smoothing scheme for the computational description of contact surfaces is outlined. Attention is focused on the two dimensional case, although the scheme is designed to be extendible to three dimensions. A numerical result illustrates the utility of the proposed scheme.

1 Introduction

In most finite element simulations of contact phenomena, contact surfaces are taken to be only C^0 continuous. Since the tangents and normals are multivalued at the nodes and suffer discontinuities there, the robustness of the contact algorithm is hampered as nodes slide across such regions. Work has been done in the past to address the lack of smoothness of the contact interface in a 2D implicit context [Eterovic and Bathe, 1991] and also in explicit dynamics calculations [Belytschko and Neal, 1991]. Some problems associated with slip motions arising out of non-smooth surface descriptions can be eliminated through the use of special integration algorithms [Agelet de Saracibar,1997]. Although these proposals treat certain types of problems satisfactorily, a fully general smoothing scheme, suitable for both two and three dimensional implicit calculations, for frictionless and frictional problems, has been lacking. This paper outlines some preliminary results in the development of such a scheme.

2 Review of Contact Formulation

We consider a two-body contact problem, involving two bodies occupying regions $\Omega^{(1)}, \Omega^{(2)} \subset \mathbb{R}^{n_{sd}}$ in their reference configurations, where n_{sd} is the

[1] Department of Civil and Environmental Engineering, Duke University, Durham, NC 27708-0287

number of spatial dimensions. The motions $\varphi^{(i)}$, $i = 1, 2$ are to be found over some time domain of interest.

In addition to the usual prescriptions of field equations and boundary conditions, we consider contact interactions by defining $\Gamma^{(i)} \subset \partial\Omega^{(i)}$, $i = 1, 2$ such that all prospective points of contact are included. Monitoring such interactions at points $\boldsymbol{X} \in \Gamma^{(1)}$, one can develop the following variational principle:

$$G(\varphi_t, \overset{*}{\varphi}) + G_c(\varphi_t, \overset{*}{\varphi}) = 0, \tag{1}$$

where φ_t and $\overset{*}{\varphi}$ are shorthand notations for the collections of the deformations $\varphi_t^{(i)}$ and variations $\overset{*}{\varphi}^{(i)}$ of the two bodies. $G(\varphi_t, \overset{*}{\varphi})$ is the usual virtual work of internal and applied forces, and $G_c(\varphi_t, \overset{*}{\varphi})$ is the virtual work due to contact, given by:

$$G_c(\varphi_t, \overset{*}{\varphi}) = \int_{\Gamma^{(1)}} [t_N \delta g + t_{T_\alpha} \delta\bar{\xi}^\alpha]\, d\Gamma^{(1)}. \tag{2}$$

In (2), $\delta(\cdot)$ denotes the variation of the indicated quantity, $g(\boldsymbol{X})$ is the *gap*, or *approach* function, $t_N(\boldsymbol{X})$ is the contact pressure, $\bar{\xi}^\alpha(\boldsymbol{X})$ is the surface coordinate of the contact point, and $t_{T_\alpha}(\boldsymbol{X})$ are the components of the frictional traction in the basis induced by the ξ^α-parameterization. Points on the surface $\Gamma^{(2)}$ are denoted as $\boldsymbol{Y}(\boldsymbol{\xi})$ for some surface parameter $\boldsymbol{\xi}$, and that point of $\Gamma^{(2)}$ in contact with $\boldsymbol{X}$ at the instant in question is written as $\bar{\boldsymbol{Y}} = \boldsymbol{Y}(\bar{\boldsymbol{\xi}})$.

The tractions in (2) are subject to contact conditions. Considering the case of Coulomb frictional contact, these can be summarized as

$$\begin{aligned} & & \Phi &:= [t_{T_\alpha} M^{\alpha\beta} t_{T_\beta}]^{1/2} - \mu t_N \le 0 \\ t_N &\ge 0 & \dot{t}_{T_\alpha} &= \epsilon_T [M_{\alpha\beta} \dot{\bar{\xi}}^\beta - \zeta \frac{t_{T_\alpha}}{[t_{T_\beta} M^{\beta\gamma} t_{T_\gamma}]^{1/2}}] \\ g &\le 0 & & \\ t_N g &= 0 & \zeta &\ge 0 \\ & & \Phi\zeta &= 0, \end{aligned} \tag{3}$$

where $M_{\alpha\beta}$, $M^{\alpha\beta}$ are the metric and inverse metric, respectively, associated with the curvilinear ξ^α coordinate system at the contact point.

3 Smoothing

In the continuum setting described thus far, the need for smoothing does not exist. However, when a finite element discretization is introduced, the parameterization $\boldsymbol{Y}(\boldsymbol{\xi})$, as well as its current position $\varphi_t^{(2)}(\boldsymbol{Y}(\boldsymbol{\xi}))$, will in general suffer slope discontinities (see Figure 1). Since the conditions (3) and are written in a coordinate system associated with this parameterization, these discontinuities result in nonphysical behaviors and in fact are likely to degrade nonlinear convergence.

To circumvent this situation, we resolve the contact interactions not in a coordinate system associated with $\Gamma^{(2)}$, but instead in a smoothed version $\hat{\varphi}_t^{(2)}$ of $\Gamma^{(2)}$. This idea may be expressed variationally by replacing (1) by the following problem statement:

Find φ_t and $\hat{\varphi}_t^{(2)}$, such that for all admissible variations $\overset{*}{\varphi}$, $\hat{\overset{*}{\varphi}}{}^{(2)}$:

$$\begin{aligned} G(\varphi_t, \overset{*}{\varphi}) + \hat{G}_c(\varphi_t, \hat{\varphi}_t^{(2)}; \overset{*}{\varphi}) &= 0, \\ H_c(\varphi_t, \hat{\varphi}_t^{(2)}; \hat{\overset{*}{\varphi}}{}^{(2)}) &= 0, \end{aligned} \tag{4}$$

where

$$\hat{G}_c(\varphi_t, \hat{\varphi}_t^{(2)}; \overset{*}{\varphi}) := \int_{\Gamma^{(1)}} [t_N \delta \hat{g} + t_{T_\alpha} \delta \hat{\bar{\xi}}^\alpha] \, d\Gamma^{(1)}, \tag{5}$$

and

$$H_c(\varphi_t, \hat{\varphi}_t^{(2)}; \hat{\overset{*}{\varphi}}{}^{(2)}) := \int_{\Gamma^{(2)}} \hat{\overset{*}{\varphi}}{}^{(2)} \cdot [\varphi_t^{(2)} - \hat{\varphi}_t^{(2)}] \, d\Gamma^{(2)}. \tag{6}$$

Both $\hat{g}$ and $\hat{\bar{\xi}}^\alpha$ are defined analogously to earlier definitions for g and $\bar{\xi}$, *but with respect to the surface defined by* $\hat{\varphi}_t^{(2)}(\Gamma^{(2)})$. One could in principle choose many different variations $\hat{\overset{*}{\varphi}}{}^{(2)}$ to enforce $(4)_2$; here, these functions are chosen to collocate $\hat{\varphi}_t^{(2)}(\Gamma^{(2)})$ with $\varphi_t^{(2)}(\Gamma^{(2)})$ at nodal points, subject also to additional slope specifications.

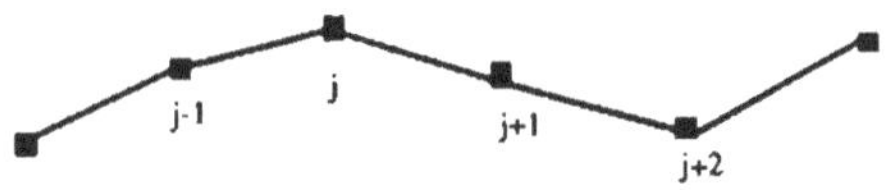

Figure 1 A discretized contact surface in two dimensions

4 A Hermite Cubic Description of $\hat{\varphi}_t^{(2)}$

One may readily define a Hermite cubic surface that suffers no discontinuities in the tangents and normals across element boundaries. Considering two adjacent nodes $\boldsymbol{Y}_j$ and $\boldsymbol{Y}_{j+1}$ on $\Gamma^{(2)}$ in a 2D problem (Figure 1), we utilize a piecewise cubic interpolation to describe the surface between them:

$$\hat{\boldsymbol{Y}}(\xi) = N_1(\xi)\boldsymbol{Y}_j + N_2(\xi)(\hat{\boldsymbol{Y}}_{,\xi})_j + N_3(\xi)\boldsymbol{Y}_{j+1} + N_4(\xi)(\hat{\boldsymbol{Y}}_{,\xi})_{j+1} \tag{7}$$

where the N_a's are Hermite cubic shape functions. $(\hat{\boldsymbol{Y}}_{,\xi})_j$ and $(\hat{\boldsymbol{Y}}_{,\xi})_{j+1}$ are constructed nodal tangent vectors, set in this work to be the tangent to the quadratic curve in ξ passing through each given node and its neighbors. Although (7) is written in the reference configuration, such a fit is done in each current configuration, so that $\hat{\varphi}_t^{(2)}$ is described by (7) as well, with the $\boldsymbol{Y}$ replaced in this instance by spatial counterparts $\boldsymbol{y}$.

5 Numerical Example

Figure 2 depicts the behavior of this approach in simulating the frictionless extrusion of an aluminum billet into a soft "die". Although contrived, the example shows that the scheme successfully eliminates much of the noise generated by the unsmoothed formulation. Additionally, the smoothing algorithm allows one to perform this simulation in about one-fourth as many time steps, indicating the enhanced robustness of the scheme.

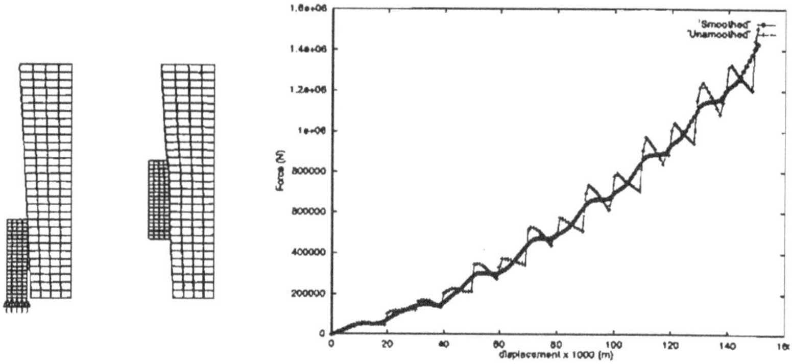

Figure 2 Numerical example: large deformation extrusion

6 Conclusion

The smoothing procedure, implemented in the frictionless 2D situation produced a marked improvement in the robustness of the solution algorithm. This benefit is expected to assume even greater significance in a frictional context. Finally, the method while already extended to regular meshes in 3D, shows promise for extension to unstructured meshes.

7 References

1. Agelet de Saracibar,C. [1997], "A New Frictional Time Integration Algorithm for Large Slip Multi-body Frictional Contact Problems," Computer Methods in Applied Mechanics and Engineering,142,pp 303-334.
2. Belytschko,T. and M.O.Neal, [1991], "Contact-Impact by Pinball Algorithm with Penalty and Lagrangian Methods," International Journal for Numerical Methods in Engineering, 31, 547-572.
3. Eterovic, A.L. and K.J. Bathe, [1991], "An Interface Interpolation Scheme for Quadratic Convergence in the Finite Element Analysis of Contact Problems," in Nonlinear Computational Mechanics- State of the Art, P. Wriggers and W. Wagner, eds , Springer-Verlag,Berlin,pp. 703-715.

Modeling Contact Forces During Impact on Flexible Systems

S. Shivaswamy[1]
H. M. Lankarani[2]

ABSTRACT Study of response of flexible systems due to impact is reported in this paper. The contact forces during impact has been modeled using modified Hertzian relation assuming quasi-static equilibrium. The model incorporates hysteresis damping to explain energy dissipation. The developed methodology was then verified with the experimental results.

1 Introduction

The energy method considering the transformation of the kinetic energy of the mass into the potential energy of deformation of the system has been used widely to study impact. An estimate of the loss in the kinetic energy during impact can be made using this approach, but the method fails to predict the instantaneous response of the system during the contact duration. Energy momentum methods are not suitable for the detailed analysis of the system during contact, specially for problems in which the contact duration is significantly large enough during which the system configurations changes from before to after impact. Timoshenko developed the integro-differential equation using the Hertzian theory of contact considering the local indentation at the impact site to model impact on beams. The concept of normalized contact force using a sine function approximation has also been followed by many.

In this paper, the response of flexible systems subjected to impact has been modeled and verified using modified Hertzian relations. The developed model incorporates the energy dissipation during impact in the form of hysteresis damping. The response of a flexible beam, undergoing gross elastic displacements when subjected to impact, is modeled. Experimental procedures were developed to verify the predicted contact forces. The presented methodology can be expanded to study the response of flexible multibody systems subjected to impact.

2 Theory

The elastic indentation during impact can be modeled using Hertz's law of

[1] Loads Dept., Cessna Aircraft Company, Wichita, KS 67277, USA
[2] Mech. Engr. Dept., WSU, Wichita, KS 67260-0133, USA (Corresponding Author).

contact. Timoshenko predicted the contact force F(t) due to a sphere of mass 'm', radius 'r' impacting with an initial velocity of 'V_0' at the center of a simply supported beam of effective mass 'M' (half of the total beam mass) and having a fundamental frequency of vibration of 'ω_1', through the integro differential equation [4].

$$\left(\frac{F(t)}{K}\right)^{\frac{1}{n}} = V_0 t - \frac{1}{m}\int_0 F(\tau)(t-\tau)d\tau - \sum_{i=1,3..}^{\infty} \frac{1}{i^2\omega_1}\frac{1}{M}\int_0 F(\tau)\sin(i^2\omega_1(t-\tau))d\tau \qquad (1)$$

The generalized stiffness parameter 'K' is a function of young's modulus and poisson's ratio of the sphere and the beam [4, 2]. The integration is carried out numerically by dividing the time 't' into small intervals during which the contact force 'F(t)' is assumed to be a constant. Then the nonlinear expression is solved for F(t) [1, 4].

The impact response as suggested by many, can be calculated using the normalized contact force functions. The contact force F(t) due to the impact of a mass 'm' on a beam of effective mass M (half of the beam mass) depends on the initial impact velocity V_0, coefficient of restitution 'e', the period of impact 'T' and is given by

$$F(t) = \frac{1}{2} m V_0 (1+e)\frac{\pi}{T}\sin\left(\frac{\pi t}{T}\right) \quad (t \le T) \qquad (2)$$

The details involved in the calculations can be obtained from reference [3].

In the Hysteresis damping model of impact where no significant permanent indentation is present, the energy dissipation during impact is due to the internal damping present in the system. The contact force is modeled as a function of the relative indentation 'α' as

$$F = K\alpha^n + \mu\dot{\alpha} \qquad (3)$$

The value of 'n' was found from the experimental results in such a way as to make the theoretical maximum contact force same as that of the experimental maximum contact force. An expression for 'μ' can be obtained by equating the kinetic energy lost by the impacting mass to the energy dissipated in the system due to the damping [1].

$$\mu = \frac{3K(1-e^2)}{4\dot{\alpha}^{(-)}}\alpha^n \qquad (4)$$

where '$\dot{\alpha}^{(-)}$' is the relative velocity of the system just before impact. The parameter 'K' is the generalized stiffness parameter and 'e' is the coefficient of restitution.

3 Experimental Setup

A schematic of the setup is shown in Figure (1). Here, a steel bar with a spherical tup (R_C = 45) is lifted up inside a teflon tube and allowed to drop freely onto a simply supported beam directly below, at its center. A beam subjected to impact from a falling weight simulates impact on a flexible system. A quartz force transducer (PCB 202A05) was placed in between the tup and the steel bar to sense the contact force during impact. The total mass of the impacting system comprising of the steel bar, tup and the transducer was 4.48 kg (≈ 10 lb). The transducer had a built in amplifier with an sensitivity of 1 volt / 4448 N (1 mV/lb) and was powered from a power supply (PCB 402B). The output from the transducer was passed to a digital storage oscilloscope (Tektronix 2204). To capture the signal during impact, an external signal from a function generator was used. The captured pulse on the oscilloscope was transferred to a computer and converted to ASCII format using the software 'GRABBER'. The ASCII file was then analyzed further using a code written in MATLAB.

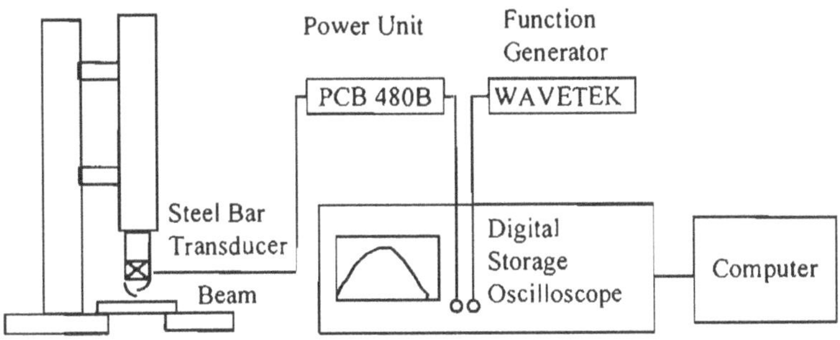

Figure 1. Apparatus Setup

4 Results and Discussion

The results obtained with a steel square hollow tube (1018 HR, $R_C = 17.5$) with outer dimensions of 38.4 mm (1.51 in) and inner dimensions of 31.34 mm (1.23 in) and a length of 126.7 mm (4.98 in) (EI = 20.865 Pa m^4, ω_1= 4.541E4 rad/s, t_p = 0.1383 ms) is presented here.
The contact force was measured and stored as an ASCII file. The ratio of the contact force to the impacting mass resulted in the acceleration of the impacting mass during impact and was integrated twice to obtain the velocity and displacement response. The coefficient of restitution 'e' was found from the ratio of the velocity at the end of impact to the velocity at the beginning of impact. The displacement of the beam during impact was evaluated numerically using the expression [1, 4]

$$y_{beam} = \sum_{i=1,3,5,...}^{\infty} \frac{1}{i^2 \omega_1 M} \int_0^t F(\tau) \sin(i^2 \omega_1^{\ 2} (t-\tau))\, d\tau \qquad (5)$$

The net relative indentation, which is the difference between the displacement of the mass and that of the beam, was calculated.
Figure (2) shows the variation of maximum contact force at different velocities for different values of 'n'. The value of 'n' in the range of 1.87 to 1.95 seems to match the theoretical results with the experimental results.

5 Conclusions

Experiments were carried out on Steel beam to simulate impact on flexible systems. The value of 'n' in the modified Hertzian relation was found to be in the range of 1.87 to 1.95 for steel-steel impact for impact velocities in the range of 1.0 to 1.6 m/s. The contact force model shows good correlation with the experimental results in the impact velocity range tested.

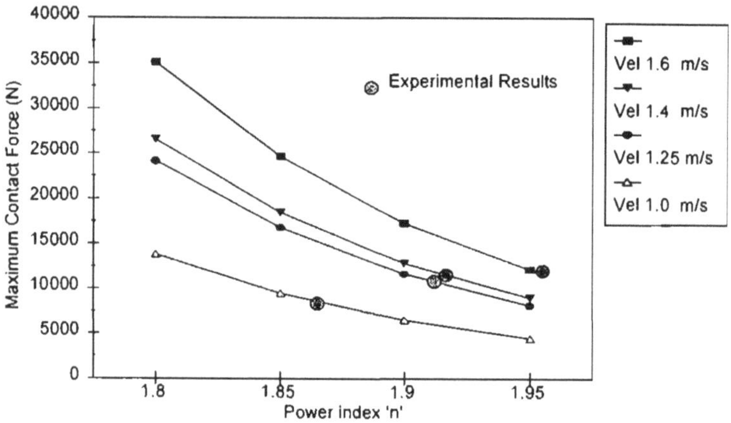

Figure 2. Variation of Maximum contact force at different impact velocities

6 References

[1] Shivaswamy, S., Modeling Contact Forces and Energy Dissipation during Impact in Mechanical Systems, Ph. D. Thesis, Wichita State University, 1997.

[2] Shivaswamy, S., Lankarani, H. M., Impact Analysis of Plates using Quasi-Static Approach, *Trans. of ASME, Jou. of Mechanical Design*, Sept. 1997, 376-381.

[3] Spotts, M. F., *Mechanical Design Analysis*, Prentice Hall, Inc., 1964, 345-366.

[4] Timoshenko, S., Zur Frage nach der Wirkung eines Stosse auf einer Balken, *Collected Papers of Timoshenko*, Mc. Graw Hill Book Company, Inc., 1953, 225-236.

Contact Stiffness and the Thermomechanical Response of Sliding Rings

C. I. Serpe[1]
G. Dargush[2]
A. Soom[1]

ABSTRACT The presence of surface texture on many engineering surfaces results in the contact being more compliant than if the interface were smooth and flat. In this paper we study a model problem, the thermoelastic contact of two steel rings sliding with friction. Contacts of this type are representative of dry clutches, brakes and seals. We illustrate the importance of including realistic contact stiffness in the thermoelastic analysis of such systems. The problem is analyzed using the ABAQUS finite element package. The stress and temperature fields change significantly as the contact stiffness is varied from rigid to softened, even when the sliding times are of the order of milliseconds in duration.

1 Introduction

Previous work on thermoelastic problems at extended sliding contacts has considered the surfaces smooth and flat. For example, Kennedy and Ling [2] have simulated thermoelastic instability and transient contact changes at the interface of an aircraft-type annular disk brake of sintered metal friction material. They have shown that, even starting from uniform contact conditions, nonuniform frictional heat generation due to velocity distribution causes an uneven localization of the contact near the outer edge of the disk.

The present investigation considers the effect of contact stiffness on the contact conditions during a transient sliding process. The role of contact compliance in determining the local static and dynamic behavior of isothermal dry sliding contacts has been studied in recent years [1]. The interface region, while stiff, is nevertheless significantly more compliant than the bulk material. The evolution of the deformation, stress and temperature field can be quite sensitive to contact stiffness. In this paper we explore by means of a simple

[1] Department of Mechanical and Aerospace Engineering, State University of New York at Buffalo
[2] Department of Civil Engineering, State University of New York at Buffalo, Buffalo, NY 14260

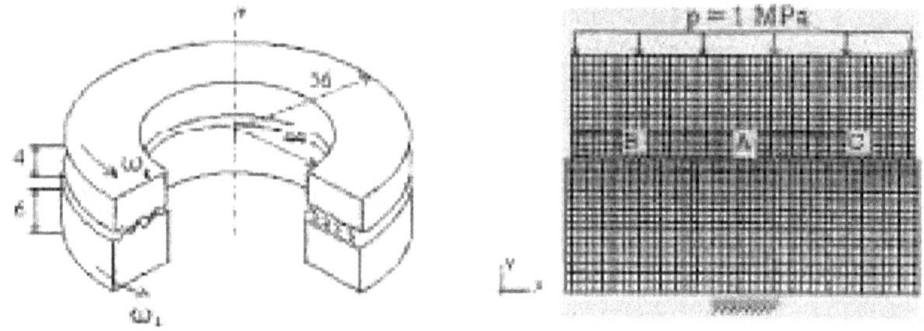

a. Sliding rings b. Finite element model

Figure 1. The problem model

example problem, the role that the contact compliance can play in thermoelastic problems. This effect has not been included in past studies. Details of individual asperity interactions, which are associated with so-called flash temperatures and are more likely to involve plastic deformations, are ignored.

2 Sliding ring problem

The sliding rings are shown in Fig. 1a. The surface roughness is represented by a set of nonlinear springs with an exponential load-deflection characteristic. An axisymmetric coupled thermomechanical analysis is performed using ABAQUS finite element package. The model is shown in Fig. 1b.

A uniform pressure loading is applied to the top of the upper ring and ramped to the full value of 1 MPa over the first 1 ms of the sliding process. The sliding time is 30 ms during which the sliding speed decreases linearly from 5000 rpm to zero. Once engagement is completed the analysis is continued for an additional 70 ms. The bottom edge of the lower ring is constrained in axial direction. A user-defined subroutine specifies the friction versus speed model and calculates the amount of frictional heat generated locally at each contact point. A linear friction-velocity model is employed at the local level. The coefficient of friction increases from 0.2 at the beginning of sliding to 0.4 at rest. All lateral surfaces are assumed to be insulated.

The contact between the surfaces may be modeled as either hard or soft. The hard contact represents contact between smooth and flat surfaces. The present thermomechanical analysis considers a hard contact (H1) and two softened contact cases (Sf2 and Sf3). The Sf2 model is representative of worn steel surfaces with an r.m.s. surface roughness on the order of 1 μm, while the Sf3 model represents a softer surface than one would find with steel surfaces. The contact stiffness values reach 0.2 N/μm/mm^2 for Sf2 and .005 N/μm/mm^2 for Sf3 at 1 MPa contact pressure. All components respond elastically, following the standard thermoelastic constitutive relationships and material properties.

Fig. 2.a presents the contact pressure distributions for the hard contact case

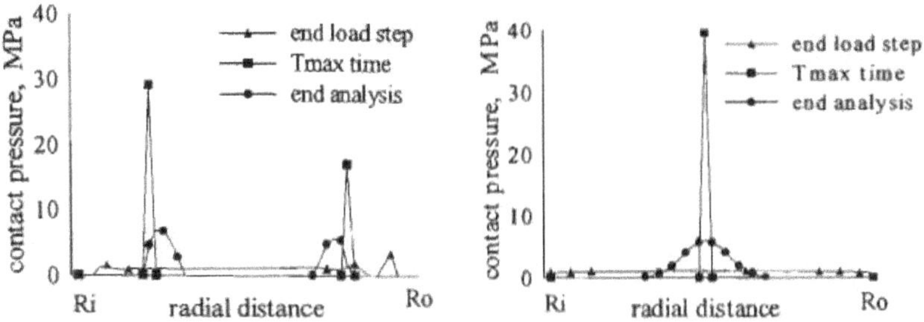

a. hard contact - H1 **b. soft contact - Sf2**

Figure 2. Contact pressure distributions during analysis

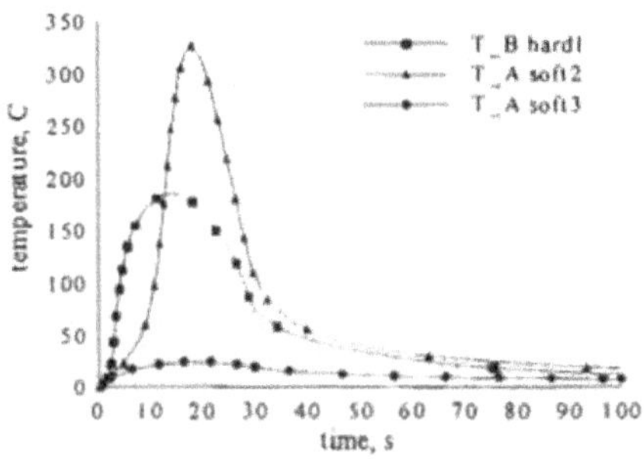

Figure 3. Temperature variation at localized contacts

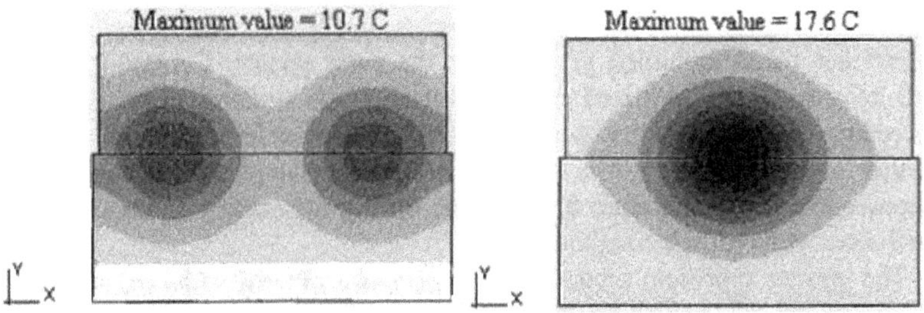

a. hard contact - H1 **b. soft contact - Sf2**

Figure 4. Temperature fields at end of analysis (t=100 ms)

at the end of the load step (t=1 ms), at maximum temperature time (t=14 ms) and at the end of analysis (t=100 ms). As a result of thermal expansion the contact pressure increases at the outer and inner contact points at the beginning of load application. When the temperature reaches its maximum value the contact is concentrated near points B and C. The contact pressure distributions for the softened case Sf2, presented in Fig. 2b, show that the contact is nearly uniform over the interface at the end of the load step. With further sliding the pressure distribution and the thermal deformations cause the localization of the contact along a line in the middle of the ring (near point A). The maximum contact pressure and temperature values are considerably higher than for the hard contact.

Fig. 3 shows the temperature versus time at the localized contacts for the three cases (point B for H1 and point A for Sf2 and Sf3 cases, respectively). The Sf2 (most realistic) softened case exhibits one localized contact with the highest temperature of 320 ^{0}C. The contact is concentrated over the smallest area, of the three cases, for a very short time. The hard contact becomes localized sooner. The maximum temperature is 190 ^{0}C and is nearly the same at the two locations B and C. The very soft case Sf3 results in a much lower temperature rise due to the uniformity of the contact pressure distribution. Throughout the process the maximum temperature reached is 24 ^{0}C, in agreement with a one-dimensional analytical solution assuming uniform pressure over the interface.

The temperature fields at the end of the analyses for H1 and Sf2 cases are shown in Fig. 4. They clearly show how the differences between the hard and Sf2 soft contact persist 70 ms after the heat input has ceased.

3 Conclusions

Local interface pressure, deformation and temperature are interdependent. The contact stiffness approach to contact modeling provides a simple way to determine localized thermomechanical response in dry sliding problems. In high speed engagements of metallic clutches and brakes the contact stiffness determines the deformation primarily in the early phases. Continued sliding will accentuate the localized contact pattern, with thermal expansion soon exceeding deflections of the "contact spring". The most important point is that neither a rigid contact nor a uniform pressure assumption can capture the pressure and the temperature distributions during the sliding of steel or other hard discs with realistic roughness.

The present approach, accounting for a realistic contact stiffness, is capable of capturing the radial localization of temperature and pressure under conditions that can only be modeled by numerical techniques.

4 References

[1] Hess D.P., Soom, A., 1992, Normal and angular motions at rough planar contact during sliding with friction, *Journal of Tribology*, Vol. 114, pp. 567-578, July 1992

[2] Kennedy, F.E.Jr., and Ling, F.F., 1974, A thermal, thermoelastic, and wear simulation of a high energy sliding contact problem, *Journal of Lu brication Technology*, July 1974, pp. 497-507

SESSION IV–(B)

Poster Session

A New Shape Memory Allow Rotary Actuator: Design and Modelling

A. Khajepour[1]
H. Dehestani[1]
F. Golnaraghi[1]

ABSTRACT In this paper we study the design and modelling of a new rotary actuator that has been developed at the University of Waterloo using shape memory alloy springs. Shape memory alloys (SMAs) have the property that they can be subjected to a large deformation when it is below martensite temperature and it recovers its original shape when it is heated over austenite temperature. In other words, a SMA can directly convert thermal energy to mechanical work! It is also interesting to note that among low and medium range actuators SMAs have the highest power to weight ratio. The designed actuator in this work is a two-way actuator and is made of NiTi SMA springs. It is modular and large rotations can be obtained using extra modules. The actuator design is first explained and then, the Ikuta's model is modified to derive a mathematical representation of the actuator. Using experimental and simulation results the mathematical model is verified.

INTRODUCTION The secret behind the SMAs behavior is in their crystalline structure. In low temperatures they have a martensite structure which in turn results in a more elastic/plastic behavior. Increasing the temperature converts the structure to austenite with much higher modulus of elasticity. As a result, the strain of a shape memory alloy under a constant tensile force changes as temperature varies. SMAs can be formed from a wide range of elements but only those that can generate significant force or can recover large strain are of commercial interest. Nitinol (NiTi) consists of nearly equal amounts of nickel, and titanium and is the most widely used SMA. In this research Nitinol springs from Mondo-Tronic Inc. are used for their high strength, large strain recovery capability and reasonable price.

SMA based actuators are becoming increasingly popular due to high power to weight ratio, smooth and silent operation, and less contamination. These features are quite important in miniaturized robotics, clean room and pharmaceutical applications. Kuribayashi in [1] studied a rotary joint actuator using two opposing SMA wires. He examined the characteristics of the SMA hysteresis when the strain was allowed to reverse in very small amounts at various points within the hysteresis. Ikuta *et al* in [2] expanded their sublayer model to include R-phase. Using solid mechanics laws they modelled mechanical properties of a SMA in its three phases, and using thermodynamics laws they modeled thermoelastic transformation. Madill *et al* in [3] improved Ikuta's model to predict minor hysteresis

[1]Department of Mechanical Engineering, University of Waterloo, Waterloo, Ontario N2L 3G1.

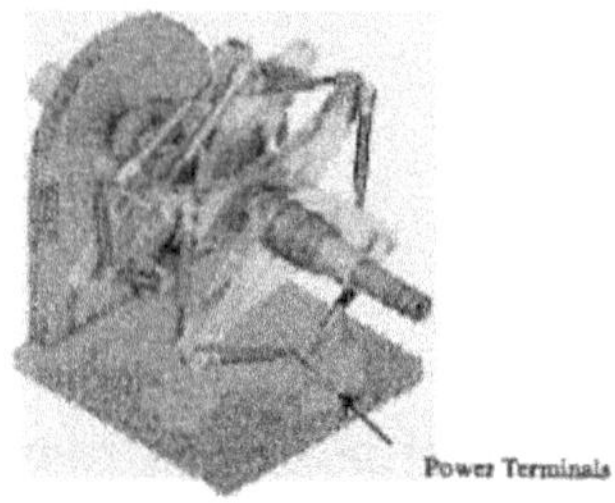

FIGURE 1. SMA Actuator

behavior. Using experimental analysis, they have verified the model and studied the closed loop stability. In [4] a linear actuator for robotic applications is studied. This actuator is made of NiTi fibers. Using SMA fibers have reduced the time constant to as low as 40 miliseconds. A variable structure control system is used to control the pulse width and current amplitude.

ACTUATOR DESIGN Figure 1 shows a 3-D model of the actuator. The front bracket in which the shaft and the power terminals are mounted, is removed for a better view of the springs arrangement. There are two sets of springs which make two spirals in opposite directions. When one of the spirals is powered, the associated springs shrink and cause a rotational motion in the output shaft. As seen in Figure 1 there are six arms to each two springs are connected. Since, each arm is connected to two springs the actuator can rotate in two opposite directions and is a two-way actuator. All arms are free to rotate about the shaft except the last one which is coupled to the output shaft via a set screw.

MATHEMATICAL MODELLING Shape memory alloys exhibit two main crystalline phases, austenite or parent phase, and martensite phase. The two phases can coexist in the material. At low temperatures martensite phase predominates while at high temperatures austenite phase predominates. Ikuta [2] proposed variable sublayer model to represent coexistence of martensite and austenite phases in a SMA wire. According to this model the stress induced in a wire is the sum of stress induced in martensite and austenite fractions. The thickness of each phase is proportional to the volumetric fraction of crystal phase in the wire that is a function of tem-

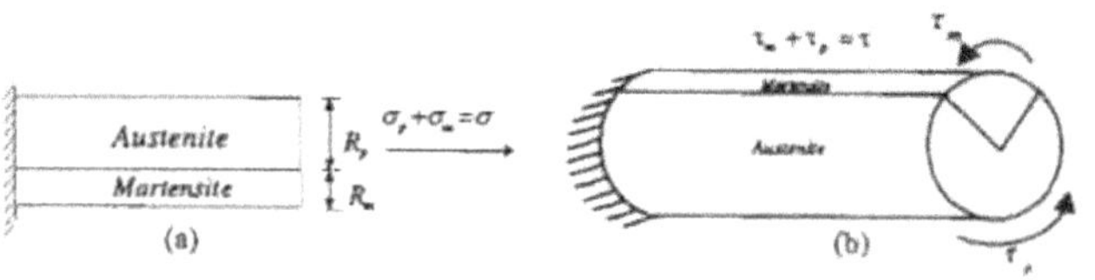

FIGURE 2. Sublayer Model (a) Tensile Stress, (b) Shear Stress

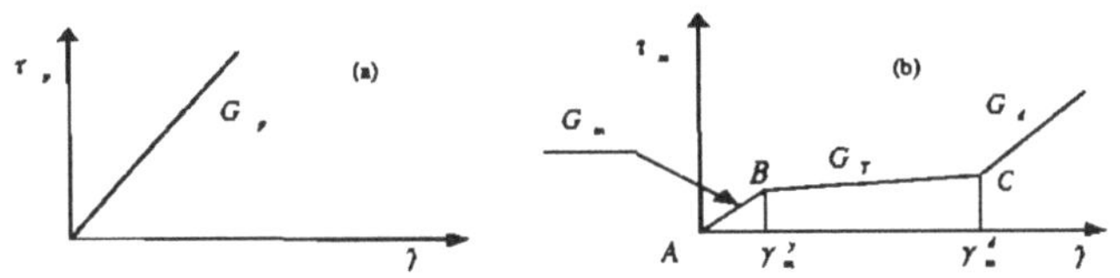

FIGURE 3. Stress-Strain Relationship (a) Austenite, (b) Martensite

perature. At high temperature the full thickness of the wire is austenite and at low temperature the full thickness is martensite. Ikuta's model is shown in Figure 2a, where R_p and R_m are the thickness of austenite and martensite, respectively. Using this model Ikuta introduced his mathematical model for a SMA wire with the main idea that the total stress in the wire is proportional to the sum of stresses in the martensite and austenite phases.

Since SMA springs are used in our prototype actuator we extend the sublayer model to shear stress. We assume a linear relation between shear stress and strain in a SMA wire. Figure 3 shows this relationship in austenite and martensite phases. The modified equation for shear stress-strain relationship in SMA wires under torsional loads is

$$\gamma = \begin{cases} \frac{\tau}{G_p-(G_p-G_m)R_m} & 0 \leq \gamma < \gamma_m^y \\ \frac{\tau+R_m(G_T-G_m)\gamma_m^y}{G_p-(G_p-G_T)R_m} & \gamma_m^y \leq \gamma < \gamma_m^d \\ \frac{\tau+R_m((G_T-G_m)\gamma_m^y+(G_d-G_T)\gamma_m^d)}{G_p-(G_p-G_d)R_m} & \gamma_m^d \leq \gamma \end{cases} \quad (1)$$

where γ and τ are the total shear strain and stress in the spring wire, and γ_m^y, γ_m^d are shear strains where detwinning starts and ends in martensite phase, respectively. The shear modulus in the austenite phase is shown with G_p, and G_m, G_T, G_d are three distinct shear moduli of martensite phase as illustrated in Figure 3b.

Since the values of shear moduli in equation (1) are not known we conducted experiments to determine them empirically. Figure 4 shows the results of the experiments in austenite and martensite phases. The shear moduli obtained from the experimental data are: $G_m = 6.33$ GP, $G_T = 1.84$

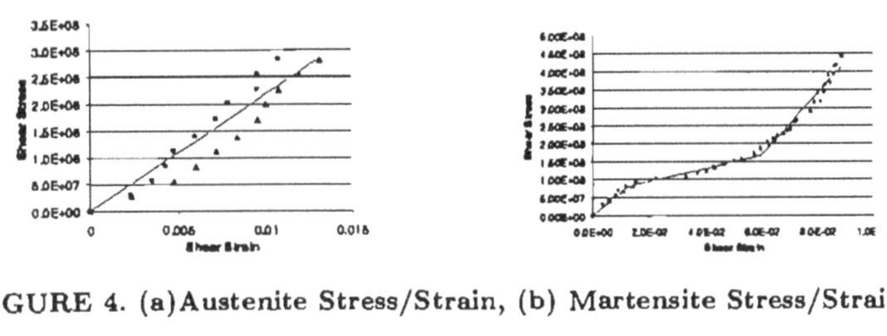

FIGURE 4. (a)Austenite Stress/Strain, (b) Martensite Stress/Strain

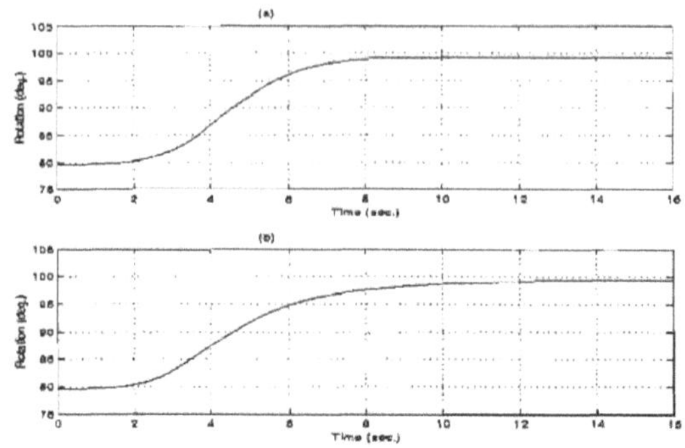

FIGURE 5. (a) Simulation Results, (b) Experimental Results

GP, $G_d = 8.55$ GP, and $G_p = 21.8$ GP.

SIMULATION AND EXPERIMENTAL STUDY Using equation (1) and the shear moduli obtained from the experiments, we verify the model. The simulated and experimental response of the actuator with one module are shown in Figure 5. In both plots a pulse width modulated current with 50% duty cycle and 4.5 A amplitude is applied to the spring. As seen in Figure 5 the simulation match the experiment. Hence, the mathematical model can be used to design a controller and can be applied to the multi-module actuator shown in Figure 1.

ACKNOWLEDGMENT The authors would like to acknowledge the National Sciences and Engineering Research Council of Canada and Control Advancements Inc. for their financial supports, and Professor D. Wang, Dr. R. Gorbet and Mr. D. Madill for their constructive discussions.

REFERENCES

[1] K. Kuribayashi, 1986, "A New Actuator of Joint Mechanism Using TiNi Alloy wire", *The International Journal of Robotics Research*", V. 4, No. 4.

[2] Ikuta, K., Hirose, S., and Tsukamoto, M., 1991, "Mathematical Model and Experimental Verification of Shape Memory Alloy for Designing Microactuator", *Proceedings of the 1991 IEEE MicroElectroMechanical Systems Conference*, pp. 103-108.

[3] Madill, D., and Wang, D., 1994, "The Modelling and L_2 Stability of a Shape Memory Alloy Position Control System", *International Conference on Robotics and Automation.*

[4] Grant, D., Hayward, V., 1997, "Variable Structure Control of Shape Memory Alloy Actuators", *IEEE Transactions on Control Systems Technology*, pp. 80-88.

Application of Graph Theoretic Modelling to Industrial Belt-Driven Gantry Robots

A. Khajepour[1]
J. Dunlop[1]

ABSTRACT In this paper we investigate the modelling of an industrial belt-driven gantry robot[1] using graph theoretic modelling (GTM). The design of the robot is such that the kinematic driver is fixed to the base frame while the end-effector is free to move in a straight line relative to the base. The power is transmitted to the end-effector using a pulley/belt arrangement. This arrangement has the benefit of radically reducing the system inertia, which permits the robot to operate at higher speed. Further increase in the operational speed of the robot is now limited by the induced vibration due to the belt flexibility. Using graph theoretic modelling we study the robot dynamics considering the base and belt flexibilities. To incorporate the pulley into the formulation, a multi-terminal model is introduced that can be used in the GTM for any simple or compound pulley with fixed or moving shaft.

INTRODUCTION The application of graph theoretic modelling, GTM, to multi-body dynamics has become well established in the past two decades. The systematic nature of the GTM has opened the door to the automation of many mechanism design and analysis tasks. A complete discussion of linear graph theory and its application to physical systems analysis is found in [1,2,3].

An integral part of any GTM is the identification of system components and their mathematical representations. In this paper we introduce a linear graph model for the kinematic and dynamic analysis of pulleys. The model is a multi-terminal element that can be used in the GTM. We apply the pulley model to a belt driven gantry robot considering the belt and support flexibilities.

PULLEY MODEL We consider a compound pulley as shown in Figure 1a with a general motion for the pulley center expressed in (X, Y) coordinate. The (X, Y) system can be an inertia or a body coordinate attached to the body at point g. Without loss of generality, we assume that the center of the pulley moves along a straight line C relative to the body. The belts are tangent to the pulley at points with relative position vectors R^n and R^p with respect to the pulley center. The positive rotation of the pulley is shown with θ in the figure. The local coordinate (x, y) is attached to the center of the pulley and has a fixed orientation with re-

[1]Department of Mechanical Engineering, University of Waterloo, Waterloo, Ontario N2L 3G1. Developed and manufactured by ATS Automation Tooling Systems Inc.

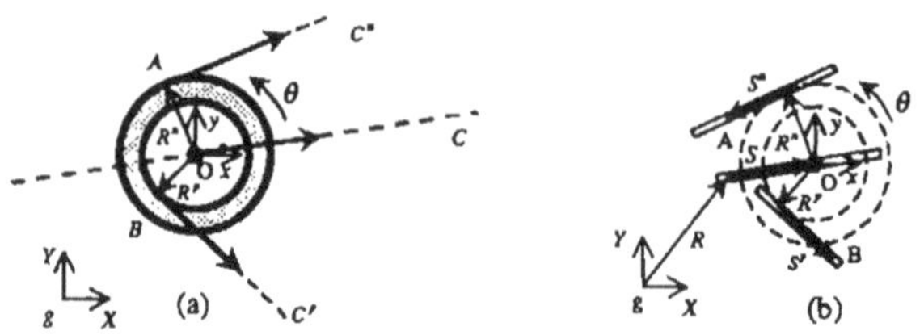

FIGURE 1. (a) Compound Pulley (b) Pulley Model

spect to the body coordinate (X, Y). At any instant, the pulley motion is in the direction C relative to the body coordinate (X, Y), and all points on the belt move along directions C^n and C^p relative to (x, y) coordinate. Hence, the pulley model can be realized using three sliders as shown in Figure 1b. Vectors S^n and S^p show the relative motion of the belt ends with respect to (x, y) which are (note that in the local coordinate (x, y), θ can be considered as a vector in z direction)

$$S^n = \theta \times R^n \quad , \quad S^p = \theta \times R^p \tag{1}$$

The position vector of the pulley center with respect to the body coordinate (X, Y) is found using the rigid arm element R and the slider element S shown in Figure 1b. Using the pulley model and equation (1) we have

$$P = R + S \tag{2a}$$
$$P^n = R^n + S^n \tag{2b}$$
$$P^p = R^p + S^p \tag{2c}$$

where P is the position vector of the pulley center with respect to the body coordinate (X, Y), and P^n, P^p are the position vectors of the belt ends at A and B relative to the local coordinate (x, y), respectively (the reason for the selection of superscripts n and p is that when the belts are parallel to the x axis, points A and B have negative and positive displacement with respect to x axis).

Combining the rigid arms (R, R^n, R^p) and slider elements (S, S^n, S^p) we define a multi-terminal element for the pulley. The translational and rotational graph model of the pulley is shown in Figure 2. In the translational graph point g is connected to the body coordinate (usually center of mass) or a ground node, and points A and B are connected to elements attached to the belt ends at A and B. If the pulley is not massless a mass element is connected to nodes O and an inertia datum. The terminal equations of P, P^n and P^p are given in equation (2). The through variable of edge P, (F_P), is the reaction force and is related to the through variables of edges P^n, (F_{P^n}) and P^p, (F_{P^p}) using the nodal equation at O. For a massless pulley the nodal equation at O takes the form

$$F_P = F_{P^n} + F_{P^p}. \tag{3}$$

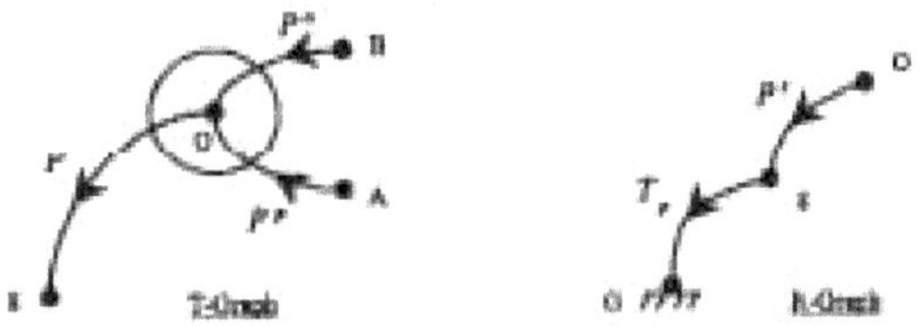

FIGURE 2. (a) Translational Graph, (b) Rotational Graph

The rotational graph in Figure 2b has two edges which P^r represents the angular rotation of the pulley with respect to the body coordinate (node g) and T_P is a dependent torque element acting on the body due to the reaction force F_P and transmitted torque from the pulley shaft. Note that when point g is an inertia datum the rotational graph is simplified and retains only edge P^r. The terminal equations for the through variable (torque) of P^r, (T_{P^r}) and T_P are

$$T_{P^r} = P^p \times F_{P^p} - P^n \times F_{P^n} \tag{4}$$
$$T_p = T_{P^r} + P \times F_P \tag{5}$$

When the moment of inertia of the pulley is not negligible it is added to the graph as a separate inertia element to nodes O and the inertia datum G in the R-graph. Other physical properties of a pulley can also be included using separate elements. For example, rotational damping can be included using an element between nodes O and g in the R-graph.

GANTRY ROBOT MODELLING The application of the pulley model is demonstrated here using an industrial robot example. The system described here is a planar mechanism with a single sliding axis for the tool plate, as depicted in Figure 3a. This mechanism can be modelled as a rigid frame supported by springs and dampers as shown in Figure 3b, upon which the sliding axis, the motor and the idler pulleys are mounted.

The system graph can now be generated from the physical model as shown in Figure 4. The rigid bodies are framed by dotted and shaded lines in order to facilitate the interpretation of the translational graph. In Figure 4a edges 1 and 2 represent the slider and base masses, respectively. Edges 3-5 are 3 rigid arms to locate the supports and slider positions. The slider mass is connected to the base by a slider element, edge 6. The pulleys are shown by multi-terminal elements 7 and 8. The slider damping is shown by edge 9. The stiffness of different parts of the belt are considered by edges

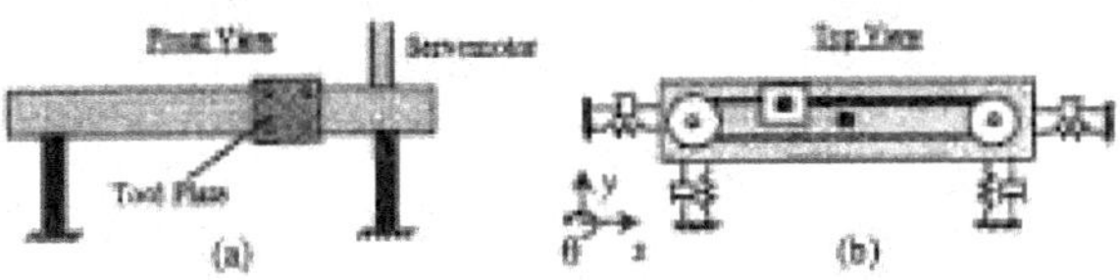

FIGURE 3. (a) Single Axis Robot, (b) Pulleys Configuration

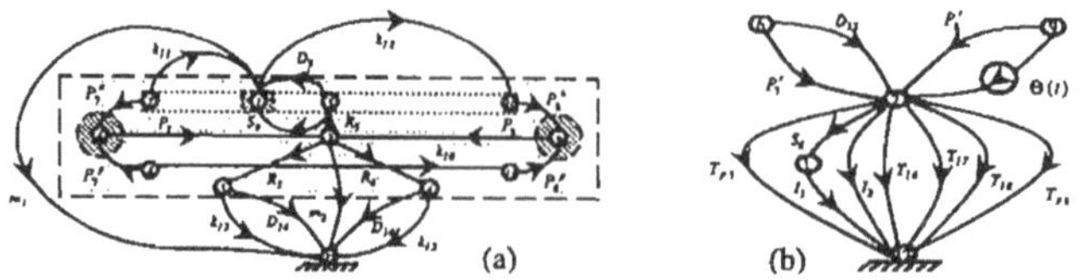

FIGURE 4. (a) Translational Graph, (b) Rotational Graph

10, 11, and 12. Furthermore, the combined supports stiffness and damping in X and Y directions are represented using edges 13 and 14, respectively.

In the rotational graph, edges 1 and 2 correspond to the rotational inertia of the two masses. Dependent torque elements, edge 16, 17 and 18 act on the base frame body to account for the torque generated by the slider and supports reaction forces, respectively. Finally, the pulley rotational edges, $P_7^r, T_{P7}, P_8^r, T_{P8}$, the rotational position driver $\Theta(t)$, and the idler pulley rotational damping D_{15} are added to complete the rotational graph.

The equations of motion using the graph model are:

$$(m_1 + m_2)\ddot{x}_2 - m_1(S_6\theta_1 + r_p)\ddot{\theta}_1 + m_1\ddot{S}_6 \tag{6a}$$
$$-2m_1\dot{S}_6\dot{\theta}_1\theta_1 - m_1(S_6 - r_p\theta_1)\dot{\theta}_1^2 + 2K_{13}^x x_2 + 2D_{14}^x\dot{x}_2 = 0$$
$$(m_1 + m_2)\ddot{y}_2 + m_1\theta_1\ddot{S}_6 + m_1(S_6 - r_p\theta_1)\ddot{\theta}_1 \tag{6b}$$
$$+2m_1\dot{S}_6\dot{\theta}_1 - m_1\dot{\theta}_1^2(S_6\theta_1 + r_p) + 2K_{13}^y y_2 + 2D_{14}^y\dot{y}_2 = 0$$
$$m_1\ddot{S}_6 + m_1\ddot{x}_2 + m_1\theta_1\ddot{y}_2 - m_1 r_p\ddot{\theta}_1 - m_1 S_6\dot{\theta}_1^2 \tag{6c}$$
$$+D_9\dot{S}_6 + K_{11}(S_6 - r_p\theta_{15}) + K_{12}(S_6 - r_p\Theta(t)) = 0$$
$$(I_1 + I_2 + r_p^2 m_1 + S_6^2 m_1)\ddot{\theta}_1 + m_1(S_6 - r_p\theta_1)\ddot{y}_2 - r_p m_1\ddot{S}_6 \tag{6d}$$
$$-m_1(S_6\theta_1 + r_p)\ddot{x}_2 + 2m_1 S_6\dot{S}_6\dot{\theta}_1 + 2R^2K_{13}^y\theta_1 + 2R^2D_{14}^y\dot{\theta}_1 = 0$$
$$D_{15}\dot{\theta}_{15} + r_p^2K_{10}(\theta_{15} - \Theta(t)) - r_pK_{11}(r_p\theta_{15} - S_6) = 0 \tag{6e}$$

where r_p is the pulleys radius and R is the distance from the supports and the pulleys shafts to the base center of mass. The stiffness and damping of the supports are shown using K_{13}^x, K_{13}^y and D_{14}^x, D_{14}^y respectively.

ACKNOWLEDGMENT The authors would like to acknowledge the National Sciences and Engineering Research Council of Canada and ATS Automation Tooling Systems Inc. for their financial supports and Professor J.J McPhee and Mr. C. Good for their constructive discussions.

REFERENCES

[1] Koenig, H.E., Tokad, Y., and Kesavan, H.K., 1967, *Analysis of Discrete Physical Systems*, McGraw-Hill.

[2] Andrews, G.C., 1975, "The Vector-Network Model: A New Approach to Vector Dynamics", *Mechanism and Machine Theory*, Vol. 10, pp. 57-75.

[3] McPhee, J.J., 1996, "On the Use of Linear Graph Theory in Multibody System Dynamics", *Nonlinear Dynamics* Vol. 9, pp. 73-90.

A Friction Failure Criterion For Two-Dimensional Granular Structures

M. A. S. Mohamed[1]

ABSTRACT Limit friction forces between structure constituents are employed to define a practical failure criterion for granular media. An analytical solution is presented to estimate local friction states built up due to given boundary forces on the structure. The question as to where failure is most likely to initiate is mainly addressed.

1 Modes of Failure of a Rhombic Cell

One of the interesting problems is the crack initiation in granular structures where compressive loads along with surface friction are the only bonding forces maintaining structure stability. Such a problem can be found in soil mechanics, aggregate stacking, powder metallurgy and the like. A review of the literature shows that failure analysis for those media has been approached using either the concept of slip surface or active-passive stress theory, or the phenomenon of structure dilatancy [1]. Other attempts to establish applicable failure criteria have been based on continuum theories and microscopic considerations [2]. The present study deals with randomly arranged rigid discs (grains) of equal radii, packed to form a rhombic structure. Figure. 1 shows a typical rhombic cell "A" in such structures. The cell consists of four discs. The centers of the four discs form a rhombus whose diagonal c_1c_3 makes an angle ϕ_A with respect to the global frame (ξ,η). At equilibrium resultant reactions due to surrounding discs form the cell's reactive or boundary forces $\{F_i, i=1,2,3,4\}$. Under a given set of $\{F_i\}$, the cell "A" may eventually fail as its grains tend to slip along the cell's diagonals following one of two possible modes, assuuming unlikely relative rotation of discs. The negative mode, $M=-1$, arises when the grains A_1 and A_3 tend to move towards o_A while A_2 and A_4 tend to move outwards. The positive mode, $M=+1$, arises when A_2 and A_4 tend to move inwards while A_1 and A_3 tend to move outwards. Therefore, when two contacting discs in the cell are isolated as one unit, their force equilibrium implies that the four total reactions at discs contacting points always form a parallelogram $k_1k_2k_3k_4$. Since all sides of the reactions parallelogram 'RP' and the rhombus of cell centers both pass the contacting points of the discs, only two friction angles $\dot{\rho}$ and $\rho^{\wedge}$ at subtended sides are sufficient to describe the cell's friction states. The two slopes of the 'RP' are given by

[1]Arab Academy for Science and Technology, Alexandria, Egypt

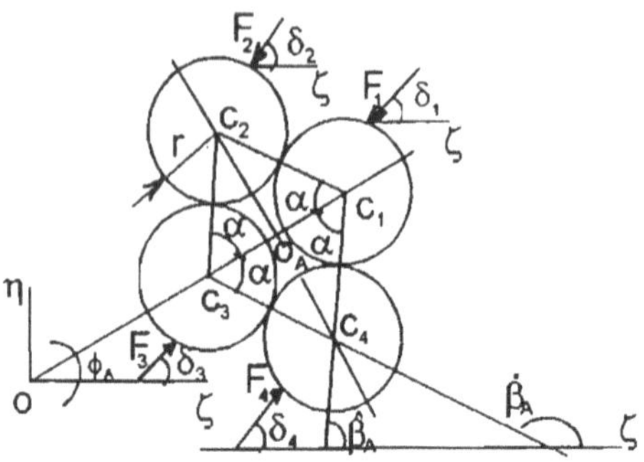

Figure.1 A single rhombic cell

$$\tan \gamma^\wedge = [F_3 \sin \delta_3 + F_4 \sin \delta_4] \;/\; [F_3 \cos \delta_3 + F_4 \cos \delta_4]$$
$$\tan \dot{\gamma} = [F_1 \sin \delta_1 + F_4 \sin \delta_4] \;/\; [F_1 \cos \delta_1 + F_4 \cos \delta_4]$$

These angles are connected to friction state angles as $\dot{\gamma} = \dot{\beta} - M\dot{\rho}$ and $\gamma^\wedge = \hat{\beta} + M\rho^\wedge$. For a stable cell, the boundary forces $\{F_i\}$ satisfy the three equations of equilibrium the moment equation of which implies that the four forces are concurrent at a certain point located somewhere inside the cell.

2 Failure of a Single Cell

The limiting conditions occur when both friction states reach the value of friction angle of disc surfaces, in which case the four total reactions intersect pairwise at points lying on the cell diagonals. In a multicellular structure, the only boundary forces are given, which are transferred through the structure core as discs reaction forces already analyzed above. The structure in Figure.2 is viewed as a number of lettered cells about to fail following conventional modes. Overall consistent motion of discs implies that each negative cell is surrounded by four positive cells and vice versa. Therefore, knowing the mode of failure of only one cell establishes the failure modes for the rest of cells. In addition, the rhombic arrangement implies that if the pairs of angles (ϕ,α) are measured for boundary cells "D", "B", and "C", a simple geometric analysis provides such pairs for successive inner cells.

3 The Network of Total Reactions

As Figure.2 shows, the two adjacent cells "D" and "B" have a common disc. Consequently, both cell rhombi intersect at the center 'c' of this disc. In addition, the two corresponding reaction parallograms 'RP's' of these two cells have one common edge point 'k'. By computing the length and slope of line 'kc', the two friction states of the cell "B" are derived, given the two friction

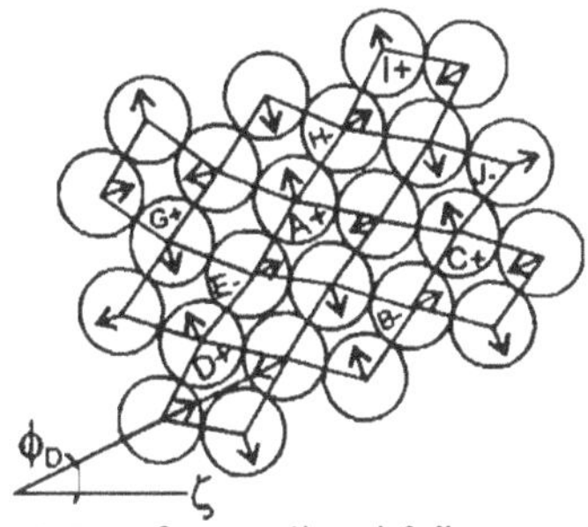

FIGURE.2 Conventional failure modes

states of the cell "D", as follows

$$\theta = U(\hat{S}_D, \hat{C}_D, \hat{T}_D, q\dot{S}_D, q\dot{C}_D, \dot{T}_D)$$

$$\hat{X}_D = V(\hat{\beta}_D, \hat{\gamma}_D, \theta) \quad \& \quad \dot{X}_D = V(\dot{\beta}_D, \dot{\gamma}_D, \theta) \quad \&$$

$$\hat{\gamma}_B = W(\hat{\beta}_B, q\dot{X}_D, \theta) \quad \& \quad \dot{\gamma}_B = W(\dot{\beta}_B, q\hat{X}_D, \theta)$$

where U,V, and W are transfer functions defined by

$$\theta = U = \tan^{-1}\{[\hat{S}\dot{T} + \dot{S}\hat{T} - (\hat{C} + \dot{C})\hat{T}\dot{T}]/[\hat{S} + \dot{S} - (\hat{C}\hat{T} + \dot{C}\dot{T})]\}$$

$$X = V = \cos(\beta - \theta) - [\sin(\beta - \theta)]/[\tan(\gamma - \theta)]$$

$$\gamma = W = \theta + \tan^{-1}\{\sin(\beta - \theta)/[\cos(\beta - \theta) - X]\}$$

and $\hat{S} = \sin\hat{\beta}$, $\hat{C} = \cos\hat{\beta}$, $\hat{T} = \tan\hat{\gamma}$.

Similarly, if we move vertically towards the cell "E" the angles $\hat{\gamma}_E$ and $\dot{\gamma}_E$ are computed, given those of cell "D". The factor 'q' controls the direction of computation such that q=+1 for horizontal transfer, e.g. D---B, and q=-1 for vertical transfer, e.g. D---E. The computed cells can then be used to compute the friction states of their neighbor cells and so on. Eventually, the friction states of all cells can be expressed in terms of only two angles $\hat{\gamma}_D$ and $\dot{\gamma}_D$, where "D" is any selected cell in the core or at the boundary.

4 Applications and Discussion

The rhombic structure in Figure.2 becomes orthogonally arranged when $\phi_I = 0$ for every cell "I". These conditions imply that either all cells have the same angle value α_I, or negative and positive cells have the same α_n and α_p, respectively. Let us discuss the case when we are given all $\alpha_I = \alpha$ and we seek to compute the friction states of boundary cells given that the core cell "A" follows a positive failure mode with two equal friction states '$\rho^{\sim}$'. Using transfer functions vertically downwards from cell "A" to "B" we get

$$\omega = \{2\tan\alpha - \tan(\alpha + \rho^{\sim})\} \quad \& \quad \hat{\gamma}_B = \pi/2 + \tan^{-1}\{-1/\omega)\} \quad \& \quad \dot{\gamma}_B = \pi/2 + \tan^{-1}\{1/\omega\}$$

Then using transfer functions horizontally rightwards from cell "B" to the corner cell "C" we get

$$\tau = \{[2\tan\alpha - \tan(\alpha + \rho^{\sim})]\tan\alpha\}/[3\tan\alpha - 2\tan(\alpha + \rho^{\sim})] \quad \&$$

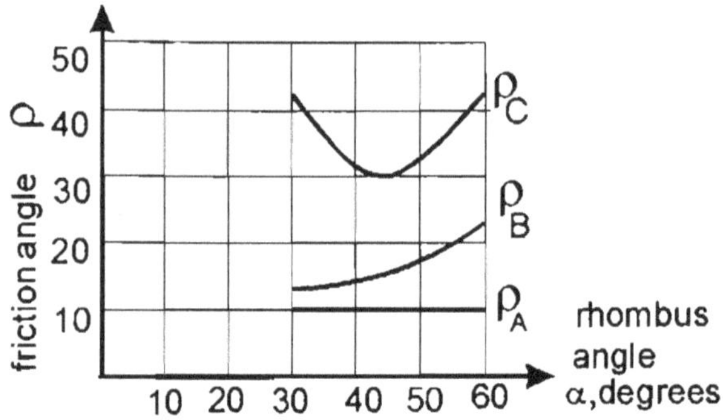

FIGURE.3 Theoretical friction angles for cells B and C when core cell A fails

$$\hat{\gamma}_C=\tan^{-1}(\tau) \quad \& \quad \dot{\gamma}_C=\tan^{-1}(-\tau).$$

We further might ask whether the cell "A" can fail, i.e. $\rho^{-}=\rho$=limit friction angle of disc surfaces, before any boundary cell. This certainly depends on the calculated friction state angles $(\hat{\rho}_B,\dot{\rho}_B)$ and $(\hat{\rho}_C,\dot{\rho}_C)$ to be all less than ρ. Replacing $\rho^{-}=\rho$ and plotting the above expressions for cells "B", negative mode, and "C" , positive mode, we get the graphs in Figure.3. For a selected $\rho=10$ degrees and different values of angle α, the figure shows that for orthogonal rhombic structures it is rather impossible to bring a core cell to fail before failure of boundary cells.

5 Conclusion

For an elementary two-dimensional granular cell with unlikely grain rotation, the friction states are fully described by only two friction angles. Accordingly, the cell is to fail when both state angles approach the limiting friction angle as the boundary forces change. By applying reactions transfer functions to multicellular random structures, the path of load transfer between grains becomes deterministic. Examples on orthogonal arrangements illustrate how to implement this analysis to predict the location of failure initiation.

6 References

[1] Bleich, H.H., On Disintegration of Bodies of Granular Materials Governed by the Coulomb Rule, *ASME J. Appl. Mech.*, pp 1-4, 1964.

[2] Jenkins, J.T., and Satake,M.,Mechanics of Granular Materials- New Models and Constitutve Equations, *Studies in App. Mech.* 7, Elsevier Publ., Amestrdam, Oxford, New York, 1983.

The Effect of Shape Memory Alloys in Vibration Suppression of a Cantilever Beam

J. O. Salichs[1]
C. Shakeri[1]
M. N. Noori[1]
H. Davoodi[1]

ABSTRACT Experiments were conducted to investigate the effect of the shape of Nitinol strips on shifting the natural frequency of a cantilever beam. Nitinol belongs to a group of metallic materials known as Shape Memory Alloys (SMA). Nitinol strips were attached to the cantilever beam in order to reduce the vibration of the beam upon activation. The results show that Nitinol strips with curved memory are more effective than straight strips in shifting the natural frequency of the system.

1 Introduction

Shape Memory Alloys (SMA) are a group of metallic materials that have the ability to return to a previous shape when subjected to heat. The SMA material used in this experiment consisted of an alloy of Nickel and Titanium known as Nitinol. This material can be deformed at lower temperatures and when subjected to higher temperatures it recovers its original memory. This memory effect is known as Shape Memory Effect (SME). The SME occurs because of the change in the crystal microstructure of the material from martensite (low temperature crystal microstructure) to austenite (high temperature crystal microstructure) (Davoodi, 1997). Shape Memory Alloys have different types of SME known as one-way and two-way memory. The one-way memory results when the material recovers its original shape only when subjected to heat. The two-way memory results when the material recovers its original shape upon heating and cooling. It is important to mention that the stiffness and elastic modulus of the material change when it is subjected to phase transformation. These property changes are responsible for the vibration mitigation abilities of the material.

[1] Department of Mechanical Engineering, Worcester Polytechnic Institute, Worcester, MA 01609

Many applications have used SMA for vibration mitigation and structural control. Both passive and active damping applications of SMA were initially proposed in 1988 (Beards 1988). Kelly and Nims at the Earthquake Engineering Research Center at Berkeley were the first to investigate the applications of SMA for seismic resistant design (Aiken, 1992). Davoodi and his colleagues (Davoodi, 1997) studied the feasibility of using SMA for controlling the amplitude and natural frequency of structures during high level of excitation. A review of applications for SMA can be found in (Shakeri, 1997).

2 Experimental Procedure

Different memories and configurations of SMA materials were investigated for the application of vibration control. Nitinol strips were attached to a cantilever beam. The dynamic response of the system was obtained for free vibration when the material was not activated and when it was activated. Electric current was passed through the strips to heat them up and thus activate them. An accelerometer was attached at the end of the cantilever beam and the response was displayed on a digital signal analyzer.

Two different memories for the Nitinol strips were used: straight and curved. By using different memories for the SMA material, the response of the system can be improved because of the change in the stiffness of the material. The natural frequency of a system can be shifted more using the same amount of material but a different shape. In addition, two different types of attachments were used: Nitinol strips attached to one side or both sides of the cantilever beam.

In order to know when the Nitinol was completely activated (austenite finish) the time response of the SMA material when subjected to heat was investigated. A 5 amps current at 10 volts was used to heat up the strips. The time to reach the austenite start temperature (29.4 C) was 3.6 seconds and the austenite finish temperature (48 C) was 14.6 seconds. These results were based on a room temperature of 26C.

A cantilever beam with a concentrated mass at the end was designed. Six SMA strips with straight memory were attached at the top of the cantilever beam in pairs. Each pair consisted of two Nitinol strips connected electrically in parallel. In addition, the pairs were connected electrically in series. The natural frequency was obtained at different temperatures, at 26C (when the material was not activated) and 58C (when the material was completely activated). Then, one pair of SMA strips was taken out of the configuration and the same procedure was repeated until just one pair of Nitinol material was at the top of the cantilever beam. The above procedure was repeated with curved memory strips.

In the second part of the experiment configuration of the Nitinol strips was changed. Two pairs of SMA strips with straight memory were attached at the bottom and another one at the top of thecantilever beam. The response of the system was obtained for free vibration. Then, one pair at the top of the beam was taken out and the same procedure was repeated until one strip was left at the top and another one at the bottom of the cantilever beam. Finally, the experiment was repeated again using SMA strips with curved memory instead of using straight memory strips. The SMA strips with curved memory were attached to the system by forming a U with respect to the beam.

3 Experimental Results

The natural frequency of the cantilever beam was shifted when the Nitinol strips were activated regardless of the shape of the strips (straight or curved) or the attachment configuration. Figure 1 shows the percentage of change of the natural frequency of the system using different number of strips. This change in natural frequency is due to the increase in the Nitinol's stiffness. As it is shown the effect of the curved memory on the natural frequency is more pronounced than the straight memory. The reason is that the Nitinol strips are acting as a pre-stressed spring. In addition, by attaching the Nitinol strips at both sides of the beam the natural frequency is shifted to a higher value than when the strips attached only at the top of the beam (no matter what memory was used). It is important to note that the effect of using different memories is more pronounce than the effect of using different attachment configurations. The maximum shift in natural frequency is 25.7%. This is obtained with 6 strips with curved memory at both sides of the beam. By using the strips with straight memory a maximum of 3.1% shift in natural frequency can be obtained using 4 strips at both sides of the beam.

4 Discussion and Future Work

The use of SMA with different memories is an efficient way to improve the vibration mitigation abilities of this material. SMA with curved memory shifted the natural frequency of the system to a higher value than the straight memory. Also higher values for the natural frequency can be obtained by attaching curved memories to the cantilever beam. It is important to note that the property that causes this shifting in the natural frequency of the system is the change in stiffness of the material.

The use of SMA in different configurations is another way to control vibrations with the same amount of material. The results showed a change in the natural frequency of the system when using SMA material at both sides of the beam rather than attaching them only to one side. However, the maximum shift in natural frequency can be obtained by using a combination of both effects: change in memory and attachment configuration.

Further research is needed to study the effect of the mass ratio of Nitinol to the structure. The result of such study will show how to get the most effect out of a specific amount of SMA material. Developing analytical models helps to gain

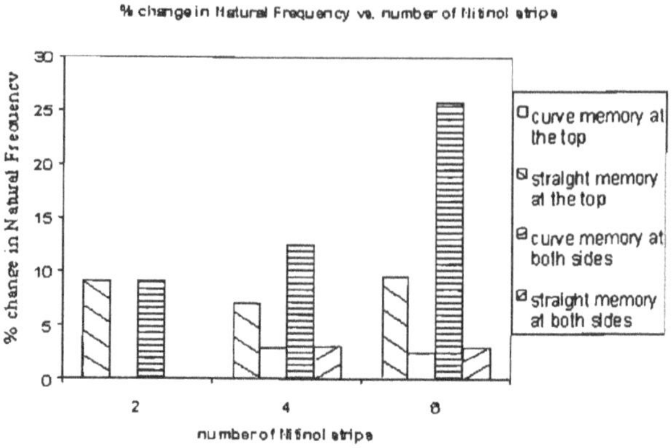

FIGURE 1 Comparison of different shapes and attachments using Nitinol

better idea of the effect of the SMA before experiment. In addition, more experiments should be done on the effect of temperature on damping ratio of SMAs in order to use it along with frequency shifting in vibration suppression.

5 Acknowledgments

Special thanks to Memry Corporation for supplying the SMA material for this investigation. Also thanks to Bob Taylor, Mr. Steve Derosie, Mr. Todd Billings, and Mr. Jared Eaton.

6 References

Aiken, I. D., Nims, D.K., Kelly, J.M., Comparative study of four passive energy dissipation systems, Bull, New Zealand National Society For Earthquake Engineering, 25(3), 1992, 175-192.

Beards, C.F., *Vibrations and Controls Systems*, Ellis Horwood Limited publishing company. First Edition, 1988.

Davoodi, H., *et al*, Vibration Control of Structures Using Shape Memory Alloys, WPI/UPRM Collaborative Project Center, Project Report, December 1996.

Davoodi, H., *et al*, Design of an Intelligent Structure, WPI/UPRM Collaborative Project Center, Project Report, May 1997.

Shakeri, C., Noori, M. N., and Hou, Z., Smart Materials and Structures: A Review, *Proceedings of the Fourth Materials Engineering Conference*, Washington, D.C., November 10-14, 1996, Vol. 2, pp. 863-876.

Nonlinear Vibrations of Moderately Thick Circular Plates

M. Sathyamoorthy[1]

ABSTRACT A nonlinear shear deformation theory is used in the present analysis to study the effects of transverse shear deformation and rotatory inertia on the large amplitude vibration behavior of isotropic circular plates. The governing equations are presented in polar coordinates making use of tracing constants to identify transverse shear and rotatory inertia effects. Based on a single-mode analysis, solutions to the governing equations are presented for moderately thick circular plates by use of Galerkin's method and the numerical Runge-Kutta procedure.

1 Introduction

Large amplitude vibrations of plates of various geometries have been studied by many researchers in recent years [1]. A review of literature indicates that there are several nonlinear theories for plates which include the effects of transverse shear deformation and rotatory inertia. However, solutions to these nonlinear formulations cannot be found in the literature. Wu and Vinson [2] derived a simplified set of nonlinear governing equations using the Berger's approximation and presented results for the large amplitude vibration of composite rectangular plates. The combined effects of transverse shear and rotatory inertia on the nonlinear vibration of rectangular plates have also been discussed by this author [3] based on a generalization of the classical von Karman plate theory [4]. From these investigations it may be concluded that the effects of transverse shear deformation and rotatory inertia are very important for moderately thick plates.

This paper is concerned with the nonlinear flexural vibration of a moderately thick isotropic circular plate with rigidly clamped edges. Nonlinear equations of motion of plates in polar coordinates expressed in terms of stress function F and lateral displacement w are used in this work. These equations are applicable for moderately thick plates and the formulation includes the effects of transverse shear deformation and rotatory inertia. Due to the complex nature of the governing nonlinear equations, only a single-mode analysis is carried out by use of Galerkin's method and the numerical Runge-Kutta procedure. A solution for w is obtained in a separable form consisting of the product of a time function and a space function. Numerical results are graphically presented for some cases

[1] Department of Mechanical and Aeronautical Engineering, Clarkson University, Potsdam, New York 13699-5725.

of isotropic circular plates including effects of transverse shear deformation and rotatory inertia. These effects are shown to be important for certain combinations of parameters of moderately thick plates.

2 Governing Equations

Consider a circular plate of radius *a* and uniform thickness *h* with isotropic material properties represented by the material constants E, ν, and G. The equations of motion for plates undergoing axisymmetric deformations can be readily derived using the strain energy and kinetic energy expressions and applying the Hamilton's principle. Such equations in terms of lateral displacement w, and stress function F are given by

$$d_1[F_{,rrrr}+(2/r)F_{,rrr}-(1/r^2)F_{,rr}+(1/r^3)F_{,r}] = -(1/r)\, w_{,r}\, w_{,rr} \tag{1}$$

$$\begin{aligned} L_1(w,F) = \{&(r - c_{10}/r)(F_{,r}\, w_{,rr} + w_{,r}\, F_{,rr}) - c_{11}(F_{,r}\, w_{,rrr} + 2F_{,rr}\, w_{,rr} + w_{,r}\, F_{,rrr}) - c_{11} \\ &r(F_{,r}\, w_{,rrrr} + 3F_{,rr}\, w_{,rrr} + 3F_{,rrr}\, w_{,rr} + w_{,r}\, F_{,rrrr}) + c_{15} R_i\, r(F_{,rtt}\, w_{,rr} + 2F_{,rt}\, w_{,rrt} + \\ &F_{,r}\, w_{,rrtt} + F_{,rr}\, w_{,rtt} + w_{,r}\, F_{,rrtt} + 2F_{,rrt}\, w_{,rt})\} + c_3(w_{,r}/r) - c_3\, w_{,rr} - 2c_4 r\, w_{,rrr} - c_4 r^2 \\ &w_{,rrrr} + d_5\, r\, w_{,rtt} + d_5 r^2\, w_{,rrtt} + \rho h(c_{10} - r^2)\, w_{,tt} - c_{15} R_i\, r^2 \rho h\, w_{,tttt} + (r^2 - c_{10})\, q \\ &(r,\theta) - c_{11}\, r(q_{,r} - r\, q_{,rr}) = 0 \end{aligned} \tag{2}$$

The coefficients c_i and d_i are functions of geometric and material parameters of the plate. Equations (1) and (2) are the two nonlinear governing equations applicable for the large amplitude axisymmetric vibration of moderately thick isotropic plates. Transverse shear and rotatory inertia effects have been included in these equations by means of tracing constants T_s and R_i. In the case of thin plates $T_s = R_i = 0$ and, therefore, the governing equations simplify considerably.

3 Method of Solution

The circular plate is assumed to be clamped and movable all along the boundary defined by r = a. Therefore, the boundary conditions are $N_r = w = w_{,r} = 0$ at r = a. A polynomial expression for w is chosen to satisfy the out-of-plane boundary conditions as $w = h\, A(\tau)\,(1 - r^2/a^2)^2$ where, $A(\tau)$ is an unknown function of nondimensional time $\tau = t(E/\rho a^2)^{1/2}$. This expression for w is substituted into the compatibility equation (1) and an expression for F is assumed as $F = \delta\,(a_0 r^2 + a_1 r^4 + a_2 r^6 + a_3 r^8)$ with $\delta = -(16 h^2 A^2/a^4)$. Substituting for F and w in equation (1), the coefficients a_i are determined as $a_1 = 1/(-8d_1 + 72d_2)$, $a_2 = -4/a^2(-24d_1 + 600d_2)$, $a_3 = 3/4(-48d_1 + 2352d_2)$. If the edges of the plate are movable in the sense that $N_r = 0$ at r = a, then $N_r = (1/r)\, F_{,r} = \delta\,(2a_0 + 4a_1 r^2 + 6a_2 r^4 + 8a_3 r^6)$. Since $N_r = 0$ at r = a, $a_0 = -a^2(2a_1 + 3a_2 a^2 + 4a_3 a^4)$. Since the coefficients a_0, a_1, a_2, a_3 are all determined, the stress function F is fully known and it is an exact solution of the compatiblity equation (1).

Having satisfied the compatibility condition and the boundary conditions, it is now required to solve the equation of motion in the lateral direction of the plate, namely equation (2). The expressions for w and F determined earlier are substituted in equation (2) to determine the error function $L_1(w,F)$. Then the following integration is performed over the entire area of the plate.

$$\int_0^a L_1\,(w,F)\,(1 - r^2/a^2)^2\,dr = 0 \tag{3}$$

Equation (3) will result in a time-differential equation in $A(\tau)$ as given below.

$$(e_1\,A + e_2\,A^3)_{,\tau\tau\tau\tau} + (A + e_3\,A^3)_{,\tau\tau} + e_4\,A + e_5\,A^3 = e_6\,q_0/E \tag{4}$$

The modal equation (4) is applicable for moderately thick isotropic clamped circular plates and the effects transverse shear deformation and rotatory inertia are included in the analysis. If these effects are ignored, as in the case of thin plates, the modal equation (4) reduces to the Duffing equation $A_{,\tau\tau} + e_4\,A + e_5\,A^3 = e_6\,q_0/E$. In the case of a circular plate subjected to a uniformly distributed lateral load q_0 undergoing large deflections, the Duffing equation will reduce further to the nonlinear algebraic equation $e_4\,A + e_5\,A^3 = e_6\,q_0/E$.

For moderately thick circular plate, the time-differential equation applicable to this case, namely equation (4) can be integrated numerically using the Runge-Kutta method. The frequency ratios, ω/ω_0 can be obtained by integrating equation (4) for various values of the amplitude of vibration of the moderatly thick circular plate. ω is the nonlinear frequency including the effects of transverse shear deformation and rotatory inertia whereas ω_0 is the linear frequency without these effects.

4 Numerical Results and Discussion

Numerical results are presented for dynamic nonlinear behavior of a circular plate in Figs. 1 and 2. In Fig. 1 the frequency ratio ω/ω_o is plotted against the nondimensional amplitude of vibration of a thin circular plate exculding the effects of transverse shear and rotatory inertia. For moderately thick circular plates undergoing large amplitude free vibration, equation (4) is integrated numerically using the Runge-Kutta integration procedure. In using this integration technique, the nondimensional time interval has been taken as 0.00001. The coefficients of the time-differential equation (4) are presented in Table 1 for moderately thick circular plates with various radius-to-thickness ratios. In Fig. 2, numerical results are presented for the nonlinear frequency ratios at various amplitudes of vibration for moderately thick isotropic circular plates with various radius-to-thickness ratios. It should be noted that the effects of transverse shear and rotatory inertia have been included in all these cases by taking $T_s = R_i = 1$. For moderately thick plates subjected to uniformly distributed loading of intensity q_o, equation (4) is applicable provided the time-dependant terms are ignored. For the limited number of cases considered in this paper, the well-known hardening type of behavior is observed between the frequency and the amplitude. The effects of transverse shear and rotatory inertia are found to decrease the frequency ratio at any amplitude of vibration.

5 References

[1] Sathyamoorthy, M., *Nonlinear Analysis of Structures*, CRC Press LLC, Boca Raton, 1998.

[2] Wu. C. I and Vinson, J. R., On the Nonlinear Oscillations of Plates Composed of Composite Materials, *Journal of Composite Materials*, 3, 548- 1969.

[3] Sathyamoorthy, M., Vibration of Plates Considering Shear and Rotatory Inertia, *AIAA Journal*, 16, 285- 1978.

[4] Herrmann, G., Influence of Large Amplitudes on Flexural Motions of Elastic Plates, *NACA TN-3578*, 1955.

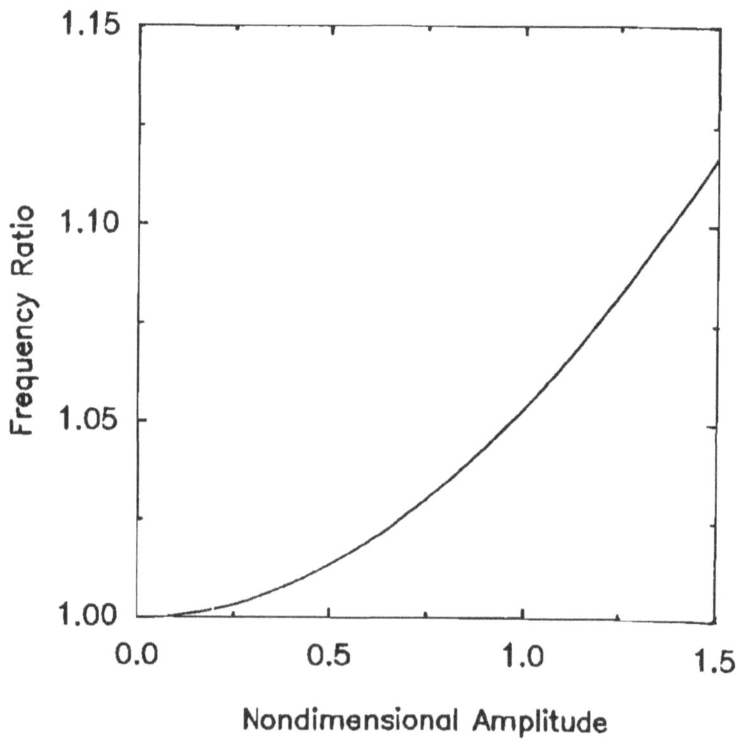

Figure 1. Amplitude -frequency curve for a thin circular plate.

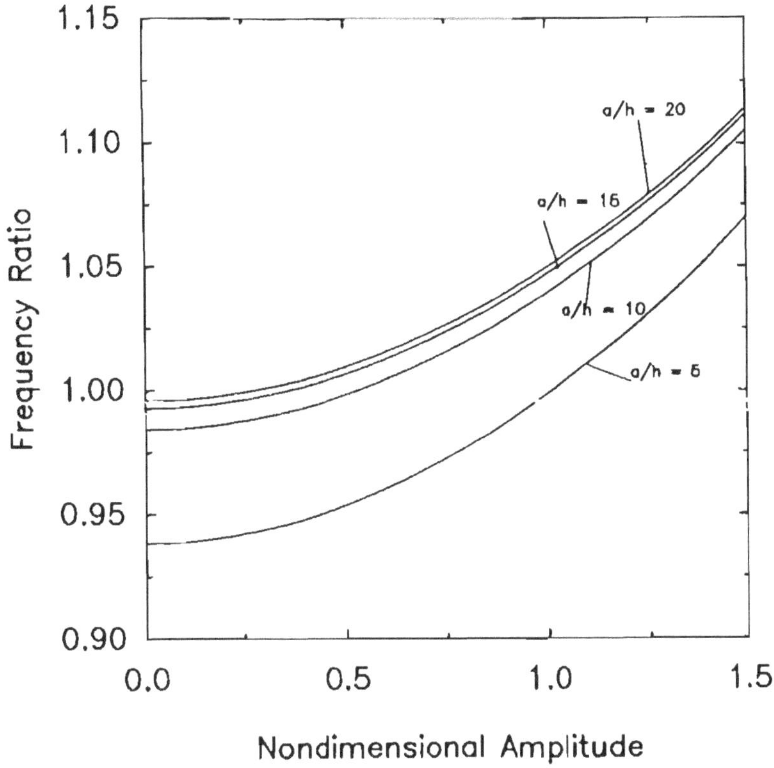

Figure 2. Amplitude -frequency curves for moderately thick circular plates.

Table 1. Coefficients of equation (4) for various values of thickness parameter, (a/h).

(a/h)	e_1	e_2	e_3	e_4	e_5	e_6
5	0.0195	0	0.0011	0.3441	0.0607	36.6964
10	0.0051	0	0.0001	0.0947	0.0145	161.5116
15	0.0023	0	0	0.0428	0.0064	369.8131
20	0.0013	0	0	0.0242	0.0036	661.4689

Mechanical Strength of Long Bones During the Healing Period

D. G. Savadkoohi[1]
K. Firoozbakhsh[2]

ABSTRACT- The purpose of this investigation was to measure the reduction in bone strength resulting from drill holes in diaphyseal bone and to investigate its mechanical strength during healing period. To determine the mechanical strength following fracture fixation, 144 fresh rabbit femurs and tibias were mechanically tested. Computed tomography (CT) and ultrasound techniques were also employed to determine local variation of bone density and modulus of elasticity. The results confirm that callous bone formation do not create strong rigidity. After a complete healing period of 6 weeks, the bone did not gain its initial strength and at least 45 days post plate removal is required for complete bone remodeling.

1. Introduction

Refracture of long bones after the removal of screw-plate device for internal fixation has been observed [5, 6]. These fractures appear to have occurred with less trauma than those which produced the initial fracture. A decreased in strength may be attributed not only to disuse osteoporosis but also to the alteration in the bone architecture following removal of screws. A void in the wall of a hollow tube will weaken the tube, for, besides decreasing the material available to withstand the applied loads, the defect also raises the local stress around the hole. Several investigators [1,3, 6] have reported the effect of implant stiffness on the stress shielding. Among techniques used to determine the mechanical properties and strength reduction resulting from drill holes in

[1] *Department of orthopaedic Surgery, Sina Hospital, Tehran, Iran*
[2] Department of Orthopaedic and Rehabilitation, Medical Center, The University of New Mexico, Albuquerque, NM 87131, USA

diaphyseal bone are mechanical testing, ultrasound and Computerized Tomography (CT) [2, 4, 7]. In this study we used all three techniques to evaluate the mechanical strength of long bones during the healing period after plate removal. Results of this study may help in choosing a safe rehabilitation program post fracture fixation.

2. Materials and Methods

Sixty, 6 months old white German rabbits were used in this study. Under anesthesia, the mid tibia and femur shafts were totally osteotomised and then fixed internally using mini plates and screws. In one group, at postoperative sequences of two weeks, 4 weeks and 6 weeks, the rabbits were sacrificed and bones were mechanically tested following plate removal. In another group, plates were removed 6 weeks postoperatively but rabbits were not sacrificed until 45 days later, when the bones were mechanically tested. Intact specimens and unosteotomised bones with the screw holes were also tested as controls. Test modes included: 4-point bending, axial compression and combined torsion-compression. Computed tomography (CT) and ultrasound techniques were also employed to determine local variation of bone density and modulus of elasticity.

3. Results

Compared to intact specimens, unosteotomised bones with screw holes were weaker *97% in* bending, 33.5% in axial compression and 23% in torsion. Their rigidities were not, however, significantly different from those of intact bones. As the fractured bones healed the bone strength progressively increased in all modes of testing. The mechanical strength at 6 weeks postoperation was higher than those of unosteotomised bone with screw holes due to callous bone formation, see Figure 2. At this stage, the strength of bone was less than intact bone (85% in bending. *83%* in compression and 96.6%, in torsion). Those bones tested 45 days post plate removal had an increased strength compared to the intact bones by 22% in bending, 12% in compression and 15% in torsion, see Figure 2. In this group the callous formation was partially resorbed and the screw holes were almost filled due to remodeling. The bones tested 6 weeks postoperation had low rigidity in all modes of loading, particularly in bending,

Figure 1. Bone stiffness at post operative sequences

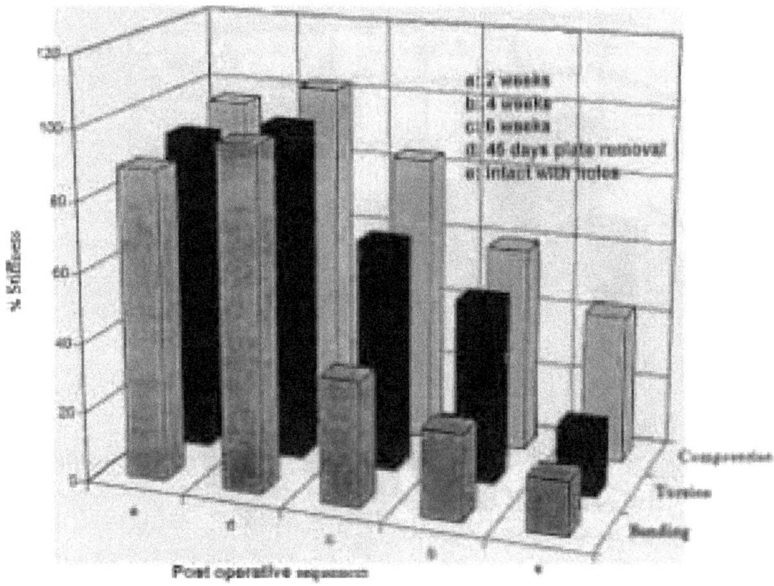

see Figure 1. Compared to those of intact bone, their rigidity was 83.5% *in* compression, 66% in torsion and only 35.5% in bending. Those specimens 45 days post plate removal, although had less callous bone formation were more rigid and their rigidity approached those of intact values, see Figure 1. Both CT scanning and ultrasound techniques confirmed the progressive nature of bone density and modulus of elasticity postoperatively.

4. Discussion

Geometric disturbances, such as holes, cause high-localized stress in an elastic body. In the application of this concept to bone, it should be noted that the risk of fracture during normal activity is significantly increased when stress risers are present. Common stress risers in cortical bone usually occurs as a result of orthopaedic surgery and includes screw holes. The results confirm that callous bone formation do not create strong rigidity. After a complete healing period of 6 weeks, the bone did not gain its initial strength and at least 45 days post plate removal is required for complete bone remodeling, see Figure 2.

Figure 2. Bone strength at post operative sequences

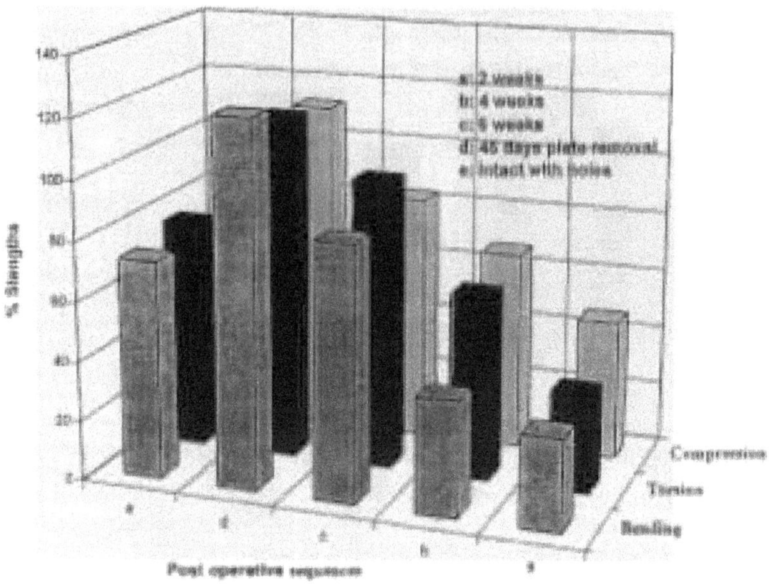

5. References

[1] Akeson, WH and Woo, SL The effects of rigidity of internal fixation plates on long bone remodeling. Acta Orthop Scand 47; pp:241-249,1976.

[2] Ashman, RB. And Cowin,SC. A continuous wave technique for measurement of the elastic properties of cortical bone. J Biomech 17(5) ; pp:344-361, 1984.

[3] Carter, Dr. the plated femur relationship between the changes in bone stress and bone loss. Acta Orthop Scand 52; pp:241-248, 1981.

[4] Rho, JY. And Ashman, RB. Young's modulus of trabecular and cortical bone material: ultrasonic and microtensile measurements. J Biomech 23(2); pp:111-119, 1993.

[5] Rosson, JW. Bone structure after removal of internal fixation plates. J Bone Joint Surg [Br] 73-B, pp:65-67, 1991

[6] Rosson, JW. Bone weakness after the removal of plates and screws. J Bone Joint Surg [Br] 78-B, pp:283-286, 1991

[7] Snyder, SM. and Schnierder , E. Estimation of mechanical properties of cortical bone by computed tomography. J Orthop Res 9(3); pp:422-431, 1991

Constitutive Description of Granular Materials

Q. S. Yang[1]

Abstract

The behaviour of granular materials is quite complicated and varies greatly under different conditions. Idealizations are therefore essential in order to develop mathematical models for description of material response. The earlier development of stress-strain relations for granular materials was strongly influenced by the well-established classical plasticity theory. However, the excessive dilation was predicted which was the result of the use of the associated flow rule. Extension of the classical theory to a non-associated form of plastic potential for granular materials was first proposed by Poorooshasb et al. (1966, 1967).

Much attention has been focused on the modeling of the granular materials behaviour under cyclic loading conditions encountered in the transient and dynamic problems. This has led in turn to the generation of nested yield surface or multisurface plasticity models, and the bounding surface model, which can be considered as a smooth extension of the nested surface models.

This paper presents a constitutive model based on the recent development of hypoplastic concept, which can incorporate the effects of the non-coaxiality. The non-coaxiality of the principal stress and principal plastic strain increment directions observed during loading involving rotation of principal stress directions. The void ratio and the inter-granular (or effective) stress tensor are used by the model to characterize the material states. It may be also considered as an extension of the critical state soil mechanics, as a compact state line is introduced to the state boundary surface in addition to the critical state line. This model has been used to simulate the deformation, stability and flow of granular materials.

[1]The Hong Kong University of Science and Technology, Hong Kong, cellan@uxmail.ust.hk

Applications of Active Noise and Vibration Control in Precision Machining

R. Ghanadan[1]

Abstract

Control and signal processing techniques developed over the past several years have shown great potential in control system design and analysis of complex and nonlinear systems. In this work we will demonstrate the use of such tools in two industrial applications. First, we consider an application of active structural control technology to the noise and vibration problem in precision machining. Our focus will be on integrated modeling of machining dynamics for performance evaluation prior and during full scale experimental tests, and real time control system design.

Our second application involves design of a DSP controlled adaptive feedback linearizer for RF power amplifiers. Our goal is to reduce harmonic distortion in intermodulation spectra over a wide bandwidth. In PCS, intermodulation distortion appears as interference in other channels. As a result, interference reduction circuitry is an essential element of any high bandwidth communication system.

[1]Bell Laboratories, Lucent Technologies, ghanadan@bell-labs.com

Index of Authors

www.ingramcontent.com/pod-product-compliance
Lightning Source LLC
LaVergne TN
LVHW051052210926
848486LV00002B/87